DATE			

Reservoir Seismology

Geophysics in Nontechnical Language

RESERVOIR SEISMOLOGY

GEOPHYSICS IN NONTECHNICAL LANGUAGE

MAMDOUH R. GADALLAH

PENNWELL PUBLISHING COMPANY
TULSA, OKLAHOMA

PennWell Publishing Company
1421 South Sheridan/P.O. Box 1260
Tulsa, Oklahoma 74101

Library of Congress Cataloging-in-Publication Data

Gadallah, Mamdouh R.
Reservoir seismology : geophysics in nontechnical language / Mamdouh R. Gadallah
p. cm.
Includes bibliographical references and index.
ISBN 0-87814-411-0
1. Seismic prospecting. I. Title.
TN269.8.G33 1994
622'.1592—dc20

93-39214
CIP

Printed in the United States of America

1 2 3 4 5 98 97 96 95 94

DEDICATION

This book is dedicated to my wife, Candida M. Gadallah, and my children, Darrell, Deena, and Dahlia. I am grateful for their extraordinary understanding which let me spend all the time needed while writing this textbook.

CONTENTS

FOREWORD

The oil industry has undergone tremendous development over the past four decades. In the early years, decisions to drill were based on surface geological anomalies and on instinct. Many large structures were discovered, but the smaller ones became increasingly difficult to find.

Today, although proven reserves are declining in the United States, worldwide reserves are staying about the same, even increasing in some areas, partly because of the increased use of advanced technology in the exploration methods.

Efforts to increase the success ratio of exploration drilling have led to great improvements in the techniques used to investigate the subsurface such as magnetic, electrical and, finally, the seismic methods. Geophysical techniques have been extended to include borehole measurements of rock parameters such as density, acoustic velocity, and other parameters.

Historically, oil companies have been structured into divisions having different responsibilities for the different stages leading to the discovery and development of a new field. Although each division has in its custody a wealth of information gained through measurements made during its particular phase of activity, the information tends to remain within the division and other divisions do not benefit from it. Though individual missions may be well executed, the overall results have sometimes fallen short of expectations because of the gap in communication between the divisions.

One of the objectives of this book is to promote team effort by acquainting nongeophysicists with the seismic method and to show how it can be used by geologists, engineers, and others to augment their own data, leading to more confidence in their decisions.

The various disciplines represented by the different divisions may be integrated by a reservoir seismologist. This is a geoscientist who integrates data on the geological and petrophysical properties of the rock with the seismic signature of the reservoir. His activities include relating the observed seismic amplitudes, velocities, and elastic properties of the rock to its lithology (rock type) or fluid content.

By mapping these properties horizontally and vertically between wells, a reservoir seismologist can delineate the lateral extent, thickness variations, subsurface topography and heterogeneity (variation in rock properties) of the reservoir, which lead to the design of improved recovery methods.

This text will provide engineers, geologists, landpeople, and managers with knowledge of the fundamentals of the seismic techniques, their applications and limitations, and the pitfalls resulting from stretching the data beyond its resolution. This knowledge will serve as a foundation for learning techniques at the leading edge of technology, such as high-resolution subsurface imaging, reservoir delineation, description of reservoir characteristics, and predicting what lies ahead of the bit. These techniques are illustrated by case histories.

The book covers a broad spectrum of applications to instruct professionals in all oil-related disciplines, as well as the managers who coordinate exploration and development programs. The material is organized in a chronological manner and includes "real-life" exercises. The text is written at a level that anyone can understand without difficulty. At the end of each chapter, you will find a list of key words that will help the reader to better understand the chapter by looking them up in the glossary at the end of the text. For those who are interested in more details, there are appendixes for some chapters that include more theory and mathematics and a very complete bibliography of references for those who want to dig deeper.

Acknowledgements

I would like to acknowledge the contributions of Dale Stone and Doyle Fouquet of Seismograph Service Corporation in reviewing some of the material in this book. Many thanks to Dr. Larry Lines and Dr. Terry Watt for their constructive suggestions on tomography and AVO. Many thanks to SEG for allowing me to use many of the illustrations. Special thanks to Dr. Yilmaz.

My gratitude to Dr. E. A. Robinson of Columbia University for sparing his precious time to read and evaluate the text.

I am grateful to Mr. Osman M. Osman of Ain Shams University, Cairo, Egypt, for gathering reference material and for his diligent efforts in proofreading the text. I am very appreciative of the time he invested carefully checking the details of the text.

My gratitude to Mr. Norman James for editing the original manuscript.

Many thanks to Dr. Norman Hyne for his encouragement and suggestions.

Many thanks to Seismograph Service Corporation, Bolt Technology Corporation, Halliburton Geophysical Services, Geophysical Press, Schlumberger, Western Geophysical, and GX Technology, who gave me permission to use illustrations and seismic sections from their literature.

Special thanks for the generosity of the authors who contributed case histories, applications, and information on new technology, and for granting permission to use their material in this book.

While I cannot possibly mention the names of all of those who contributed to this text, I thank all of them for their time and efforts.

M. R. Gadallah, Petroleum Consultant
4-D International
Tulsa, Oklahoma

CHAPTER 1

INTRODUCTION

An adage in the oilfield is "oil is where you find it," which implies that a drill bit is the primary exploration tool. This was undoubtedly true in the past, and the result was a low success ratio. However, the development of the earth sciences has changed all that. Exploration techniques now involve most aspects of modern technology, techniques that could only be dreamed of 50 years ago. Yet, the full potential of these methods is not being realized because of a lack of knowledge and understanding on the part of everyone involved in the various phases of exploration and development.

Techniques used in exploration and development of fossil hydrocarbons are not all routine. As geologists, we dream. As geophysicists, we share those dreams, and as engineers we predict and estimate it all in the grey zone. Although the geologist uses some subsurface control and analogy to identify a lead, he still dreams. It has been said that "oil is found in the brain of man."

Then comes the geophysicist to find and detail the dream from the surface. Do the two disciplines work together and change ideas to optimize their effort to generate a geologically and geophysically sound (and economically justified) prospect? Did the engineer have any input about the possible reservoir quality for his prediction of future production before drilling the wildcat? Did the engineer have any input about the economic factors of the play, such as reservoir thickness, porosity, saturations, recovery factor, and other parameters? Normally, a wildcat effort is the responsibility of the exploration department. Even within the exploration department, the communication between geologist and geophysicist was not always frequent. Often, two interpretations were done independently. Until recently, only a few oil companies had adopted the team concept for better communication and integration of efforts.

Now, let us discuss the operation side; that is, the drilling of a well. The geologist has decided to drill the high shot point, and perhaps the geophysicist with his seismic data participated in the decision. They gave the location and the desired log suites to run, cores to cut, and other information to the drilling engineer, who designed a sound drilling program, applicable to the geological setting of the area in general.

On the seismic section, some anomalies at a certain projected depth may suggest the possibility of abnormally high pressure zones. The engineer must be prepared to control this high-pressure zone, either by setting casing or increasing mud weight. Many high-amplitude anomalies on the seismic section in the shallow part of the stratigraphic column of the well have been overlooked until a blowout occurred.

If the geologist and the engineer are not certain of the type of anomaly, the geophysicist can run a VSP before the drilling bit reaches this zone; this will aid him in predicting what is ahead of the bit and the depth of a high-pressure zone. What is the downtime to run such a survey? Whatever it is, it is definitely cheaper than losing the well. In addition, the geologist and geophysicist can get better information about the next seismic marker and check their well prognosis.

So far, all parties should benefit from the communication and the exchange of ideas. Now, the well has been drilled, and we have two alternatives: a dry hole or a marginal producer to be plugged, or our efforts have succeeded and we will gear up to complete the well and develop the field.

Normally, at this point the exploration department's mission would be completed, and it would set out to find another wildcat play. The geologist would go back to his logs, and the geophysicist would go back to his sections and wiggles. The development engineer and development geologist, or the reservoir geologist and engineer, would remap the field of the wildcat discovery, run calculations, and decide on the spacing and number of wells to develop the field.

The drilling engineer would start his job and, more than likely, there wouldn't be communication between the operation and development departments except the daily drilling report and occasional casual communications. Besides the routine work and usual day-to-day problems, no honest effort would be made to conduct intelligent communication between the disciplines to optimize the activities and exchange ideas for better solutions.

In a routine scenario if the confirmation well is a producer, the rig is moved to another location and then to another, until a dry hole is drilled and we define the reservoir boundaries and, probably, overdrill the reservoir. If the confirmation well is a dry hole, the result is bad news and long faces. Then somebody gets the blame, probably the exploration department who did not do their homework. We drilled without a good, detailed seismic grid, and we drilled on the downside of the fault. After the fact, we will run more detailed seismic data to delineate the structure and (we hope) determine whether there are more surprises or define the next drilling location.

Did we consider a tighter 2-D seismic grid or a 3-D seismic survey? Why is 3-D better than 2-D? Or will a 2-D survey using a tighter grid serve the purpose? How tight must the seismic grid be in order to detail the expected feature? Have the geophysicist and the exploration geologist given any input to the development department since the transfer of all data to that department? The answer, more than likely, is "no," simply because it is not the exploration department's responsibility at that point. Its responsibility is to find another potentially successful wildcat play. Do we need to have a geophysicist in the development department to carry on what the geophysicist in the exploration department started? The answer is probably "yes" if we want intelligent communications and an exchange of ideas between the disciplines.

Now, back to the flow of events. The field has been developed, the primary production is declining, and we are ready for an enhanced oil recovery (EOR) project. We now need to collect more information about the lateral changes of permeability, saturations, and other properties. It would also be helpful if we could get some more information in the vertical sense.

How can we get this information from 5 or 10 wells in the field that are spaced 100 meters apart? Shall we interpolate between wells? Extrapolate between wells? Use the best guess between wells? Or can we find other ways to get better imaging of the vertical and horizontal variations of the reservoir properties?

Every day we come to realize that the reservoir rock is heterogeneous. It is risky to assume homogeneity in the reservoir properties, so we definitely need to image the reservoir. We need to image the fine details, the thin beds, their geometries, and their reservoir characteristics. With the large number of wells drilled, a great amount of oil can be recovered if we can obtain better resolution of the reservoir. This can be accomplished if we insert an energy source and receiver down the holes and inject and record high-frequency energy to detail the thin beds and describe their properties. We can accomplish much if we can relate seismic velocities to reservoir properties by using the inversion process or tomography, and if we all communicate together and have one goal in mind: to find the *right* answer regardless of who has the right answer. The key is integrating the knowledge of the disciplines by exchanging ideas and discussing alternatives.

For all that modern science adds to petroleum technology, the exploration and development of hydrocarbons still involves considerable art as well as applied science. A prospect may begin as a geologist's dream, but economic success may depend on the contributions of the geophysicist and the engineer, as well as the ability of each to interpret properly the data at his disposal. The skill and experience of the interpreter will influence the utility of the data. Failure of any of these people to apply correctly his own knowledge, or to integrate the expertise of one of the others with his own, may result in economic disaster. How can we (as earth scientists) prevent such a disaster? By communicating with each other to take full advantage of the expertise of each discipline. We must learn each other's terminology and educate one another so that we can understand the value, potential, and limitations of all the disciplines and can communicate effectively. Then we can all work together toward one goal—to discover and recover more hydrocarbons.

With today's unstable oil prices and the low level of wildcatting, minimizing risk in exploration is a must. Reservoir geophysics will play an important role in helping oil producers to enhance their methods of recovery by better describing the reservoir properties using borehole seismic measurements.

The value of seismic surveying to exploration geology is well known. Engineers, however, have been skeptical that seismic data have sufficient resolution to aid in reservoir exploitation. This may have been true at one time, but advancements such as Vertical Seismic Profiling, 3-D Seismic Surveys, Tomography, and vastly improved data processing are changing this notion. Monitoring of enhanced recovery is one important application of seismic-to-oil production.

Seismology, like any other tool we apply in the oil business, has its limitations. Other information and tools must be used in conjunction with the seismic data to increase the degree of confidence and minimize the risk involved. All the tools are area dependent; one tool may work well in one geologic setting and fail in another.

The future of the oil industry will lie not only in discovering new reserves but also in recovering the most from the present reserves. All of the geological, geophysical, and engineering information should be integrated to recover all possible barrels or "reserves" from the ground. The key is to understand each other's terminology and establish intelligent communication to close the gap between these disciplines.

Let us study a seismogram, Figure 1–1. This "seismic section" is the acoustic image of a slice of the earth. The vertical axis is scaled in milliseconds of time; the total time is usually on the order of a few seconds. If we know the formation velocities, the scale can be converted to depth. The horizontal axis is in feet, sampled at typical intervals of about 200 feet (60 m).

Conventionally, the black areas in the seismic trace represent the increase in the velocity and density of the rock and give a positive kick (i.e., swing). The higher the swing of the signal, the greater the velocity and density changes. On the contrary, soft rock such as sandstone will

FIGURE 1–1. A seismic section (courtesy of Seismograph Service Corporation)

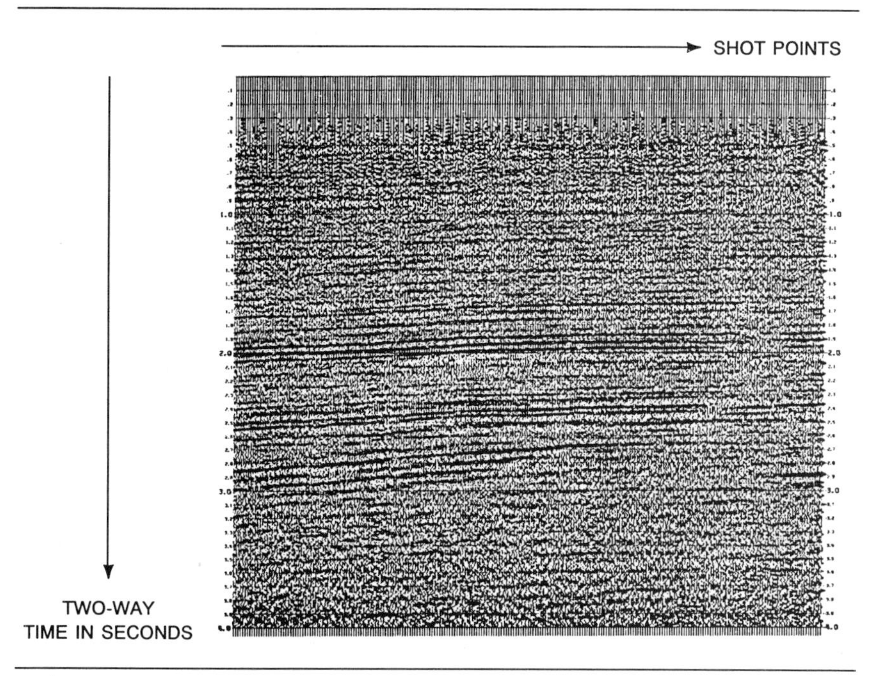

show a decrease in velocity and density. This will give a negative kick and is colored white.

The sampling rate, which may be at 1, 2, or 4 ms (or greater intervals), is usually selected not to limit the vertical resolution or the ability to separate the top and the bottom of a thin layer. We are usually able to see changes in dip, faults, anticlinal features, unconformities, some lateral changes, and other rock properties. With any image, just as in medicine, the meaning of what we see is often a little ambiguous. An interpreter uses his knowledge and experience to put meaning to these images.

To learn the meaning of this seismogram and to evaluate the utility of these images, let us next review the basic physics of wave propagation.

KEY WORDS

Abnormally high pressure zones
Amplitude
Anomaly
Blowout
Casing
Cores
Down time
Exploitation

Extrapolation
Fault
Frequency
Formation
Heterogenous
Homogenous
Hydrocarbons
Interpolation
Inversion
Limestone
Logs
Mud weight
Permeability
Polarity
Porosity
Recovery factor
Reservoir
Resolution
Saturation
Seismic marker
Seismic velocity
Shotpoint
Stratigraphic column
Tomograph
Unconformity
VSP
Well prognosis
Wildcat (well)

CHAPTER 2

OVERVIEW OF GEOPHYSICAL TECHNIQUES

INTRODUCTION

A broad division of geophysical surveying methods can be used either on land or offshore. Each of these methods measures a parameter that relates to a physical property of the rocks. Table 2–1 lists the different methods, the parameters they measure, and the related rock properties.

TABLE 2–1. Geophysical surveying methods

METHOD	MEASURED PARAMETER	PHYSICAL PROPERTY DERIVED
SEISMIC	TRAVEL TIME OF REFLECTED/ REFRACTED SEISMIC WAVES	DENSITY AND ELASTIC MODULI, WHICH DETERMINE THE PROPAGATION VELOCITY OF SEISMIC DATA
GRAVITY	SPATIAL VARIATIONS IN THE STRENGTH OF THE EARTH'S GRAVITATIONAL FIELD	DENSITY
MAGNETIC	SPATIAL VARIATIONS IN THE STRENGH OF THE GEOMAGNETIC FIELD	MAGNETIC SUSCEPTIBILITY AND RESONANCE
ELECTRICAL RESISTIVITY	EARTH RESISTANCE	ELECTRICAL CONDUCTIVITY
INDUCED POLARIZATION	FREQUENCY-DEPENDENT GROUND RESISTANCE	ELECTRICAL CAPACITANCE
SELF-POTENTIAL	ELECTRICAL POTENTIAL	ELECTRICAL CONDUCTIVITY
ELECTRO-MAGNETIC	RESPONSE TO ELECTROMAGNETIC RADIATION	ELECTRICAL CONDUCTIVITY AND INDUCTANCE

One can see from the table that the physical property to which a particular method responds determines the applicability of that method. For example, the magnetic method is very suitable for locating buried magnetic ore bodies. Similarly, seismic or electrical methods are suitable for the location of the subsurface water table, since saturated rock may be distinguished from dry rock by its higher seismic velocity and higher electrical conductivity.

Geophysical methods are often used in combination. For example, the initial search for hydrocarbons in the continental shelf area often includes simultaneous gravity, magnetic, and seismic reconnaissance surveys.

THE SEISMIC METHOD

At the present time, the most elaborate, expensive, and effective exploration tool is the seismic method.

As previously discussed, the gravity and magnetic methods are used as reconnaissance surveys to delineate an area of interest, whereas the seismic method is a tool to determine local subsurface geology in detail. Seismic surveys are usually done by either the refraction method or the reflection method.

The refraction method has the advantage of supplying data that enables the interpreter to identify a rock unit, provided the acoustic velocity is known. Refraction shooting has recently become important as a method of detailing rock structure of deep, high-velocity carbonate and evaporite sediments, where reflection shooting fails to deliver data of adequate quality. This method will be reviewed.

The reflection method has been the most successful seismic method for identifying formation tops, and it is the most widely used. This method will be discussed in some detail in the following sections.

SEISMIC RECORD

Before discussing basic seismic principles, let us first become acquainted with the *seismic trace, seismic record,* and *seismic section.*

A seismic trace, or “wiggle trace” in Figure 2–1, is the response of a single seismic detector to the earth’s movement due to seismic energy. Each part of the wiggle trace has some meaning; it’s either reflected or refracted energy from a layer of rock in the subsurface or some kind of noise. Excursions of the trace from the central line appear as peaks and troughs; the peaks represent “positive” signal voltages, and the troughs represent “negative” signal voltages.

A seismic record, or common-shot record, is a side-by-side display of all the wiggle traces that were recorded simultaneously from a number of detectors for a single shot point. The “peaks” are toward the right

FIGURE 2–1. A seismic record

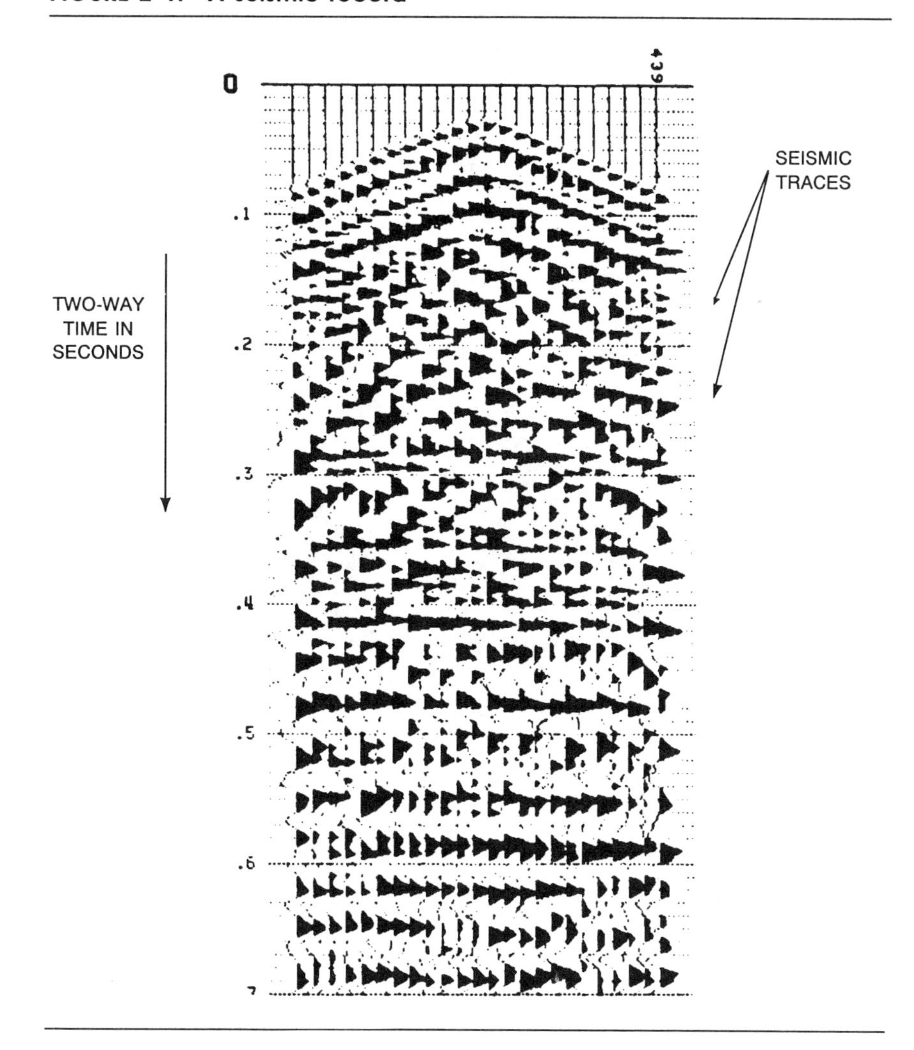

side of the display and are filled in with black ink to make patterns more visible. Zero time is at the top of the record, with time increasing downward (see Figure 2–1). This display is a raw image of the subsurface over a limited area, and it contains noise and other signal distortions.

A seismic survey generates a large number of shot records to cover the area under study. Many steps of processing are applied to the data to enhance the signal, to minimize noise, and to increase resolution. All the traces corresponding to *a particular depth point* in the subsurface are combined into a single trace, called a *common-depth-point stack.*

FIGURE 2–2. A seismic section (Reprinted from O. Yilmaz, *Seismic Data Processing*, 1987, courtesy of the Society of Exploration Geophysicists)

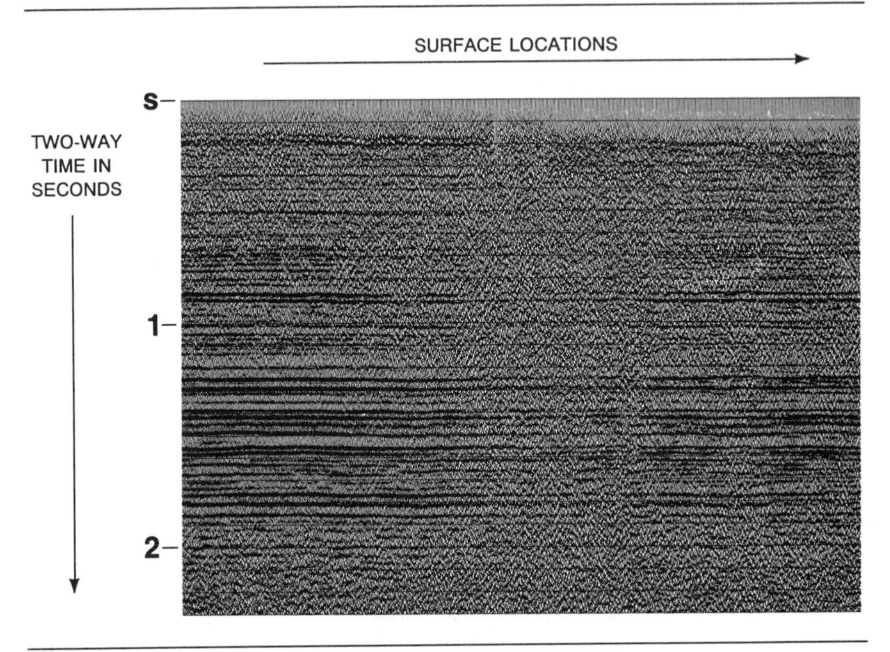

SEISMIC SECTION

When processing is complete, all the common-depth-point stacks are displayed side by side to make a seismic section, which is the final output of the seismic survey. The section is an image of the subsurface, which can be used to plan drilling and development programs. The section in Figure 2–2 shows many rock beds and a potentially hydrocarbon-bearing structure.

SUMMARY AND DISCUSSION

Chapter 2 is a quick review of the geophysical methods used in hydrocarbon exploration and development.

Gravity and magnetic methods may be used for reconnaissance surveys to delineate focused areas of interest. They should be conducted before (or in conjunction with) the seismic method.

With the increasing demand for high-resolution seismic data, and the massive volume of information acquired by 3-D seismic surveys, some recording systems can gather as much as 1,024 channels. Manufacturers are working on a recording system that can record up to 19,000 data channels.

The seismic industry is continuously developing. The need to find more hydrocarbons to replace reserves, and the subtle nature of the reservoirs to be discovered, demand more accurate information so that the fine details of a reservoir can be studied.

Key Words

Carbonate
CDP
CMP
Common shot point
Electrical conductivity
Electrical method
Evaporite
Gravity method
Magnetic method
Noise
Reconnaissance survey
Reflection method
Refraction method
Seismic detector
Seismic record
Seismic trace
Seismic section
Signal
Stacking

CHAPTER 3

BASIC SEISMIC PRINCIPLES

SEISMIC WAVE PROPOGATION

Following a seismic pulse through the earth is a very difficult and perhaps impossible task because the problem is so complex. Instead, we follow pulses through greatly simplified earth models. Even when using simple models, we face problems that can make the correlation task a difficult one. Therefore, field techniques and processing approaches are developed on the bases of these simple models. To learn how to follow a seismic pulse, let us establish some guiding principles and discuss some simplified models of the earth.

The principle of sound propagation, even though it can be a very complex phenomenon, is something with which we are fairly familiar. At some time, everyone has dropped a pebble in still water. When the stone hits the water's surface, we observe ripples propagating away from the center in definite circular patterns, which get progressively larger in diameter. If we look closely, we see that the water particles do not physically travel away from the shock center, but only displace vertically the adjacent particles and return to their original positions. This continuous and progressive displacement of adjacent water particles is the means by which the shock is propagated. We can visualize the same process in the vertical plane; thus, the wave propagation is a three-dimensional phenomenon.

The principle of seismic wave propagation is essentially the same as the process just discussed. As we induce seismic energy into the ground by means of an explosive source or a vibrating source, energy starts to propagate in ever-expanding spherical shells through the earth media. If at any time we could take a snapshot of the motion of the traveling waves, we could observe the waves moving away from the center. The leading edge of the energy is termed the *wave front.* Seismic waves propagate in three dimensions by means of these wave fronts.

WAVE FRONTS AND RAYS

If we begin at the source and connect equivalent points on successive wave fronts by perpendicular lines, we have the directional description of the wave propagation. The connecting lines form a *ray*, which is a simple representation of a three-dimensional phenomenon. Remember, when we use a ray diagram we are referring to the wave propagation in that particular direction; that is, the wave fronts are perpendicular to the ray at all points (see Figure 3–1).

WAVE THEORY

Wave theory explains events, travel time, form and size. As it applies to seismic energy in the earth, wave theory is a very complicated subject. It is applied to simplified models so that we can understand the phenomena.

The simplest model represents the earth as a homogeneous, infinite, elastic solid, which is made up of infinitesimal cubes. Forces and deformations are studied for each cube, and their transfer from cube to cube is governed by the wave equation.

FIGURE 3–1. Wave fronts and rays

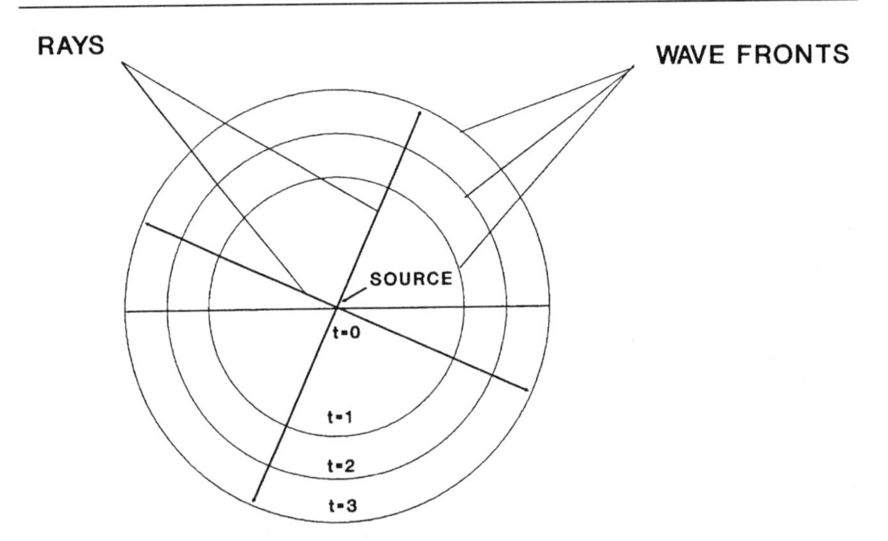

WAVEFRONTS ARE THE EXPANDING SPHERES OF ENERGY EMANATING FROM THE SOURCE.
RAYS ARE LINES THAT REPRESENT THE DIRECTION OF PROPAGATION OF THE WAVEFRONTS, AND ARE PERPENDICULAR TO THE WAVEFRONTS.

Many types of wave motion are admitted by the wave equation. The primary reflection events that we would like to detect are *compressional waves* (P-waves) in which particle motion in the medium is in the direction of wave propagation. *Shear waves*, which are also important for the information they contain, propagate through particle motion transverse to the direction of travel, much like the ripples on the surface of water. Figure 3–2 illustrates the characteristics of compressional and shear waves.

At boundaries between media of different characteristics, a more complex type of propagation occurs. One type of behavior is the *surface wave*, which largely confines itself to the boundary region. One kind of surface wave, the *Rayleigh wave*, exhibits a retrograde elliptical particle motion. It can be recorded on the seismic section, and it is known as *ground roll* (see Figure 3–3).

FIGURE 3–2. Compressional and shear waves

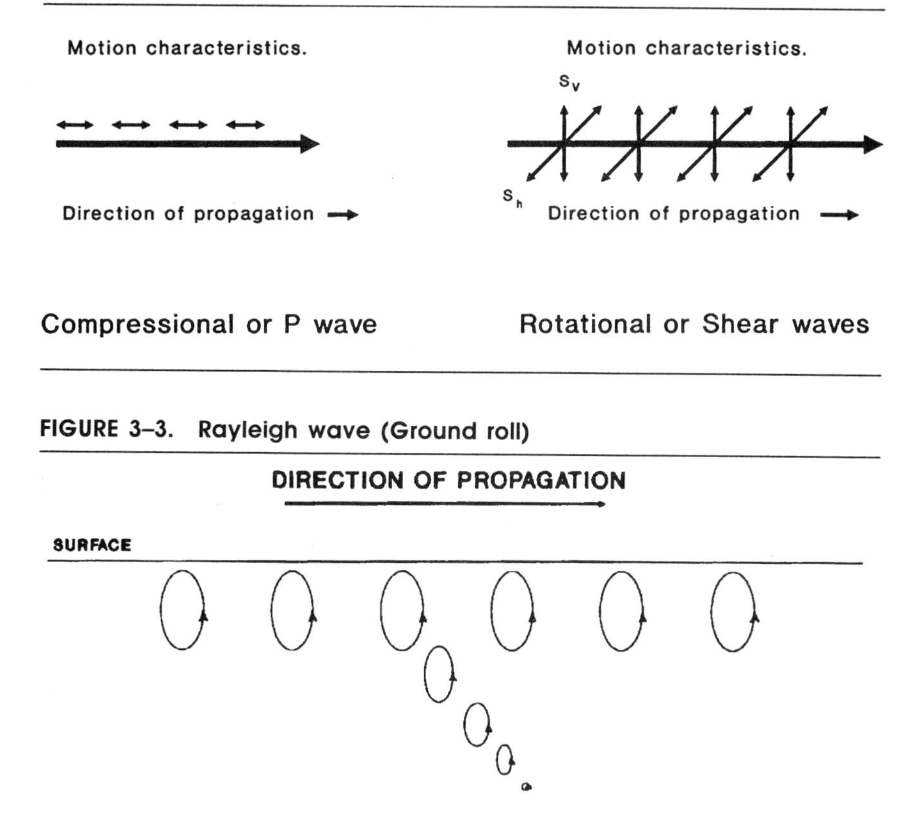

FIGURE 3–3. Rayleigh wave (Ground roll)

PARTICLE MOTION IS ELLIPTICAL AND RETROGRADE, THAT IS, THE MOTION AT THE TOP OF THE ELLIPSE IS TOWARD THE SOURCE. THE MAGNITUDE OF THE MOTION DECREASES WITH DEPTH.

REFLECTION AND REFRACTION

There are basic principles that govern the propagation of wave fronts through a complex medium such as layered earth. When an incident compressional wave strikes a boundary between two media having different velocities of wave propagation, part of the energy is reflected from the boundary and the remainder is transmitted into the next layer. Although the energy of the incident wave is divided into various components, the sum is equal to the incident energy.

FERMAT'S PRINCIPLE

A seismic pulse traveling in a medium follows some physically connected path from the source to a particular receiver. Fermat's principle admits the possibility of multiple travel paths, or more than one primary reflection event. The classical *buried focus* ("bow tie") *effect* illustrates Fermat's principle (see Figure 3–4).

FIGURE 3–4. Fermat's principle and buried focus

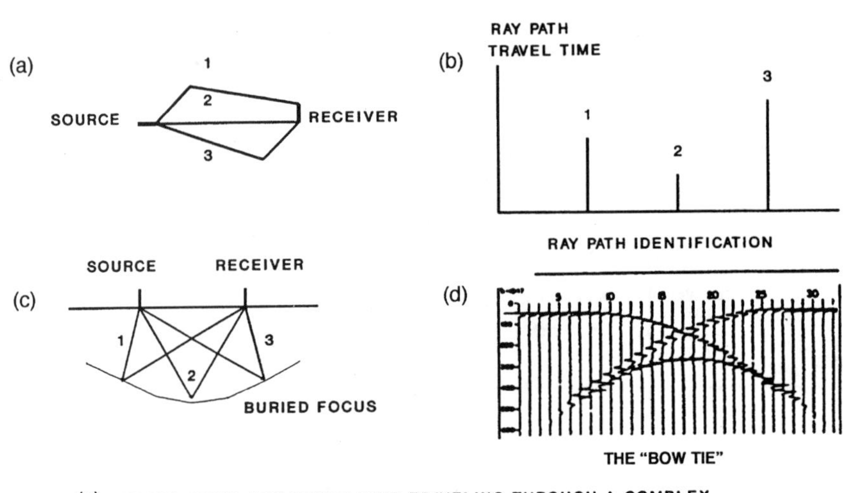

(a) TRAVEL PATHS FOR THREE RAYS TRAVELING THROUGH A COMPLEX INHOMOGENEOUS MEDIUM SUCH AS THIS SYNCLINAL REFLECTOR.
(b) TRAVEL TIME FOR EACH RAY PATH.
(c) WITH ONE GEOPHONE BEING IN POSITION TO RECEIVE INFORMATION FROM TWO OR EVEN THREE PARTS FROM THE SYNCLINE, THE RAY PATHS CROSS ON THE WAY DOWN.
(d) THE RECORD SECTION SHOWS TWO CROSSING LINEUPS OF ENERGY WITH A "BOW TIE" SHAPE. THIS IS CALLED THE "BURIED FOCUS" EFFECT.

REFLECTION AND TRANSMISSION COEFFICIENTS—ZOEPPRITZ'S EQUATION

The relative portions of the energy transmitted and reflected are determined by the contrast in the *acoustic impedances* of the rocks on each side of the interface. It is difficult to precisely relate acoustic impedance to tangible rock properties but, in general, the harder the rock the higher the acoustic impedance.

The acoustic impedance of a rock is the product of its density and the velocity of a compressional seismic wave through it, ρV, designated Z.

Consider a P-ray of amplitude A_0, normally incident on an interface between two media of differing velocities and densities (see Figure 3–5a). A transmitted ray of amplitude A_2 travels on through the interface in the same direction as the incident ray, and a reflected ray of amplitude A_1 returns to the source along the path of the incident ray.

The reflection coefficient R is the ratio of the amplitude A_1 of the reflected ray to the amplitude A_0 of the incident ray. In terms of the acoustic impedances of the two media, for a normally incident ray

Zeoppritz's equation states:

$$R = (Z_2 - Z_1) / (Z_2 + Z_1)$$

where Z_1 and Z_2 are the acoustic impedances of the first and second layers, respectively.

FIGURE 3–5. Zoeppritz's equations

(a) NORMAL INCIDENCE

INCIDENT WAVE AMPLITUDE Ao

REFLECTED WAVE AMPLITUDE A1

V1, ρ1

V2, ρ2

ρ2•V2 • ρ1•V1

TRANSMITTED WAVE AMPLITUDE A2

(b) P AND S WAVES

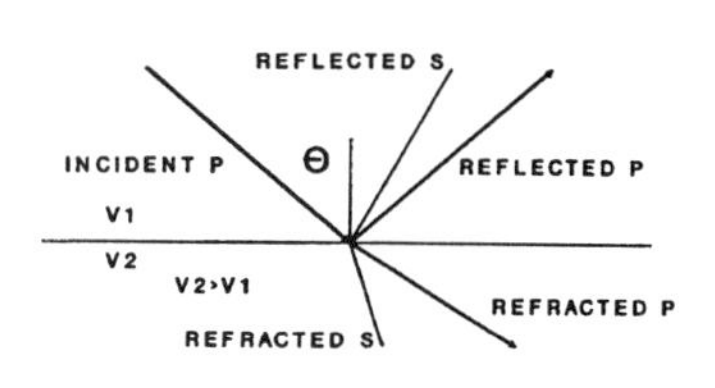

$R = A1/A0$

$R = (\rho 2 \cdot V2 - \rho 1 \cdot V1)/(\rho 2 \cdot V2 + \rho 1 \cdot V1)$

IF WE SET $Z = \rho V$, THEN

$R = (Z2-Z1)/(Z1+Z2)$

$T = A2/A0 = 1 - R^2$

$T = 4Z1Z2 / (Z2+Z1)^2$

(a) REFLECTED AND TRANSMITTED WAVES ASSOCIATED WITH A WAVE NORMALLY INCIDENT ON AN INTERFACE OF ACOUSTIC IMPEDANCE CONTRAST

(b) REFLECTED AND REFRACTED P- AND S- WAVES GENERATED BY A P-WAVE OBLIQUELY INCIDENT ON AN INTERFACE OF ACOUSTIC IMPEDENCE CONTRAST

Typical values of reflection coefficients for near-surface reflectors and some good subsurface reflectors are as follows:

NEAR-SURFACE REFLECTORS:

Soft ocean bottom (sand/shale)	.33
Hard ocean bottom	.67
Base of weathered layer	.63

GOOD SUBSURFACE REFLECTORS:

Sand/shale versus limestone at 4,000 ft.	.21
Shale versus basement at 12,000 ft.	.29
Gas sand versus shale at 4,000 ft.	.23
Gas sand versus shale at 12,000 ft.	.125

You can see that a soft, muddy ocean bottom reflects only about one-third of the incident energy, while a hard bottom reflects about two-thirds of the energy.

The transmission coefficient is the ratio of the amplitude transmitted to the incident amplitude:

$$T = A_2/A_0$$

When a P-ray strikes an interface at an angle, both reflected and transmitted P-rays are generated as in the case of normal incidence. However, some of the incident compressional energy is converted to reflected and transmitted shear rays (see Figure 3–5b), which are polarized in the vertical plane. Zoeppritz's equations give the amplitudes of the four components as a function of the angle of incidence. The converted rays contain information that can help in identifying fractured zones in the reservoir rocks. In this text, however, we shall discuss compressional waves only.

SNELL'S LAW

Snell's Law, originally applied to light and optics, applies equally well to seismic waves and the earth. For a reflected ray, Snell's Law states that the angle θ between the reflected ray and the normal to the reflecting surface is equal to the angle between the incident ray and the normal to reflecting surface. In seismology, of course, the reflecting surface is the boundary between two layers having different acoustic impedances.

The portion of the incident energy that is not reflected is transmitted through the boundary and into the second layer. The transmitted ray travels through the second layer with changed direction of propagation, and it is referred to as a *refracted ray*.

Snell's law of refraction states that the ratio of sine of the angle θ to the velocity is a constant. For a refracted P-ray:

$$(\sin\theta_1)/V_1 = (\sin\theta_2)/V_2$$

then

$$\sin\theta_1/\sin\theta_2 = V_1/V_2$$

where the subscripts refer to layer 1 and layer 2, respectively. These relationships are illustrated in Figure 3–6a.

CRITICAL ANGLE AND HEAD WAVES

When the velocity is higher in the underlying layer, there is a particular angle of incidence, known as the *critical angle*, θ_c, for which the angle of refraction is 90°.

This gives rise to a critically refracted ray that travels along the interface at the higher velocity V_2:

$$(\sin\theta_c)/V_1 = (\sin 90°)/V_2 = 1/V_2$$

$$\theta_c = \sin^{-1}(V_1/V_2)$$

This wave, known as a *head wave*, passes up obliquely through the upper layer toward the surface, as shown in Figure 3–6b.

FIGURE 3–6. Snell's law and critical angle

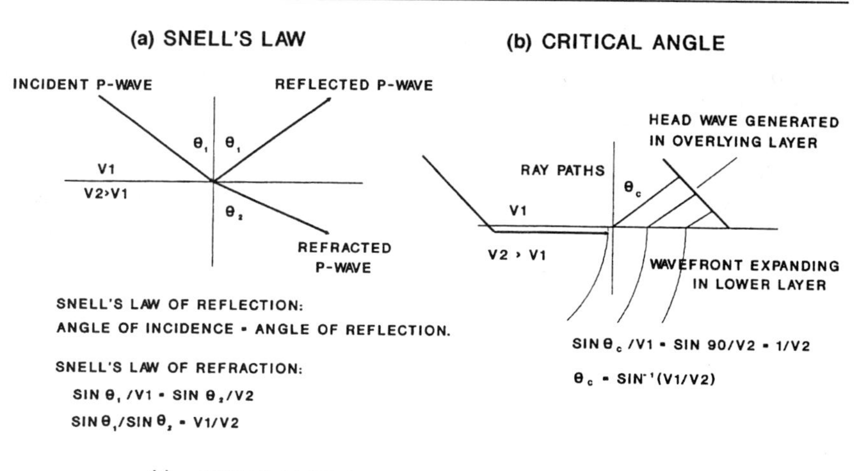

(a) PART OF AN OBLIQUELY INCIDENT RAY IS REFLECTED AT THE ANGLE OF INCIDENCE, AND PART IS TRANSMITTED AT AN ANGLE THAT DEPENDS ON THE RATIO OF THE VELOCITIES IN THE TWO LAYERS.

(b) A HEAD WAVE IS GENERATED IN THE UPPER LAYER BY A WAVE PROPAGATING THROUGH THE LOWER LAYER ALONG THE BOUNDARY.

FIGURE 3–7. Huygens' principle

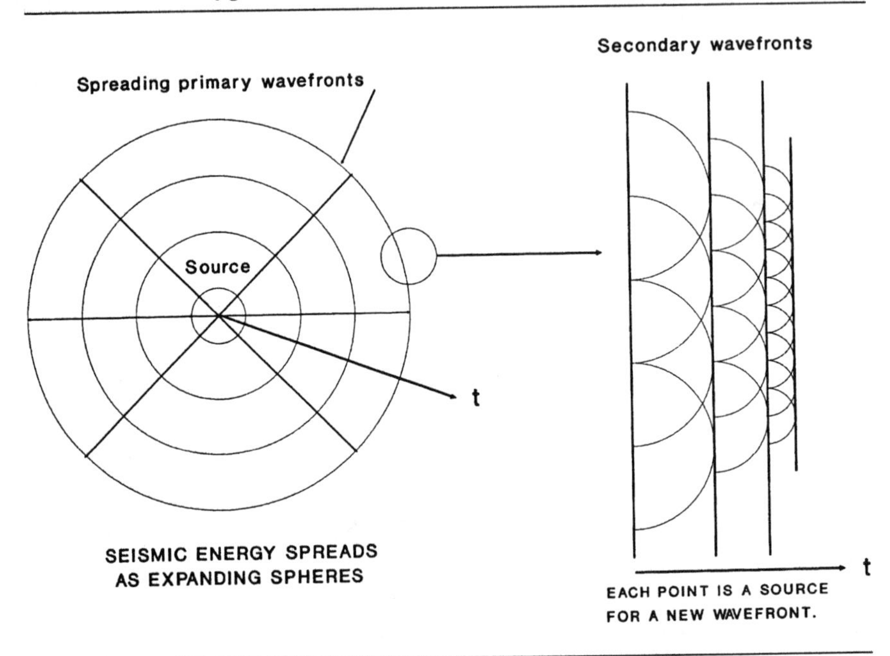

HUYGENS' PRINCIPLE

This principle states that every point on the primary wave front surface is a source of secondary wavelets. The position of the wave front at a later instant then is found by constructing a surface tangent to all secondary wavelets. This concept is a very powerful tool for understanding all types of wave propagation, from electromagnetic waves to seismic waves.

Huygens' Principle, illustrated in Figure 3–7, regards each point on the advancing subsurface wave as a source that generates new wave fronts, which radiate in all directions. It explains one of the most important mechanisms by which a propagating seismic pulse loses energy with depth.

PROPAGATION MODEL FOR EXPLORATION SEISMOLOGY

Exploration seismology was developed for exploring the sedimentary basins which have gentle dip, horizontal continuity over a large area, and layered structure. It follows, then, that propagating seismic pulses in simple models that embody these essentials are of great value in understanding and interpretating practical seismology.

The model that we shall consider assumes that the seismic energy propagates along paths to a multiplicity of receivers, using a multiplicity of source positions. From the following propagation models, it will become clear that redundancy will allow estimation of the velocity information. Since velocity is one of the most important parameters of our geological model, propagation characteristics that allow velocity estimates are considered in our tracking of the seismic pulse.

The *dipping reflector* is one of the most important models used in the development of present-day exploration approaches, whether applied in the field or in the data processing center. (See Figure 3–8, right.)

The appearance of the dipping reflector in time sections depends upon which traces have been assembled to make the time section. A *common source gather*, for example, is a collection of traces recorded for a single source discharge, where the trace arrangement is ordered by the distance from the source to the receiver. The reflection appears as a hyperbola with its apex offset from the source position.

When the reflector is horizontal (see Figure 3–8, left), the apex of the hyperbola is at $X = 0$ and is symmetrical about this offset.

FIGURE 3–8. Dipping and horizontal reflector models

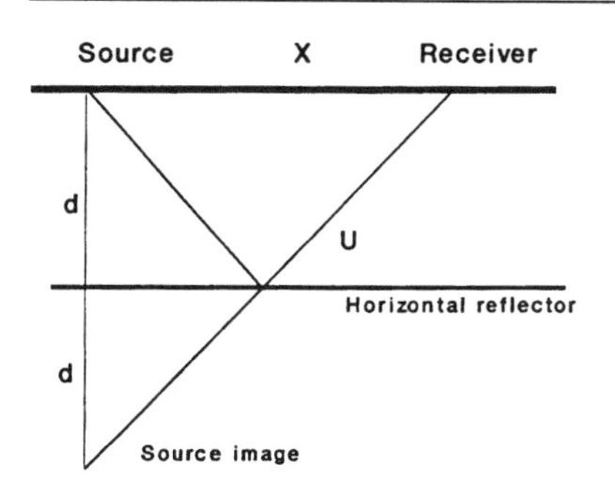

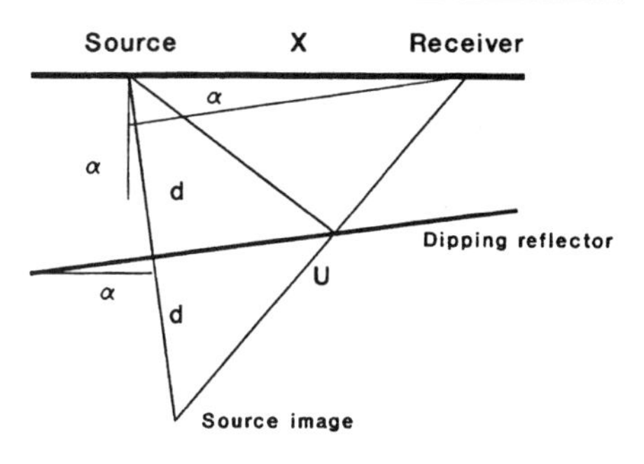

For a horizontal reflector the relation is simplified to:

$$T_x^2 = T_0^2 + (X/V)^2$$

This is the equation of a hyperbola that is symmetrical about $X = 0$

In the case of a dipping reflector in a homogeneous medium at depth d from the source and velocity V, the relation between travel time from source to receiver, vertical travel time at $X = 0$, and dip angle α is expressed by:

$$T_x^2 = T_0^2 - 2d\sin\alpha/V + (X/V)^2$$

This is the equation of a hyperbola, shifted horizontally by the middle term.

SUMMARY AND DISCUSSION

Seismic wave propagation is a three-dimensional phenomenon; it is a difficult task to follow a seismic pulse in the earth. Instead, we follow pulses through greatly simplified earth models.

As the seismic energy induced in the ground using an explosive source such as dynamite or a vibrating source such as the Vibroseis®* system, energy starts to propagate in expanding spheres through the earth. A wavefront is the surface of equal travel time of the seismic wave.

The ray path is perpendicular to the wavefront; thus, remember that when we use a ray diagram in that particular direction that the wavefronts are perpendicular to the ray at all points.

A very important theory was introduced "Fermat's principle," which admits the possibility of multiple travel paths, or more than one primary reflection event. The Seismic method applies many theories from optics such as Snell's law of reflection and refraction, which is the fundamental theory of how the seismic energy travels in the earth.

KEY WORDS

Acoustic impedance
Compressional wave
Critical angle
Density
Fermat's principle
Ground roll
Huygens' principle
Ray
Shear wave
Snell's law
Wave equation
Wave front

EXERCISES

1. Name and describe the three types of seismic waves.
2. Define the following terms:
   a. Acoustic impedance
   b. Snell's law of refraction
   c. Critical angle
3. Two layers are separated by an interface. The upper layer has a velocity of 3.3 km/sec and thickness of 0.5 km. The lower layer has a velocity of 6.6 km/sec. If a ray travels downward through the top layer at an angle of incidence of 20°, then at what angle will the ray travel in the lower layer? What is the critical angle? What is the traveltime of the head wave for an offset of 20 km?

*Vibroseis® is registered trade mark of Conoco

4. With $\theta_1 = 30°$, use Snell's law to compute θ_2 for the following conditions:
   a. V_1 = 1,600 m/sec, V_2 = 2,100 m/sec
   b. V_1 = 2,000 m/sec, V_2 = 1,200 m/sec
   c. V_1 = 3,000 m/sec, V_2 = 4,000 m/sec
5. Most of the energy in a seismic wavelet is contained in a band of frequencies centered about the dominant frequency. The dominant period can be defined as the time between two major crests. The dominant frequency is the reciprocal of the dominant period. The equation for wavelength, λ, is:

$$\lambda = \text{velocity/frequency.}$$

   Calculate wavelengths for the following cases:
   a. Shallow rocks: 2,000 m/sec, 50 Hz
   b. Deep rocks: 6,000 m/sec, 25 Hz

BIBLIOGRAPHY

Bath, M. *Introduction to Seismology.* Basel-Stuttgart: Birkhauser Verlag, 1973.

Birch, F. "Compressibility, Elastic Constants." S. P. Clark, ed., *Geological Society of America Memoir* 97 (1966): 97–173.

Dix, C. H. "Seismic Velocities from Surface Measurements." *Geophysics* 20 (1955): 68–86.

Ewing, M., W. Jardetzky, and F. Press. *Elastic Waves in Layered Media.* New York: McGraw-Hill, 1957.

Faust, L. Y. "Seismic Velocity as a Function of Depth and Geological Time." *Geophysics* 16 (1951): 192–196.

Gardner, G. H. F., L. W. Gardner, and A. R. Gregory. "Formation Velocity and Density—the Diagnostic Basis for Stratigraphic Traps." *Geophysics* 39 (1974): 770–780.

Koefoed, O. "Reflection and Transmission Coefficients for Plane Longitudinal Incident Waves." *Geophysics Prospect* 10 (1962): 304–351.

Muskat, M., and M. W. Meres. "Reflection and Transmission Coefficients for Plane Waves in Elastic Media." *Geophysics* 5 (1940): 115–148.

Sharma, P. V. *Geophysical Methods in Geology.* Amsterdam: Elsevier, 1976.

Sheriff, R. E. "Addendum to Glossary of Terms used in Geophysical Exploration." *Geophysics* 34 (1969): 255–270.

Telford, W. M., L. P. Geldart, R. E. Sheriff, and D. A. Keys. *Applied Geophysics.* Cambridge: Cambridge Univ. Press, 1976.

Trorey, A. W. "A Simple Theory for Seismic Diffractions." *Geophysics* 35 (1970): 762–764.

CHAPTER 4

ACQUISITION OF SEISMIC DATA

FIELD SEISMIC DATA ACQUISITION

Acquisition of seismic data is often left to the experience of the contractors, without regard to its ultimate role in producing a geologically sound section. Inappropriate or poorly designed parameters can severely limit the quality and utility of the seismic data. Properly designed parameters, based on knowledge of the area and the exploration target, normally lead to greatly enhanced and interpretable seismic sections. The material that follows is a review of the techniques, procedures and equipment from which seismic surveys are designed.

DESIGN OF FIELD PARAMETERS

The well-designed seismic survey begins with a clear knowledge of the survey objectives in general terms. Several factors merit consideration in the design of ultimate field parameters, including economics, time of the survey, type of energy source, type of geophones, and their patterns.

TRANSLATION OF EXPLORATION OBJECTIVES INTO FIELD PARAMETERS

The choice of an appropriate parameter set depends upon the general and specific survey objectives, as well as on the operational and environmental circumstances. One environmental condition that exerts an extraordinary degree of control over the nature of the survey is where the survey is to be conducted—on land or offshore. Other elements that affect the nature of the survey are inhospitable terrain, excessive cultural or natural background noise of acoustic or electrical nature, and the presence of man-made structures.

Some parameters of a seismic acquisition program are:

- Maximum offset: distance from the source to most remote receiver.
- Minimum offset: distance from the source to nearest receiver.
- Group interval: distance between geophone arrays. Constant for a survey.
- Shot interval: distance between two shots.
- Fold coverage: number of times a subsurface point is surveyed by different sources and detectors.
- Sample interval: the time interval between digital samples of the signal. Varies from less than 1 ms to 4 ms. This sample rate is chosen not to limit the vertical resolution and to record the desired maximum frequencies.
- Choice of source and geophone arrays.
- Number of recording channels.

ACQUISITION PARAMETERS AND NOISE

Any signal other than the reflections is considered noise. In order to obtain an interpretable, geologically sound seismic section, we need to enhance the signal and/or suppress the noise. Depending on the economic factors, some noise patterns can be attenuated or even excluded by designing an optimum group interval or by selecting an appropriate geophone pattern. It can also be accomplished by cutting the low-frequency end of the signals, provided that the noise does not have the same frequency as the desired data.

Figure 4–1 is a schematic diagram comparing the desired information and recorded information.

Figure 4–2 is a schematic seismic section showing the major signal and noise features. It has been corrected for *normal moveout* (NMO), which is the time correction applied to each trace to account for its offset. The first arrivals at the top of the record are labeled P-waves; these are typically refracted waves from near-surface formations. These are followed by two coherent noise patterns. The first one, characterized by low frequency and a velocity that varies from 3,500 to 5,500 ft/sec, is the *surface wave* or *ground roll.* The second one is the *air wave,* which is energy traveling from the source to the detectors through the air having a high-frequency component and low velocity of 1,100 ft/sec. The third type of noise is the *multiple,* which is a repeat reflection from the same interface. It may be either a simple multiple from one of the shallow reflectors, or an intrabed multiple bounced between two reflectors and back to the surface, or one of several other types in which the reflected ray reverses direction at some point.

FIGURE 4–1. Desired and recorded seismic data

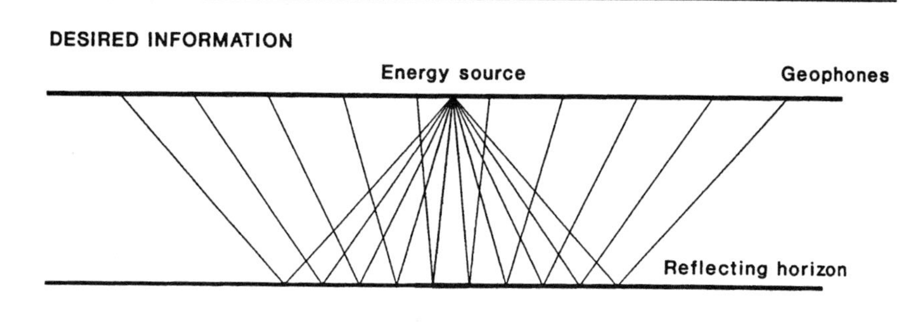

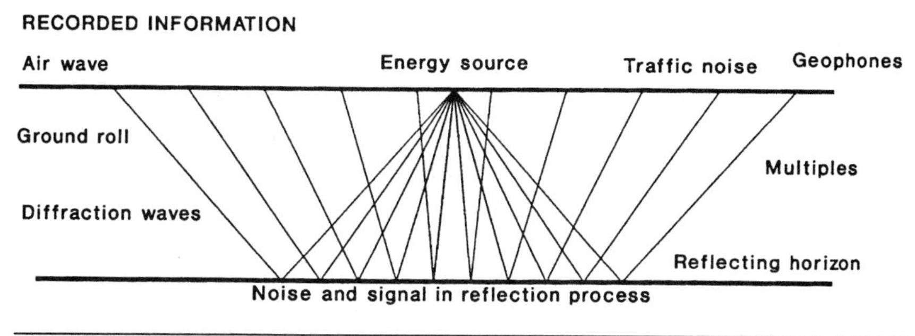

FIGURE 4–2. Schematic seismic record—normal moveout applied

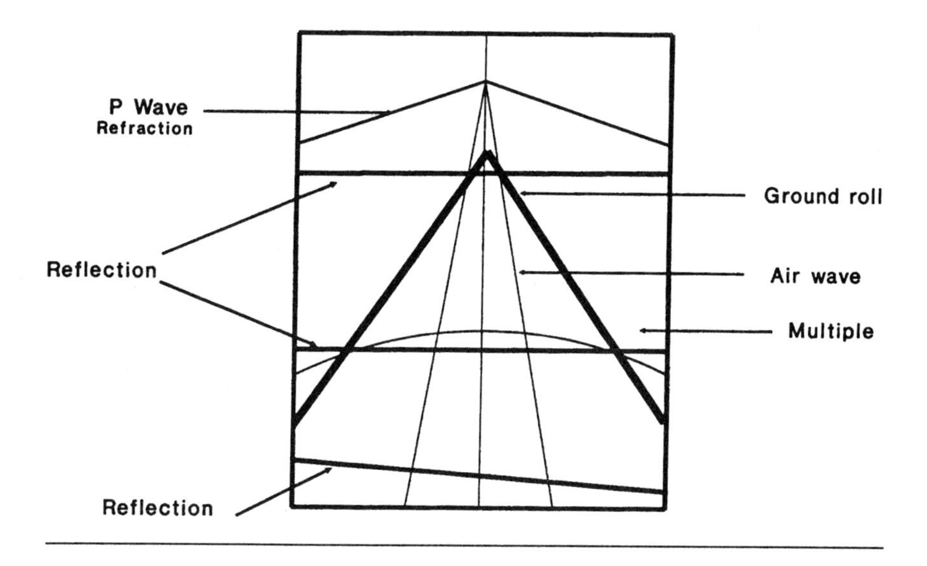

Keep in mind that data processing techniques cannot add any frequencies that were not recorded nor enhance information outside the bandwidth of the seismic data acquired in the field.

FIELD NOISE TEST

In every new area, it is advisable to conduct a noise test to analyze and study the predominant noise patterns in the area, and then design the field parameters to enhance the signal and attenuate noise.

Figure 4–3 illustrates a noise test showing the types of noise patterns. Two refraction patterns were identified at velocities of 9,700 ft/sec and 15,300 ft/sec. This was followed by a ground roll of a velocity of 5,400 ft/sec and wave length varying from 108 to 173 feet. An air wave, characterized by higher frequency and a velocity of 1,120 ft/sec, is traced from 100 ms in time down the record. Notice at the right-hand of the section the hyperbolic pattern from the reflection with the velocity of 5,200 ft/sec.

FIGURE 4–3. Noise spread (modified after Brasel, 1973)

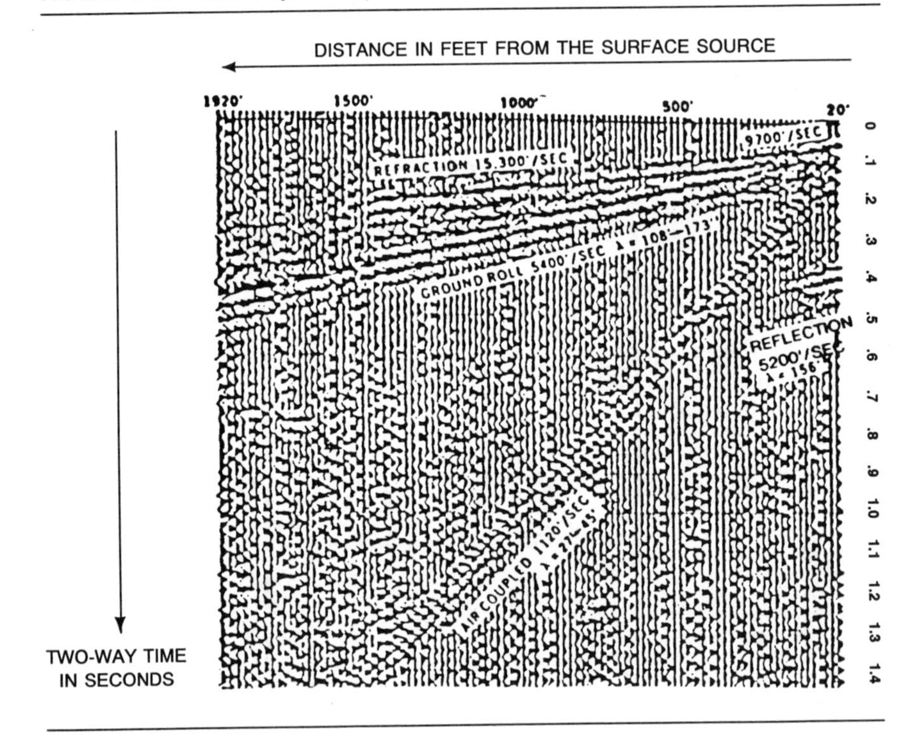

ELEMENTS OF A DATA ACQUISITION SYSTEM

The elements of a seismic acquisition system are:

- Sources and arrays.
- Detectors and arrays.
- Instrumentation.
- Field geometry or survey configuration.
- Surveying, positioning, and navigation.

Seismic Sources

There are a number of types of seismic sources used in exploration. They all fall within one of two principal categories: (1) Impulsive, such as explosives and (2) Distributed or diffuse in nature, such as vibrators.

Vibrating sources may cause less environmental impact. However, because of their lower energy, compensating treatment such as using multiple vibrators in source arrays is required. An alternative is to revibrate at the same location several times and then vertically sum the sweeps; that is, the data recorded are added together to form a composite record. Vertical summing enhances record quality by canceling random noise.

1. Operating factors

   Generating seismic energy at frequencies between 5–80 Hz or higher can be accomplished in a variety of ways. The various approaches differ in the following ways:
   - operating cost,
   - energy output,
   - operational speed, and
   - wavelet produced.

   All the previous factors will come to mind when we need to study a certain variable in the data, such as the use of seismic to make a stratigraphic interpretation. Frequency bandwidth plays the key role in selecting the energy source.
2. Multiple-Source Arrays

   Source arrays can be developed that can be discharged in concert or with appropriate delays to achieve particular objectives, such as beam forming.

   In marine surveys, arrays are used to increase energy output and to improve waveform characteristics.

   For on-land operations (in addition to increasing energy output and improving waveform characteristics), arrays can be designed to at-

FIGURE 4–4. Source arrays

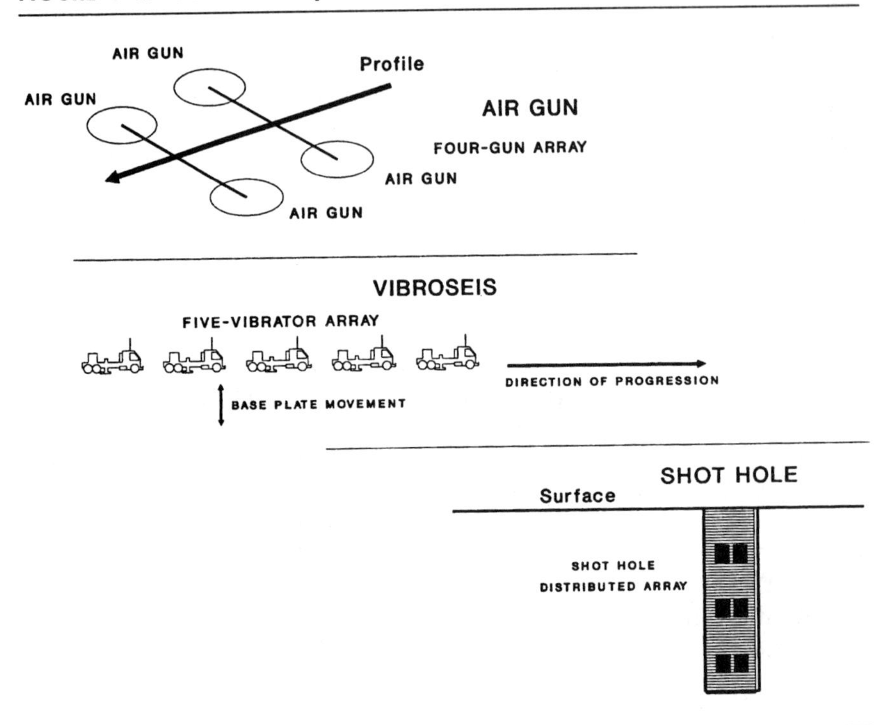

tenuate dispersive near-surface waves, such as ground roll. Vertical arrays in shot holes may be used to attenuate ghost reflections. (See Figure 4–4.)

SEISMIC DETECTORS

A variety of receiver types are available for detecting seismic waves. On land, we may use geophones, which respond to either vertical displacement or rotational motion. Vertical-displacement geophones are commonly used in land seismic data acquisition. They measure the rate of change of displacement, or velocity. In fact, they can measure the second derivative of displacement, which is acceleration.

Displacement can be measured along three principal axes. Such information can help identify surface waves and other complex wave motions. Figure 4–5 illustrates the types of receiver elements.

In marine surveys, pressure-sensing hydrophones are used. Multiple receivers can be connected in arrays to enhance the signal and to reduce noise. Types of arrays used include:

FIGURE 4–5. The receiver element

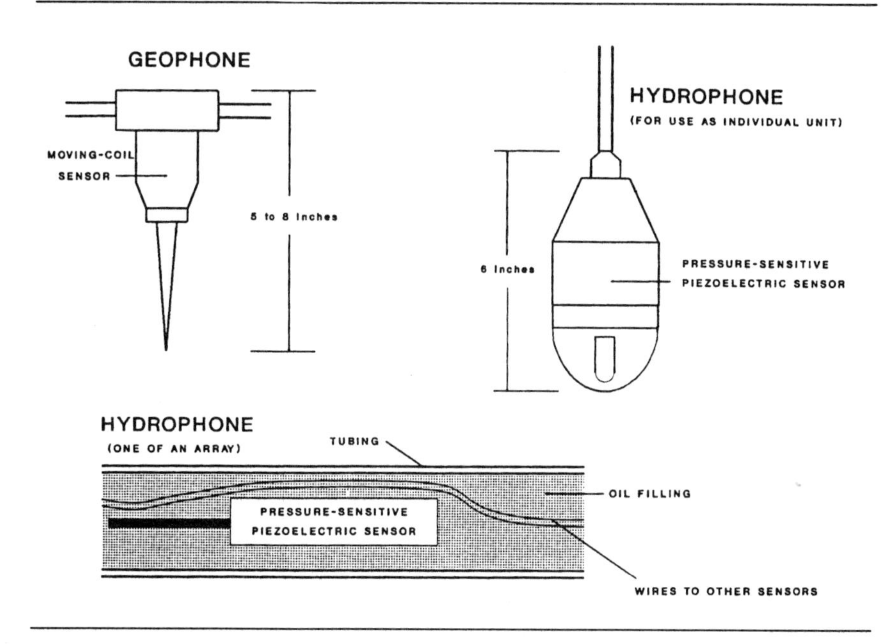

1. Linear array: a line of single geophones.
2. Weighted or tapered array: a line in which the number of geophones at each position varies so that the outer elements have the smallest number of geophones (less weight) and the center element has the greatest number of geophones (most weight). The change in number of geophones from position to position is the *taper* of the array.

Figure 4–6 shows the theoretical response of a linear array and illustrates the signal-to-noise improvement as a function of wavelength. On the graph, the lower the position of the line, the greater the signal-to-noise ratio. For frequencies below 70 Hz, the noise level is at least 13 dB below the signal. Figure 4–7 shows the theoretical response of a strongly tapered, weighted array. Here the signal-to-noise ratio is at least 38 dB from 20 to 70 Hz.

In practice, the enhancement is lesser due to the geophones plants accuracy in the field.

INSTRUMENTATION

Many of the improvements in seismic data quality are due to advances in instrumentation, binary gain recording, and (its successor) instantaneous floating-point recording. These represent great advancement in the re-

FIGURE 4–6. Typical geophone array response linear array (modified after Brasel, 1973)

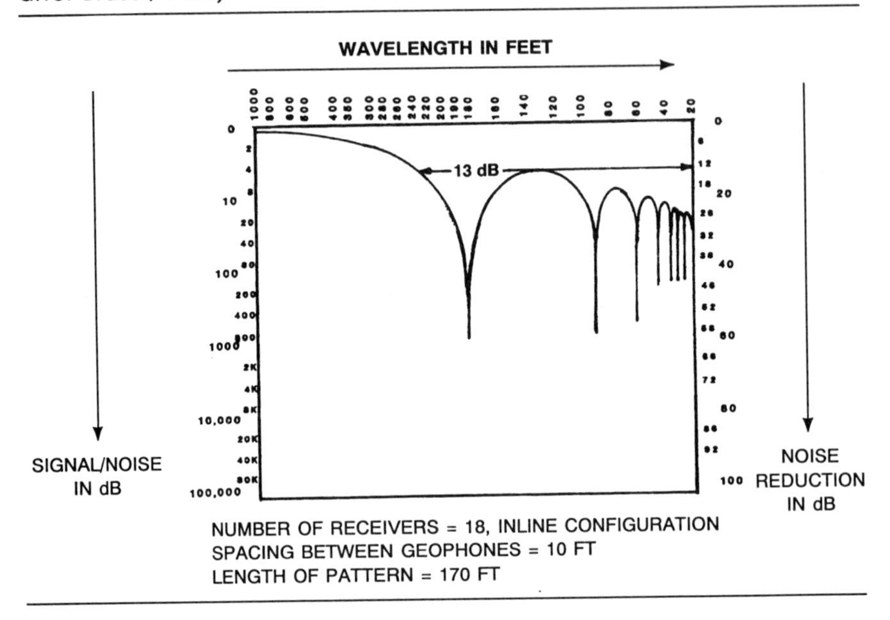

FIGURE 4–7. Geophone array response strong taper (modified after Brasel, 1973)

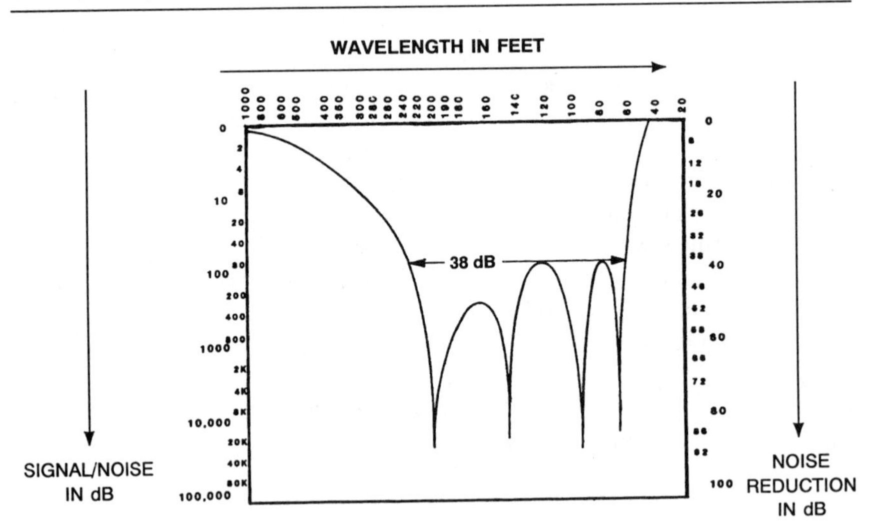

FIGURE 4–8. Instrumentation

— AMPLIFIER-FILTER-RECORDER
  Common mode rejection
  Narrow or broad band.

— ALIAS FILTER
  Alias filter frequency = 0.6 Alias frequency.
  72 dB+ of amplitude.
  Discrimination from 0.6FA – 1.0 FA

— DIGITIZER
  14 Bits + sign
  Resolution dB = 6N + 8.8

— GAIN RANGING AND CONTROL
  Ganged AGC with recording.
  Binary gain with gain coding.
  Instantaneous floating point (gain ganged by 4)

— SAMPLE-AND-HOLD APERTURE
  Time uncertainty in sample time.

cording of seismic signals without loss of information and with the ability to recover field amplitudes.

Alias filters assure us that our digital sampling is properly accomplished and that high-frequency noise components do not masquerade as contributions of lower frequency by *aliasing*. Aliasing is a property of sampling systems in which an input signal of one frequency can yield the same sample values as a signal of another frequency. The alias filter frequency is equal to 0.6 of the alias frequency; it should have at least 72 dB[1] rejection and discrimination from 0.6 to 1.0 FA.

The *digitizer* converts the analog electrical signals from the geophones into discrete samples. In order to accomplish such conversion, it is necessary to hold a short portion of the analog signal for conversion to the digital equivalent. This is the function of the *sample-and-hold* element.

Figure 4–8 illustrates typical instrumentation.

[1]The decibel (dB) expresses the *ratio* between two values or amplitudes, regardless of their magnitudes. For example, a ratio of 2:1 is 6 dB, and the ratio of 200:100 is also 6 dB. A ratio of 3:1 is 10 dB, and a ratio of 10:1 is 20 dB. The numbers of decibels for the ratio of two values A_1 and A_2 is calculated from: $dB = 20 \bullet \log_{10}(A_1/A_2)$

FIELD GEOMETRY—COMMON DEPTH POINT

A *common depth point* (CDP) gather is a group of seismic traces that represent a single point on a flat reflector. A *common midpoint* (CMP) gather is a gather of traces from a *dipping* reflector.

The traces of a common depth point (or common midpoint) gather are put together in one family. Each trace is from a different source and different receiver, but represent the same subsurface point. This reordering of the traces and *stacking* (summing them together) enhances the signal-to-noise ratio by attenuating random effects and undesired events, such as multiple reflections, whose variation with offset differs from that of the primary reflections.

Seismic data may be sorted in CMP or CDP for further analyses such as velocity analysis, static corrections, and so forth. These applications will be discussed later in the text.

Figure 4–9 illustrates the ray path geometry for six shots taken and twelve geophones in the line. The near offset is four stations from the source, and this geometry stays constant each time the source is moved one location. One can see that from shot location #1 (S_1) that we obtained twelve subsurface points, indicated by the short vertical lines labeled Tr. 1 through Tr. 12.

The source was then moved one group interval to location S_2, and the receivers were moved one station by means of the CDP switch in the

FIGURE 4–9. Common depth point field geometry

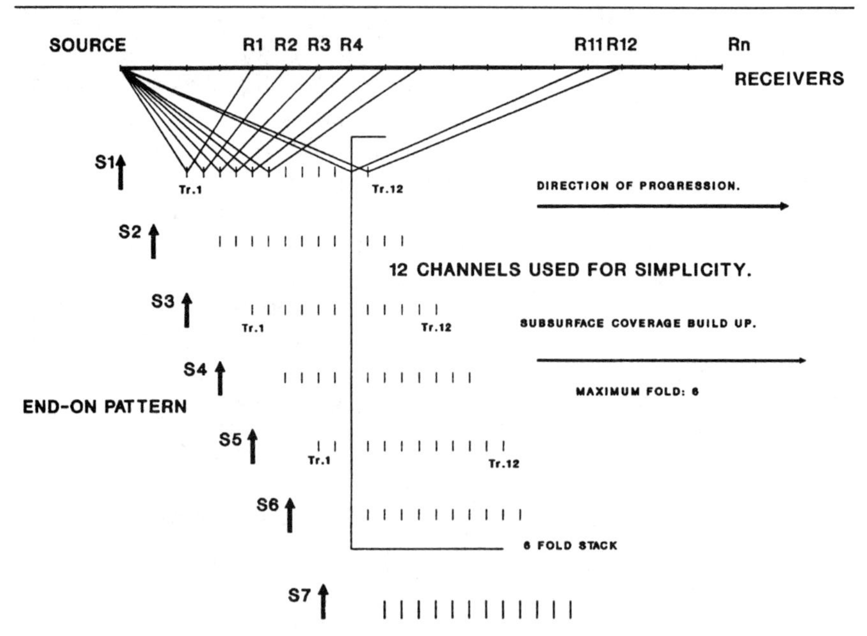

recording truck. Another coverage of twelve subsurface points was obtained.

By progressively moving the sources and the receivers, traces are obtained for a given subsurface point. These traces were obtained using different sources and receivers. The number of traces obtained for a common depth point is called the *fold*. In this case, the maximum fold is six. This will stay constant until the end of the line unless there is some surface obstruction that prevents shooting at certain locations. At surface location R_4, you can observe that the common-depth subsurface point does have six traces. This type of shooting pattern is called *end-on*. If Trace 1 is the one closest to the source, the shooting configuration is called *pushing the cable*; if Trace 1 is the farthest from the source it is called *pulling the cable*.

Another type of shooting pattern is split spread or *straddle shooting*, in which the source is in the middle, the spread is symmetrically split, and the two banks of the spread are the same length. For better imaging of dipping structures, an asymmetrical spread can be designed to get the required subsurface details.

A technique called *shooting through the cable* may be used to place depth points as close to the line termination as possible. In this method, the geophone spread is stationary and the source is moved along the line.

Sometimes it is difficult to maintain consistency in the geometry of the line because of surface obstructions. Accordingly, some of the shots will be skipped and the fold multiplicity will drop. One can build up this missing coverage by *undershooting* (placing the shots on one side of the obstruction and the geophone line on the other).

NOISE REDUCTION THROUGH MULTIPLICITY

The use of multiple receivers enhances signal-to-noise ratio. Summing, or stacking, two or more traces that correspond to a single point in the subsurface reinforces the "real" signal by canceling random noise signals. The more traces that are summed, the greater the improvement.

Multiplicity can be obtained by increasing the number of shots at the source locations or the number of geophones at the receiver location, or both. It is commonly obtained by shooting in such a way that the traces corresponding to a given point in the subsurface are from different sources and different receivers but represent the same subsurface point. This technique is called common depth-point (CDP) or in case of dipping reflectors is called common mid-point (CMP).

MARINE DATA ACQUISITION

Marine crews are sent out to gather data on a wide range of information, from simple water-bottom profiling with a fathometer to activities that include seismic, gravity, magnetic and electrical surveys.

Marine vessels cruise at sea from a week to several months. For economic reasons, they have contact with land only through radio, supply vessels, or helicopters because the cost of having a vessel idle can run into thousands of dollars per day.

Decisions concerning the field parameters to be used while shooting, (e.g., where to shoot, how many miles or kilometers, what shooting order, the type of information to gather, etc.) are made before the vessel goes out.

MARINE EQUIPMENT

Equipment carried by a seismic vessel includes a seismic recording system, which is a cable equipped with seismic transducers to be towed behind the moving vessel. It includes recording fathometer, which is a system for determining precisely the position of the vessel and the cable at any given time. It also includes an energy source to generate seismic energy such as air gun, water gun, or sparker.

MARINE CABLE

There are many types of cables and they vary in details, but a cable generally consists of one section per trace. Each section has several piezoelectric crystals that react to changes in pressure. These sections are connected together to form one long cable called a *streamer cable.* The cable is connected to the vessel by a stretch section, which is designed to absorb the shocks of cable being towed through the water.

The cable depth affects the signal-to-noise ratio, so it is essential to maintain a uniform depth. At every few sections on the cable, there is a device called a *depth transducer,* which measures the depth of the cable in water. If the cable is riding too shallow or too deep, the weight is changed or the towing speed is adjusted.

The cable depth can be changed by a small device called a *"bird."* Several birds are placed along the cable at regular intervals. A buoy is placed at the end of the cable for three purposes. First, it shows the end of the cable in the water. Secondly, it keeps the end of the cable from sinking straight down. Third, it's there for streamer positioning on 3-D surveys. (i.e., for determining the coordinates of the end of the streamer). Figure 4–10 illustrates a marine seismic cable layout.

ELECTRONIC SURVEYING SYSTEM

One of the major problems in marine exploration is to know the exact location in the sea. Obviously, land survey methods as well as normal navigating methods are useless because they cannot provide the accuracy demanded in seismic exploration.

FIGURE 4–10. Marine seismic cable layout

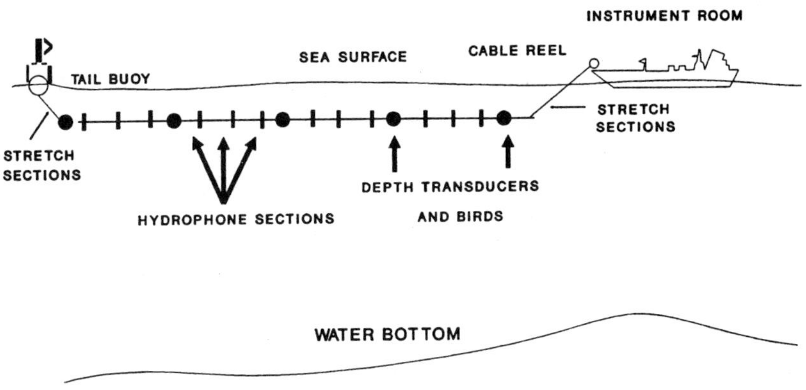

The LORAC system is a system of radio location that provides continuous position coordinates derived from signals from groups of shore-based transmitters.

A dramatic breakthrough in electronic positioning occurred when the U.S. Navy TRANSIT satellite navigation system was released for public use. This system provides position fixes from orbiting satellites anywhere in the world. With the independent of shore-based transmitters, the SAT/NAV system is a time-saver because it reduces the number of runs made onshore to "check points" for checking positioning. These systems have since been upgraded with GPS (Global Positioning System) and DGPS (Differentiate Global Positioning System), which provides greater accuracy.

Two different positioning systems are automatically integrated so that each system enhances the accuracy of the other. Systems can be integrated together such as SAT/NAV or LORAC/NAV.

ENERGY SOURCE

There are quite a few energy sources available for marine seismic data acquisition. A few of them will be discussed briefly. (A number of published papers are listed at the end of the chapter for readers interested in knowing more about this topic.)

The two most popular energy sources in use today are the *air guns* and *sleeve guns*. The first type uses compressed air in a cylinder called an air gun. The second type explodes compressed air in a rubber sleeve, causing the sleeve to expand and contract.

In both cases, the energy released is a short burst. The energy released with each "pop" (or shot) is much less than a dynamite charge, so many guns (typically 16–32) are fired simultaneously. Data from several compressed air shots may be added together by the computer in a process called *vertical stacking* to enhance the signal.

The air gun releases a bubble of air that acts as a secondary source and appears as if a second shot was fired. The data are debubbled by one of several means available.

This energy source is popular and can be used for 24 hours a day. It causes no hazard to marine life, and little objection is raised by environmentalists.

Another type of energy source used for shallow high-frequency marine data acquisition is the sparker, which creates a spark between two electrodes in the water. In this type of source, a high voltage is built up in a capacitor then discharged in oil. This results in a high-frequency seismic signal that gives high resolution. However, the signal has poor penetration into the subsurface, so this type of gun is used for high-resolution, shallow seismic surveys with short cable length. Figure 4–11 is a photo of an air gun.

SHOOTING CONFIGURATION

The most simple and obvious surveying method is with a cable of hydrophones pulled by a single vessel. If the energy source is air guns and there are 48 hydrophones, every "pop" per hydrophone station will give 24 fold CMP. However, the cable may have 72 or 120 stations, and there may be 2, 3, 4 or some variable number of pops between stations.

The variable number is often the case, since a constant ship speed is impossible and shots are usually fired at fixed intervals. As the vessel speeds up, fewer shots are taken between stations, while more shots will be obtained when the vessel slows down. Gun control systems are attached to the ship navigation system.

Recently, a mini cable section was towed between the vessel and the streamer. The hydrophone intervals were shorter on this cable than on the streamer cable. The purpose of this arrangement was to increase the fold and reliability of the shallow reflections.

The mini-cable is rarely used on standard seismic surveys today, due to the higher channel capacity of modern recording systems.

Marine 3-D seismic surveys utilize multiple streamers and multiple sources. The operation is more complicated than the 2-D surveys.

MARINE SHOOTING VERSUS LAND SHOOTING

There are many advantages to marine shooting over land shooting, some of these advantages are:

- The marine shooting environment is nearly constant as in the recording of the data.

FIGURE 4–11. Marine air guns (a, courtesy Bolt Technology Corporation. b, courtesy Halliburton Geophysical Services)

a AIR GUN

b SLEEVE GUN

- The areas to be shot are easy to permit.
- Very high fold data are recorded with relative ease.
- Recording operations can continue for days without interruption and for 24-hour periods.
- Grids are usually shot so that lines can be tied together and the whole area can be processed and checked as a unit.
- Marine shooting is much faster and therefore much cheaper. One day of marine surveying may produce as much seismic data as one month of moderately difficult land recording.

SUMMARY AND DISCUSSION

It is crucial to acquire reliable seismic data in the field that will contain the required resolution and the geological information.

The effort to record a good signal starts in the field operation. However, efforts to attenuate the various types of noise should be done with economic limits in mind.

These types of noise may be of random nature, such as wind, traffic, etc., or coherent noise due to predominant near-surface seismic waves. Boundary waves such as ground roll or air waves or even some converted waves could be recorded with the compressional waves.

Weighted source and receiver arrays are used to attenuate such noise patterns. The common-midpoint technique is effective in attenuating random noise and source/receiver noise patterns.

There are two different schools of thought concerning the approach to designing the optimum field parameters for land data acquisition. The first is to conduct a noise test and then design the field parameters. This can be costly because the field crew may be idle while the test is being processed and analyzed, which may take few days.

The other school is to modify the parameters while acquiring data in the field. Data is processed on site and parameters are modified as needed.

Whichever method is used to select the optimum parameters for acquiring seismic data, economic analysis should be done to decide if the extra cost needed to enhance the data quality is justified.

The cost per mile to acquire seismic data is dependent on the nature of the survey (land or marine), the geographic location, and the season. It is dependent on the field configuration, such as the number of channels, record length, fold, sampling interval, noise problems in the area which may need special geophone array, near-surface problems which may need a certain type of energy source, and certain weighted array. Line preparation and cleanup add more cost per mile.

Land seismic data acquisition normally requires more attention to design economically sound field parameters. However, marine seismic data is certainly not free from problems. Generally, in most areas it is easier to acquire decent marine seismic data economically.

It is difficult to venture an estimate of the cost per mile to acquire seismic data. As you have seen, there are many variables involved. Add to these variables the question of the availability of a seismic crew that will do for you a specific job in your time frame.

KEY WORDS

Aliasing (Alias filter)
Bandwidth
Binary gain
Digitizer
Dispersive wave
Floating-point recording
Fold
Ghost
Multiple
Noise pattern

Noise test
Random noise
Straddle-shooting (Split-spread)
Vertical stacking
Wave length

EXERCISES

1. Describe and compare the following seismic sources. Include the advantages and disadvantages of each.
   a. Dynamite
   b. Vibroseis
   c. Air gun
2. Name three types of undesirable noise patterns.
3. Draw a 50-Hz sine wave traveling at 2,000 m/sec in time domain. What are the period and wavelength?
4. A crew is operating with a recording truck that has a 24-trace recording system. The line is an east-west line, with the first station at 101 and the last station at 160. The group interval is 100 meters. The vibrators cannot operate at stations 127 and 128 because of a tank battery, nor at stations 147, 148, and 149 because of a gas compressor station. That is, these five stations cannot be used as source stations.

   For a split-spread configuration with 3 stations near offset (5 gaps):
   a. Draw a surface and subsurface diagram representing the field geometry for 12-fold coverage. Use cross-section paper for uniform geometry.
   b. Program to maintain 12-fold coverage under the five vibrator stations that were skipped because of the surface obstructions.
   c. How far back would you start vibrating the ground to have 12-fold coverage at station 101?

   If you have difficulty in starting this exercise, refer to the CDP method in the text.

BIBLIOGRAPHY

Courtier, W. H. and H. L. Mendenhall. "Experiences with Multiple Coverage Seismic Methods." *Geophysics* 32 (1967): 230–258.

Dix, C. H. "Seismic Velocities from Surface Measurements." *Geophysics* 20 (1955): 68–86.

Dobrin, M. B. *Introduction to Geophysical Prospecting.* New York: McGraw-Hill, 1960.

Gardner, D. H. "Measurement of Relative Ground Motion in Reflection Recording." *Geophysics* 3 (1938): 40–45.

Giles, B. F. "Pneumatic Acoustic Energy Source." *Geophysics Prospect* 16 (1968): 21–53.

Godfrey, L. M., J. D. Stewart, and F. Schweiger. "Application of Dinoseis in Canada." *Geophysics* 33 (1968): 65–77.

Griffiths, D. H. and R. F. King. *Applied Geophysics for Engineers and Geologists.* London: Pergamon, 1965.

Marr, J. D. and E. F. Zagst. "Exploration Horizons from New Seismic Concepts of CDP and Digital Processing." *Geophysics* 32 (1967): 207–224.

Mayne, W. H. "Common Reflection Point Horizontal Stacking Techniques." *Geophysics* 27 (1962): 927–938.

Mayne, W. H. "Practical Considerations in the Use of Common Reflection Point Technique." *Geophysics* 32 (1967): 225–229.

Mayne, W. H. and R. G. Quay. "Seismic Signatures of Large Air Guns." *Geophysics* 36 (1971): 1162–1173.

Neitzel, E. B. "Seismic Reflection Records Obtained by Dropping a Weight." *Geophysics* 23 (1958): 58–80.

Nettleton, L. L. *Geophysical Prospecting for Oil.* New York: McGraw-Hill, 1940.

Parr, Jr., J. O. and W. H. Mayne. "A New Method of Pattern Shooting." *Geophysics* 20 (1955): 539–564.

Peacock, R. B. and D. M. Nash, Jr. "Thumping Technique Using Full Spread of Geophones." *Geophysics* 27 (1962): 952–965.

Poulter, T. C. "The Poulter Seismic Method of Geophysical Exploration." *Geophysics* 15 (1950): 181–207.

Shultze-Gatterman, R. "Physical Aspects of the Air Pulser as a Seismic Energy Source." *Geophysics Prospect* 20 (1972): 155–192.

Chapter 5

Seismic Data Processing

INTRODUCTION

It is very important for the interpreter to be aware of all the problems encountered in seismic data processing. The geophysicist must know and understand the particulars of each processing step. In addition, a high level of experience is required for quality control at each step to ensure its validity before proceeding to the next step. Engineers and geologists, however, in order to have a better appreciation for the applications and limitations of seismic methods, should at least understand the physical meanings of the terms.

The final interpretation is only as good as the quality of the processing of the seismic data. Special attention and care should be given to the stratigraphic applications of seismic data in areas of subtle traps and of rapid lithology and facies changes.

The process of converting field recordings into a meaningful seismic section involves many steps of data manipulation. Before a CMP gather is corrected for normal moveout and stacked, the data should be corrected for near-surface time delays. These adjustments are called the *static corrections*, or simply *statics*.

In addition, various deconvolution and filter tests are done, then parameters are designed to enhance signal-to-noise ratio and increase vertical resolution. Finally, we want to convert our seismic reflections into a picture representing the true subsurface geology. This is accomplished by the process called *migration*.

There is no unique processing sequence or cookbook routine to follow in processing the data. Each geologic setting stands on its own. Extensive testing must be done to study the problems involved and to design the optimum parameters for each step of the data-processing flow.

It is important to have a good idea about the regional geology of the basin and specific problems in the area where the seismic data was acquired.

In all cases except virgin exploration, communication between the interpreter and the processing geophysicists should be clear and frequent. It is possible, sometimes, for the interpreter to give the processor some starting points for the velocity range expected or to warn him about known multiples or other noise in the particular area.

DEMULTIPLEXING AND GAIN

Raw data are recorded on field tapes in *multiplex* form, where sample #1 from every channel is recorded, followed by sample #2 from each channel, and so on. The first step in data processing is to reorder the data so that the complete signal for each receiver/shot combination is reconstituted. The *demultiplex* rearranges this configuration to have all the samples of channel #1 in sequence from time 0 to the end of the record, and then all samples for channel 2, and so on. The signal consists of amplitude values, spaced at a certain sample interval (usually 4 ms) to obtain a certain resolution. In high-resolution work, a shorter sample period (1 or 2 ms) is often used so that higher frequencies can be recorded without aliasing.

The output is in trace sequential form. Every record has as many channels as there are geophones in the field geometry for each shot point. The signal from each channel is called a *trace*, and the traces are arranged and numbered in the order of their distances from the source.

Figure 5–1 illustrates the multiplexing and demultiplexing processes.

Because the amplitudes of the reflections decay with depth due to spherical divergence, the data normally is recorded in the field with a

FIGURE 5–1. Demultiplex

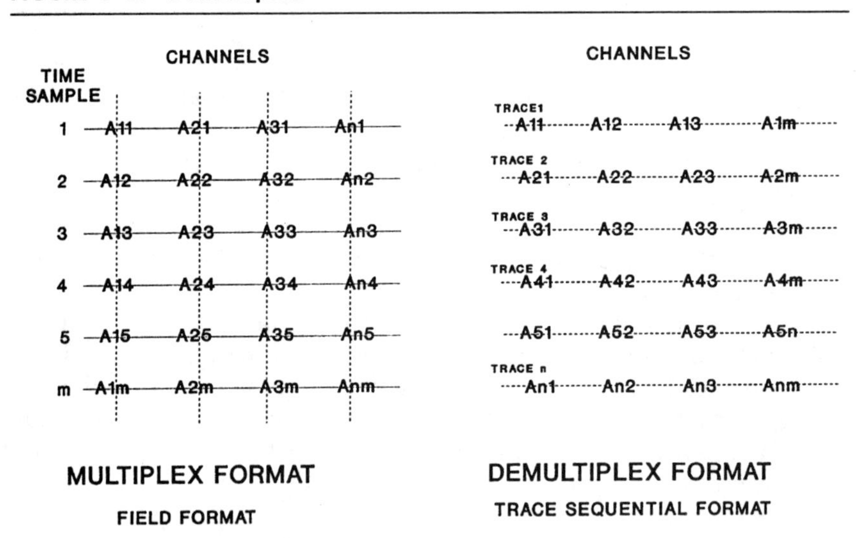

binary gain, a multiplier which amplifies all samples to the same level. This multiplier is removed in the processing, and an exponential gain-recovery curve, which amplifies the signal to a lower reference level, is applied. Other gain-related scalers, such as for attenuation due to absorption, are also applied to the data.

AMPLITUDE RECOVERY AND GAIN

Since it is known that lateral variations in the amplitudes of the reflections carry much useful information, it has become common practice to avoid artificial tampering with the recorded amplitude values. However, it is very important to correct for the gradual decrease in amplitude down the record due to the spherical spreading and absorption of the source wave.

During the demultiplexing stage, the binary gain is removed and a gain curve is applied to compensate for energy loss with depth.

Figure 5–2 is a comparison between field records before and after correcting for geometric spreading. The amplitude has been restored at late times, but, unfortunately, ambient noise has also been strengthened.

TYPES OF GAIN

Various types of gain are used in practice, based on the desired criteria. A gain function is derived from the data and applied to trace amplitudes at each time sample.

Three types of gain are commonly used: programmed gain, surface-consistent gain, and automatic gain control (AGC). These are explained in the paragraphs to follow.

PROGRAMMED GAIN CONTROL (PGC)

PGC is the simplest form of gain. A gain curve can be defined by interpolating between gain values specified at particular time samples. Naturally, larger gain values would be applied to later time samples. Normally, a single PGC function is applied to all traces in the gather or stacked section to preserve the relative true amplitude variations in the lateral direction.

Figure 5–3 illustrates the method of recovering amplitude by the PGC method. The gain function is estimated at time samples indicated by solid circles, and it is interpolated between these samples.

SURFACE-CONSISTENT GAIN

This is the same as PGC gain except that the gain is computed for each trace within the CMP. A gain curve for every source and receiver is derived and applied, giving more accurate compensation for the amplitude decay due to spherical divergence.

FIGURE 5–2. Correction for geometric spreading (Reprinted from O. Yilmaz, *Seismic Data Processing*, 1987, courtesy of the Society of Exploration Geophysicists.)

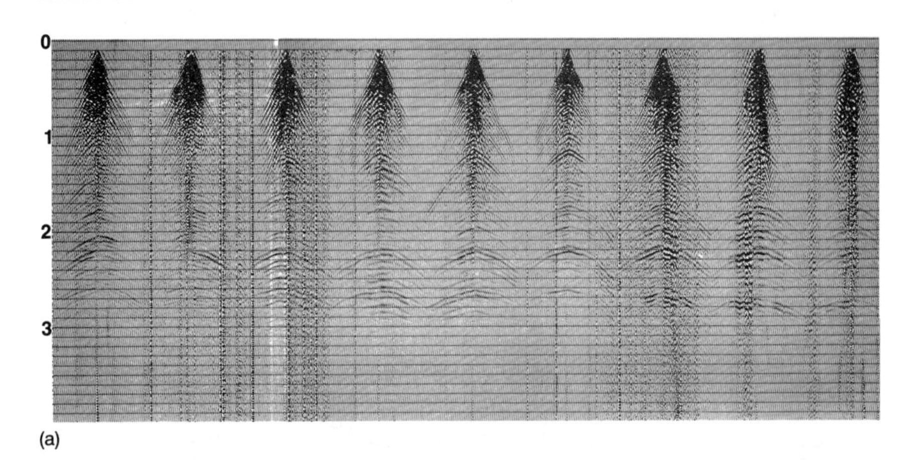

(a)

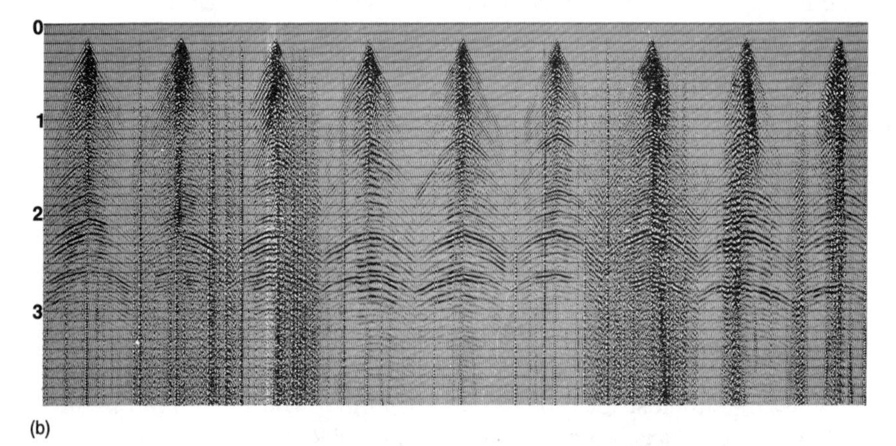

(b)

a) RAW RECORDS FROM LAND SURVEY. NOTE THE RAPID DECAY IN AMPLITUDE AT LATE TIMES.
b) THE SAME FIELD RECORDS AS IN (a) AFTER CORRECTING FOR GEOMETRIC SPREADING. THE AMPLITUDES HAVE BEEN RESTORED AT LATER TIMES, BUT NOISE IS STRENGTHENED.

This type of gain correction is applied within the CMP for every trace to preserve the variation of amplitude with *offset*, the lateral distance between source and receiver. In other words, it preserves the amplitude variation with the angle of incidence. (See Figure 5–4.)

This technique is used in the Amplitude Versus Offset (AVO) analysis technique.

FIGURE 5–3. Programmed gain control (Reprinted from O. Yilmaz, *Seismic Data Processing*, 1987, courtesy of the Society of Exploration Geophysicists.)

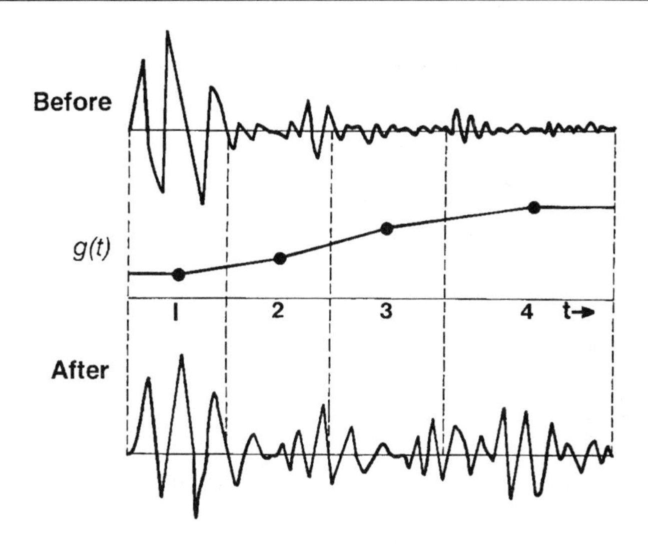

FIGURE 5–4. Surface consistent gain (courtesy of Seismograph Service Corporation)

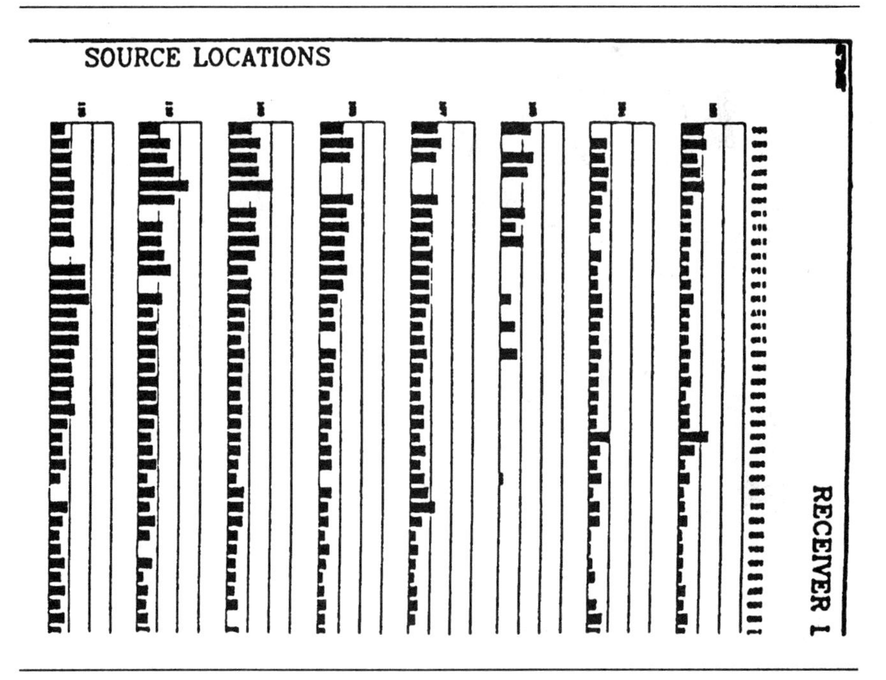

AMPLITUDE AGC

This is the most commonly used type of gain.

Automatic gain control is a process in which the gain in each channel (trace) is controlled automatically and independently from the other channels. The output level controls the gain to keep the output level at a desired value within the entire limits of the defined gate (time windows). It is applied to every sample in the trace. No interpolation is needed to define this gain function.

Figure 5–5 illustrates the AGC for gates of 64 ms, 128 ms, 512 ms, and 1,024 ms.

The shorter the gate, the higher the amplitude; the longer the gate, the lower the amplitude.

In summary, gain is applied to the seismic data for a number of reasons:

1. Geometric spreading correction to compensate for amplitude loss with distance from the source.
2. AGC-type gain functions are applied to seismic data to bring up weak signals. This gain function is time variant.
3. Program gain control can be used to preserve the relative true amplitude relationship.
4. Gain must be used with care, since it can destroy the signal character, as seen in the short gate AGC display.

FIGURE 5–5. Automatic Gain Control (AGC) (Reprinted from O. Yilmaz, *Seismic Data Processing*, 1987, courtesy of the Society of Exploration Geophysicists.)

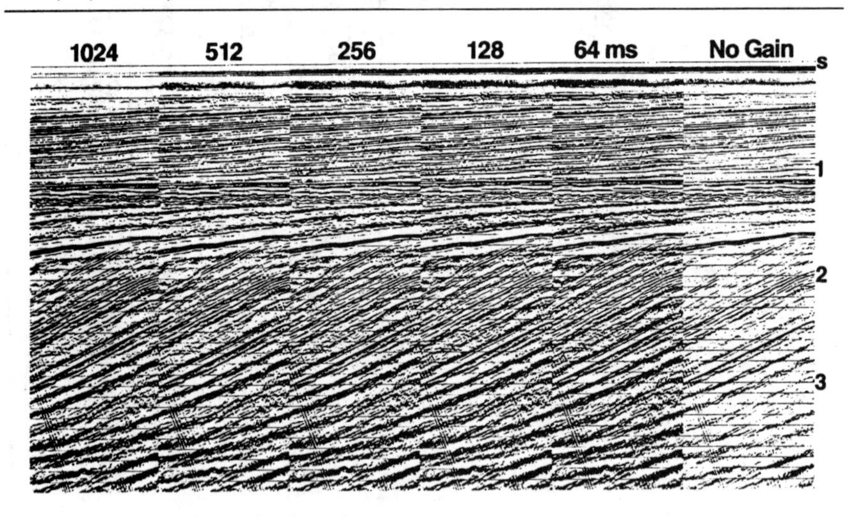

SIGNAL THEORY

Digital seismic processing uses techniques from signal theory, which were originally developed to improve radar and radio signals. In those applications, the objective is to clear up the message received by enhancing the signal and reducing interfering noise.

These techniques involve combining two lists of numbers to obtain a third list, which bears some desired relationship to the first two. This is done by multiplying a set of numbers in one list by those in the other list. Many applications can be performed using these techniques. For example, one list can be reversed before shifting past the other. Or, a specially designed "operator" list, which is usually shorter than the other, is combined with the first list to produce the third.

These lists of numbers are numerical representations of the data received; they are the seismic traces in digital form, which actually is the form in which they are recorded. The lists may be plotted as curves, forming an analog representation of the data. These operations applying signal theory, such as correlation, convolution, and deconvolution, are some of the means used to improve seismic data in digital processing.

CORRELATION

Correlation is one of the applications of signal theory to seismic data processing.

An analog trace in a record section is a wiggly line that moves to the right and left of a central line. This is a close representation of the movement of the earth due to the seismic wave. In digital form, it is a list of numbers sampled from the data every one, two, or four milliseconds. Positive and negative numbers represent the right and left excursions (peaks and troughs) of the wiggle trace.

When we put two traces side by side in digital form and multiply the first number of the first trace by the first number of the second trace, then the second number of the first trace by the second number of the second trace, and so on, we obtain a column of products. Adding all these products together yields a single number that is a measure of how much alike the two traces are. The larger the number, the more they are alike. Since each pair of numbers is multiplied together, matching troughs as well as matching peaks yield positive products.

A peak matched with a trough produces a negative number, as the product of a positive number and a negative number is negative. Therefore, a good match is positive and a mismatch is negative.

TIME ADVANCE AND DELAY

At this point we should define the meaning of *time advance* and *time delay*. Consider this example. If a meeting was scheduled at 4:00 but we advanced the meeting time to 3:00, we have shifted the meeting in the time direction. By convention, time runs to the right, so an advance is represented by a shift to the left. *A* time delay, of course, is a negative advance, and is therefore a shift to the right.

Now let us go back to our discussion of correlation. Figure 5–6 illustrates the *cross-correlation* procedure of two wavelets in digital form. Consider two wavelets, each represented by a list of three numbers:

$$A = (2, -1, 1) \text{ and } B = (2, 1, -1)$$

Let wavelet *B* be held in place while wavelet *A* is advanced with respect to it. In the upper-right corner of the figure is a table of numbers representing the positions of the wavelets as wavelet *A* is advanced from –3 to +3. The position of wavelet *B* is shown at the next-to-bottom line of the table.

To advance wavelet *A* by –1 time unit, we shift it to the right one notch. We now have:

$$A \text{ (advanced } -1 \text{ unit)}: (2, -1, 1) \text{ and } B: (2, 1, -1)$$

The sum of the products is

$$(2)(1) + (-1)(-1) + (1)(0) = 3$$

Since the product of two numbers of like sign is positive, both matching peaks and matching troughs yield positive numbers. This in-

FIGURE 5–6. Cross-correlation (modified after E.A. Robinson)

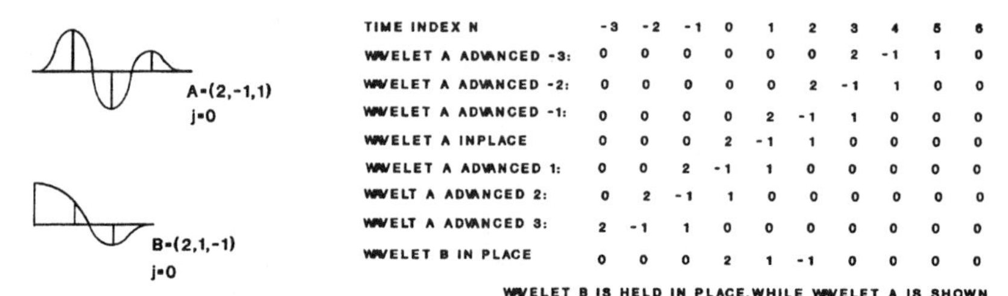

TIME INDEX N	-3	-2	-1	0	1	2	3	4	5	6
WAVELET A ADVANCED -3:	0	0	0	0	0	0	2	-1	1	0
WAVELET A ADVANCED -2:	0	0	0	0	0	2	-1	1	0	0
WAVELET A ADVANCED -1:	0	0	0	0	2	-1	1	0	0	0
WAVELET A INPLACE	0	0	0	2	-1	1	0	0	0	0
WAVELET A ADVANCED 1:	0	0	2	-1	1	0	0	0	0	0
WAVELT A ADVANCED 2:	0	2	-1	1	0	0	0	0	0	0
WAVELT A ADVANCED 3:	2	-1	1	0	0	0	0	0	0	0
WAVELET B IN PLACE	0	0	0	2	1	-1	0	0	0	0

WAVELET B IS HELD IN PLACE,WHILE WAVELET A IS SHOWN IN POSITIONS FROM AN ADVANCE OF -3 TO AN ADVANCE OF +3. AN ADVANCE IS DEFINED AS A SHIFT AGAINST THE DIRECTION OF TIME.SINCE TIME RUNS TO THE RIGHT,A POSTIVE ADVANCE REPRESENTS A SHIFT TO THE LEFT,WHEREAS A NEGATIVE ADVANCE REPRESENTS A SHIFT TO THE RIGHT.

THE CROSS-CORRELATION OF A AND B IS THE FUNCTION MADE UP OF THE SUM OF CROSS PRODUCTS OF A WAVELET A ADVANCED AND WAVELET B IN PLACE.

A ADVANCED-3, B IN PLACE	2(0) + (-1)(0) + (1)(0) =	0
A ADVANCED-2, B IN PLACE	2(-1) + (-1)(0) + (1)(0) =	-2
A ADVANCED-1, B IN PLACE	2(1) + (-1)(-1) + (1)(0) =	3
A IN PLACE AND B IN PLACE	2(2) + (-1)(1) + (1)(-1) =	2
A ADVANCED 1, B IN PLACE	2(0) + (-1)(2) + (1)(1) =	-1
A ADVANCED 2, B IN PLACE	2(0) + (-1)(0) + (1)(2) =	2
A ADVANCED 3 B IN PLACE	2(0) + (-1)(0) + (1)(0) =	0

COMPUTION OF THE CROSS-CORRELATION VALUE FOR EACH TIME SHIFT.

creases the value of the sum. The sum is the cross-correlation for the time shift (advance of –1) and in this case has a large value, 3.

Let us now advance wavelet *A* by +1 time unit; that is, we shift it to the left by one time period. We have:

$$A: (2, -1, 1), \text{ and } B: (2, 1, -1)$$

The cross-correlation for a time advance of +1 is

$$(2)(0) + (-1)(2) + (1)(1) = -1$$

The lining up of the trough (–1) on wavelet *A* with a peak (2) on *B* produces a negative number of –2, which decreases the sum, and thus the waveforms in this position are negatively correlated.

Now we repeat the process of shifting, multiplying, and adding. The sums are the cross-correlation values for each time shift. The cross-correlation of wavelets *A* and *B* is the function made up of the sum of the cross-products of wavelet *A* advanced and wavelet *B* in place.

Plotting the advances versus the cross-correlation values forms a third wavelet, which is called the *cross-correlation function.*

In the case of vibrator data, another cross-correlation process is done during the demultiplexing stage. This extracts the data by correlating the sweep with the total data sample and outputs the listen time, which is the actual record length. Demultiplexed data are normally displayed every n^{th} record for quality control, or even every record in case of short seismic land lines or problem lines. A good approach is to output all the sweeps on the line and plot them side by side to check that all sweeps are in phase and for editing purposes in case of bad sweeps or short records.

AUTOCORRELATION

Autocorrelation is a special case of cross-correlation. Instead of correlating two different waveforms, autocorrelation consists of correlating a waveform with itself. The two rows of numbers used in the process are made up of the same digital form. The autocorrelation is a symmetrical, zero-phase function, while a cross-correlation is not necessarily symmetrical.

Figure 5–7 shows an autocorrelation function and a cross-correlation function obtained from two different waveforms.

CONVOLUTION

Convolution is a process similar to correlation, with one important difference. When we correlate two digital waveforms, we place one above the other, shift one to the right or left, multiply vertical entries, and then add the products. The resulting sum is the correlation value of the time shift in question.

FIGURE 5–7. Autocorrelation and cross-correlation

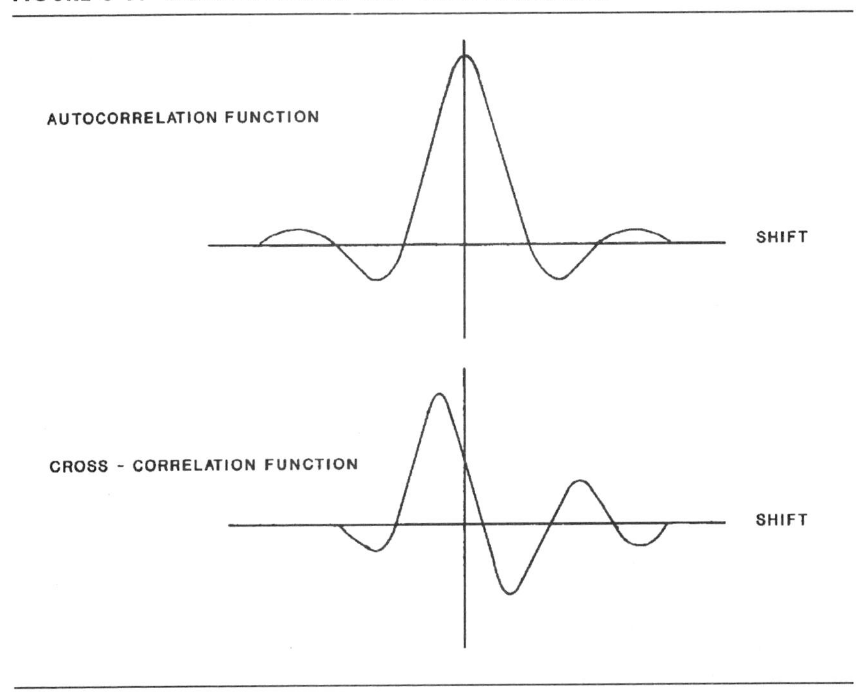

To convolve two digital waveforms, we first *reverse* the direction of one so that its values are in the opposite order with respect to time. We then carry out the same correlation procedure as discussed above. Figure 5–8 illustrates this process.

One of the application of the convolution process is to apply band-pass filter to seismic data to attenuate some undesired frequency range. This can be done by obtaining a display of a seismic record filtered with different filter pass bands. Figure 5–9 illustrates a filter test.

DECONVOLUTION

Deconvolution (decon) is a process that improves the vertical resolution of seismic data by compressing the basic wavelet. It is called "inverse filtering."

Decon is normally applied before stack (DBS), but is sometimes applied after stack (DAS).

The earth is composed of layers of rocks with different lithologies and physical properties. Seismically, rock layers are defined by their densities and the velocities at which seismic waves propagate through them. The product of density and velocity is called *acoustic impedance*. It is the impedance contrast between layers that causes the reflections that are recorded along a surface profile.

FIGURE 5–8. Convolution

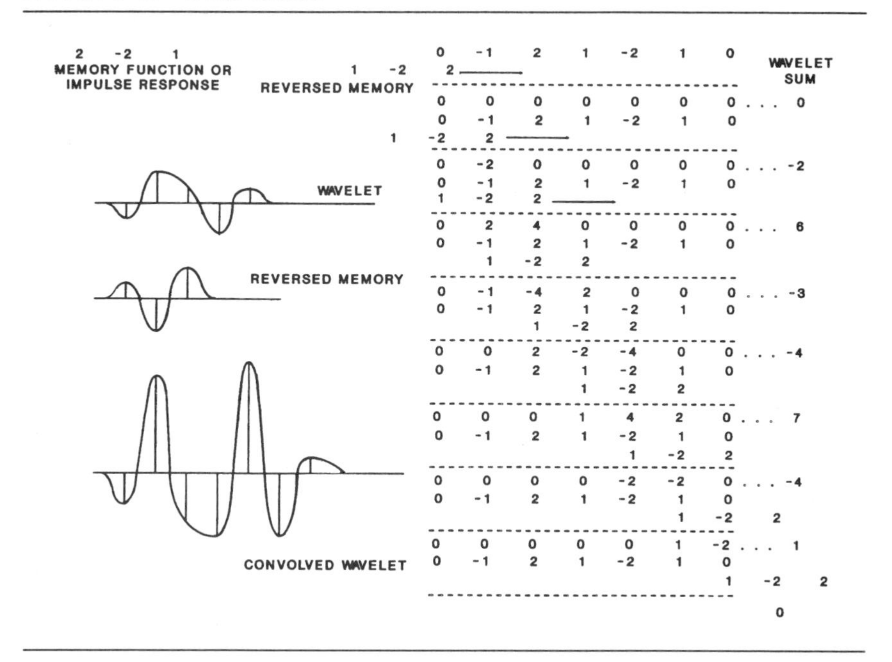

FIGURE 5–9. Filter test (courtesy of Seismograph Service Corporation)

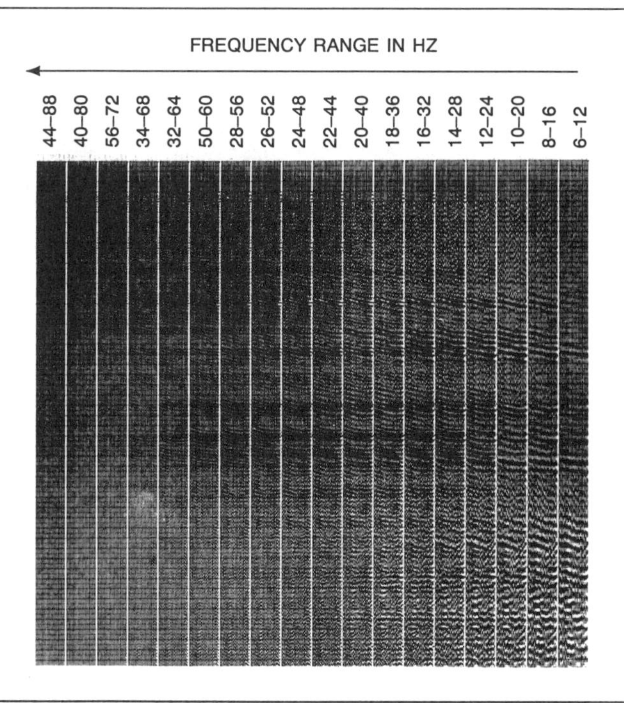

A seismic trace can be modeled as the convolution of the input signal with the reflectivity function or earth impulse response. The recorded seismic trace has many components, including source signature, recording filter, surface reflections, and geophone response. It also has primary reflections (reflectivity series), multiples, and all types of noise.

Ideally, decon should compress the wavelet components and attenuate multiples, leaving only the earth reflectivity on the seismic trace.

Basically, if the inverse filter is convolved with a seismic wavelet, it will convert to a spike when applied to the seismogram. The inverse filter should yield the earth's impulse response. This is called *spiking decon.*

Figure 5–10 shows how deconvolution is performed on a seismic wavelet. Note that the side lobes represent the undesired signal, such as some or all of the noise discussed before. By deconvolving the seismic wavelet with the optimum decon parameters, we hope to attenuate these undesired signals. One method of quality control is to run an autocorrelation function on the deconvolved trace to check if the side lobes are attenuated.

In order to select the optimum parameters to deconvolve seismic data, a decon test should be run before and/or after stack. The decon

FIGURE 5–10. Deconvolution

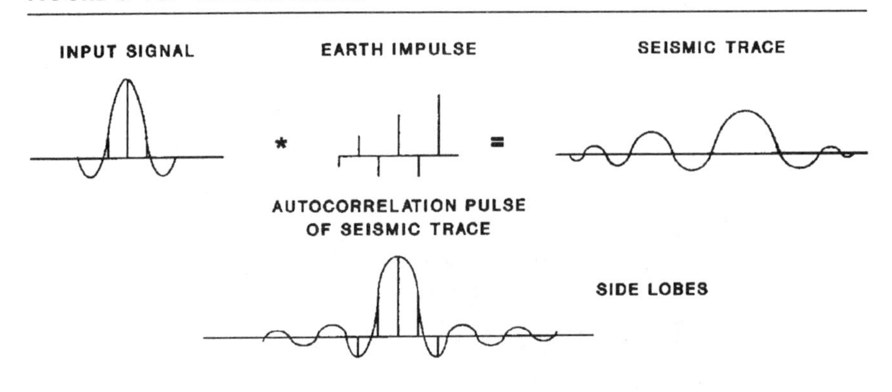

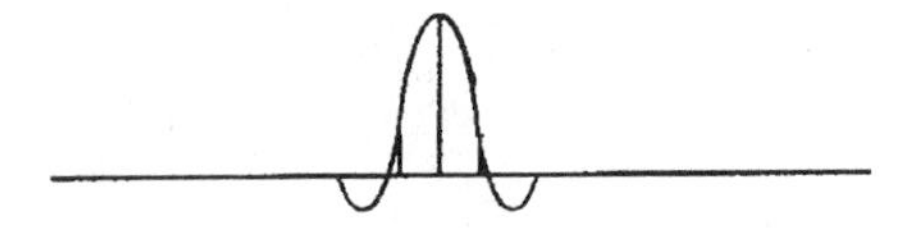

DECONVOLUTION ATTENUATES THE SIDE LOBES AND COMPRESSES THE WAVELET SHAPE.

FIGURE 5–11. Deconvolution test (courtesy of Seismograph Service Corporation)

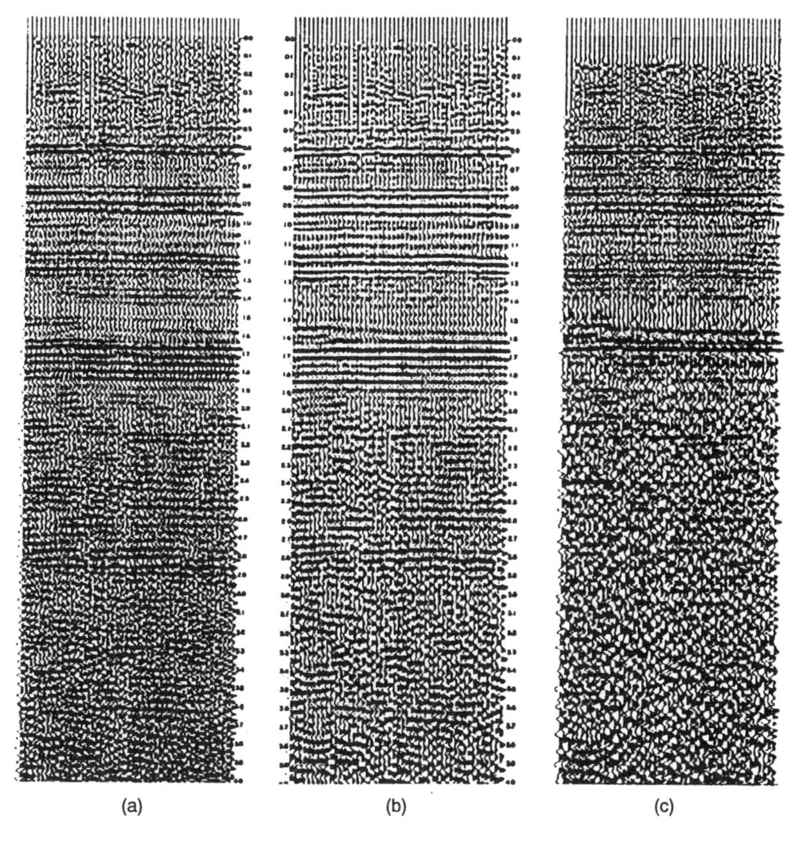

(a) PART OF A STACK SECTION—NO DECONVOLUTION
(b) PASS BAND FILTER APPLIED ON THE STACK
(c) DECONVOLUTION APPLIED BEFORE STACK

before the stack test can be seen on Figure 5–11 with different decon parameters.

Figure 5–12 illustrates a comparison between nondeconvolved and deconvolved stacked sections. One can see better separation between reflection and multiple attenuation on the deconvolved one.

DBS is normally applied in order to attenuate the short-period multiples. On the other hand, a long decon operator may be used after stack to attenuate the long-period multiples.

DAS should not be implemented if it does not help. It may be omitted to avoid tampering with good data. Figure 5–13 illustrates a deconvolution after stack. This decon section had a post-stack filter.

FIGURE 5–12. Deconvolution comparison (Reprinted from O. Yilmaz, *Seismic Data Processing*, 1987, courtesy of the Society of Exploration Geophysicists.)

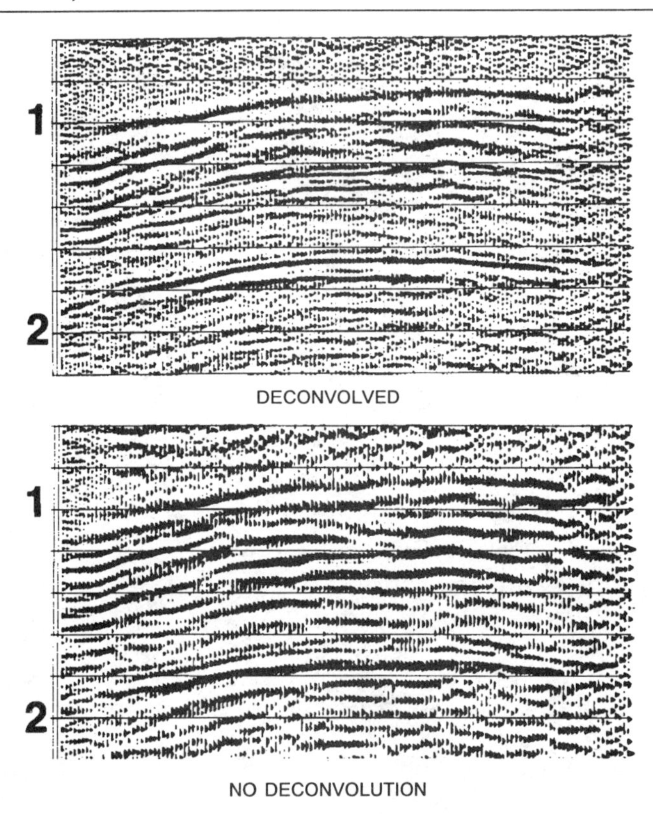

SUMMARY OF CONVOLUTION PROCESSES

Convolution, cross-correlation, and summing (stacking) are three basic processes that constitute practically the entire range of digital processing in the time domain.

Stacking of traces is simply the process of adding corresponding samples from two or more time series in order to enhance the S/N ratio.

Cross-correlation and convolution both follow the same operational procedure of cross-multiplying corresponding samples of two shifted time series and then summing the products to obtain a single output point. One time series will be shifted and the operation is repeated to obtain the second output point, and so on. The only operational difference between cross-correlation and convolution is that in convolution one time series is reversed end-for-end before any computation is conducted. Deconvolution

FIGURE 5–13. Deconvolution comparison (Reprinted from O. Yilmaz, *Seismic Data Processing*, 1987, courtesy of the Society of Exploration Geophysicists.)

is an inverse filter, and it is used to retrieve the shape of the original seismic pulse by attenuating the undesirable signals.

NORMAL MOVEOUT

Normal moveout (NMO) is the procedure that removes the time shift due to the offset between the source and the receiver and corrects all traces to zero offset; that is, with the source and receiver at the same surface point, which is at the mid-point between the actual source and receiver.

NORMAL MOVEOUT FOR A NONDIPPING HORIZON

Figure 5–14 shows a simple case of normal moveout geometry for a single, horizontal reflector. At a given midpoint *M*, the travel time along the ray path SDG is $t(x)$, where x is the offset from source to receiver position. If V is the velocity of the medium above the reflecting horizon, $t(0)$ is twice the travel time along the vertical path MD.

$$t(x) = \text{SDG}$$

define $t(0) = 2\text{MD}$

$$t(x)^2 = t(0)^2 + x^2/v^2$$ which is the equation of a hyperbola

NORMAL MOVEOUT IN MULTIPLE HORIZONTAL REFLECTORS

Consider a medium consisting of horizontal velocity layers, as seen in Figure 5–15. Each layer has a certain thickness that can be defined in

FIGURE 5–14. Normal moveout for flat reflector

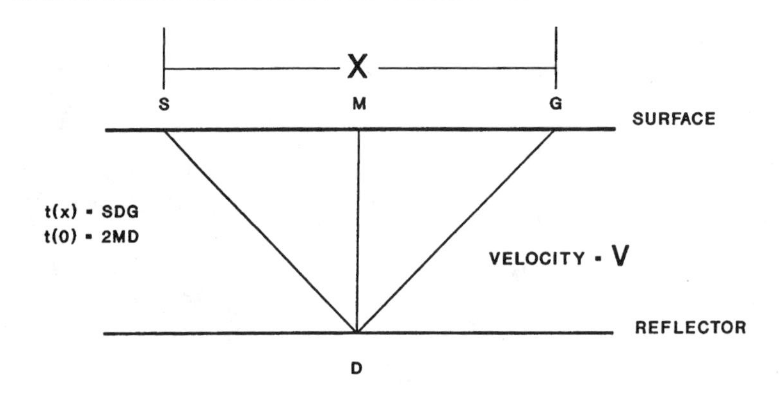

SIMPLE CASE OF NORMAL MOVEOUT GEOMETRY FOR A SINGLE HORIZONL REFLECTOR. AT A GIVEN MIDPOINT M THE TRAVEL TIME ALONG RAYPATH SDG IS t(x)

X IS OFFSET FROM SOURCE TO RECEIVER POSTION.
V IS VELOCITY OF THE MEDIUM ABOVE THE REFLECTING INTERFACE.
t(0) IS TWICE THE TRAVEL TIME ALONG VERTICAL PATH MD.

$$t(x)^2 = t(0)^2 + X^2/V^2$$ EQUATION OF A HYPERBOLA

FIGURE 5–15. NMO in horizontally-stratified earth

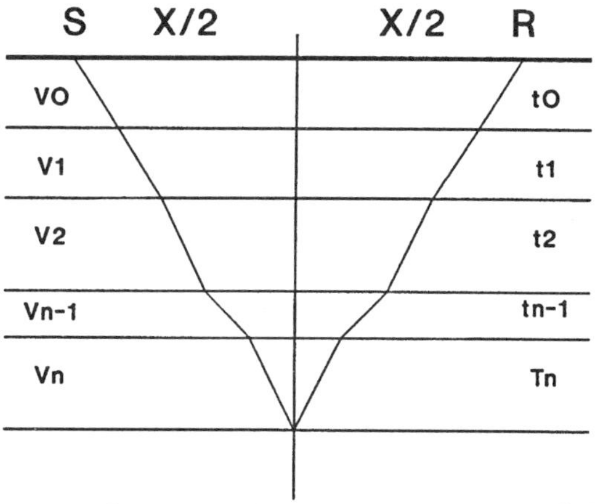

S X/2 X/2 R

Vo to
V1 t1
V2 t2
Vn-1 tn-1
Vn Tn

$$Vrms^2 = \left(\frac{Vo^2 * to + V1^2 * t1 + V2^2 * t2 + \ldots + Vn^2 * tn}{to + t1 + t2 + \ldots + tn} \right)^{0.5} \qquad (1)$$

$$t(x)^2 = t(0)^2 + X^2/Vrms^2 \qquad (2)$$

terms of two-way zero offset time. The layers have interval velocities V_0, V_1, V_2, . . . Vn, where n is the number of layers.

A study of this relationship was done by Dix (1955), and Taner and Koehler (1969) have derived the relationship between RMS velocity and interval velocities, as indicated in Eq. (1) of Figure 5–15.

The normal moveout equation for the multi-layer case is given in Eq. (2). This is similar to the NMO equation for a single horizontal layer except for the velocity, which is the RMS velocity.

VELOCITY ANALYSIS

Acoustic well logs provide direct measurement of formation velocity as a function of depth. Seismic data, on the other hand, provides an indirect measurement of the velocity. By using both types of information, the explorationist can derive a large number of different types of velocity, such as interval, apparent, average, RMS, instantaneous, phase, NMO, stacking, migration, and so forth.

The basic objective is to measure "true" interval velocities; this is often difficult, or even impossible, with the data available. The potential of seismic data as an exploration tool depends significantly on the use of velocity information by the geologist, geophysicist, and the engineer.

Velocity may be the most under-utilized tool available. It is an extremely powerful tool if its role is understood and if its applications are utilized and implemented.

VELOCITY TERMINOLOGY

Velocity, in seismic work, is simply the rate of travel of a seismic wave through a medium with respect to time. However, as we have seen, there are several kinds of velocity. Velocity may be defined by the type of seismic wave motion or by the method used to determine it. Before going into the specifics of velocity analysis, we must understand the terms. Figure 5–16 illustrates a simple layered-earth model. Assume that the seismic energy is initiated at the surface at time t_0 and passes through different reflectors at depths h_1, h_2, and h_3, which have velocities of V_1, V_2, and V_3, respectively. Within h_1 there some reflectors at times t_1, t_2, t_3, and t_4.

1. Interval velocity: the velocity measured between two reflectors.

$$V_1 = h_1/(t_4 - t_0)$$

FIGURE 5–16. Velocity terminology

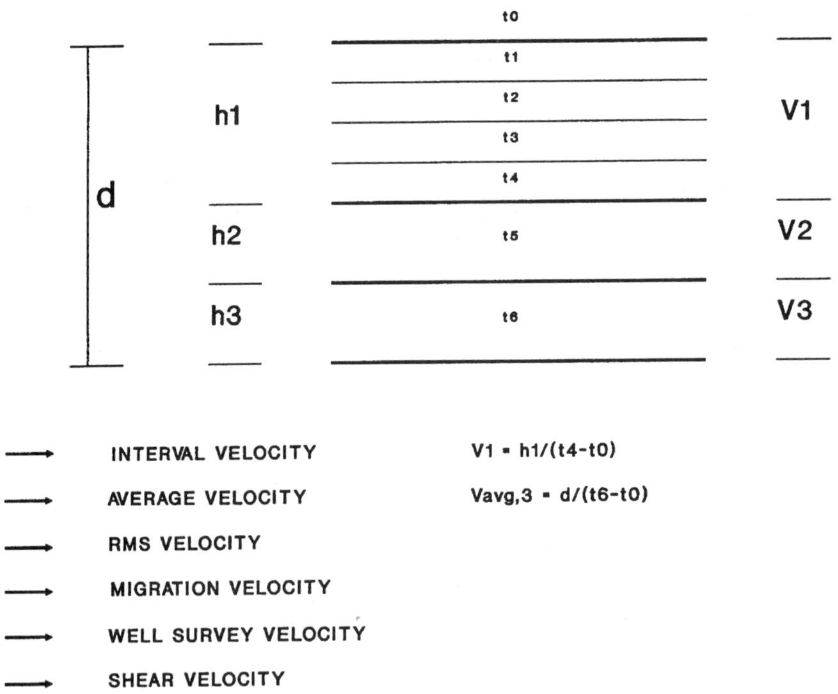

t0
t1
h1
t2
V1
t3
t4
d
h2
t5
V2
h3
t6
V3

INTERVAL VELOCITY V1 = h1/(t4-t0)
AVERAGE VELOCITY Vavg,3 = d/(t6-t0)
RMS VELOCITY
MIGRATION VELOCITY
WELL SURVEY VELOCITY
SHEAR VELOCITY

2. Average velocity: depth divided by seismic travel time to that depth.

$$V_{\mathrm{AVG}}\left[\text{at horizon 3}\right] = d/\left(t_6 - t_0\right)$$

3. RMS (Root-Mean Square) Velocity

$$V^2_{\mathrm{RMS},n} = \sum_{i=1}^{n}\left(V_i^2 \Delta t_i\right) / \sum_{i=1}^{n} \Delta t_i$$

Example:

$$V^2_{\mathrm{RMS},2} = V_1^2\left(t_4 - t_0\right) + V_2^2\left(t_5 - t_4\right)/t_5$$

RMS velocity is derived from seismic data acquired by CDP method.

4. Stacking Velocity: measured directly from CDP seismic data.
5. Migration Velocity: measured from seismic wavefront.
6. Well Survey Velocity: determined using borehole geophone.
7. Shear velocity: the velocity of waves with particle motion perpendicular to the direction of propagation.

FACTORS AFFECTING VELOCITY ESTIMATES

There are several factors that affect the estimation of velocity using the common depth point family of seismic data:

- Field geometry—long offset.
- Multiplicity—Stacking fold.
- Signal-to-noise ratio.
- Front-end mute.
- Time gate length for velocity estimates.
- Velocity increment—sampling.
- Coherency measures.
- Data frequency—bandwidth of the data.
- Statics—True departure from hyperbolic moveout.

Spread length is important, as a longer spread length helps give better NMO values and thereby obtains better discrimination between primary reflectors and multiples.

Depending on the record quality in an area, the fold, can affect signal-to-noise ratio. Normally, the higher the fold the better the signal-to-noise ratio, as higher multiplicity attenuates random noise.

"Mute" is an operation that affects the quality of the velocity estimates. Front-end mute is used to eliminate noise trains, such as body waves. Care must be taken in selecting the mute and protecting (not muting) as many short traces (close to the source) in order to enhance the coherency of the shallow markers. *Surgical mute*, applied inside the record, is normally used in case of severe coherent noise patterns such as airwaves and some dispersive waves. Digital filtering cannot be applied in such cases because the noise train frequency is equal to the addressed data frequency at depth on the seismic trace.

Resolution of the velocity estimates depends on the time gate length to conduct the normal moveout calculations in the vertical sense. This gate is normally a 20 ms window of time. In the horizontal sense are the velocity increment and velocity range. Both factors affect the resolution of the estimates and the run time on the computer.

The frequency content of the data affects the velocity details, especially shallow in the section. It affects the coherency and the lateral and vertical resolution. The *velocity spectrum* method distinguishes the signal along hyperbolic paths even with a high level of random noise. This is because of the power of cross-correlation in measuring coherency. The accuracy of the velocity spectrum is limited, however, if S/N is poor.

The velocity spectrum is computed along a hyperbolic search path for a range of constant velocity value, or constant ΔNMO. The hyperbolic path spans a certain two-way time gate at zero offset. If the gate chosen is too small, it costs a lot of computational time; if it is too large, the spectrum will suffer from lack of vertical resolution.

STACKING VELOCITY AND VELOCITY ANALYSIS

Normal moveout is the basis for determining velocities from seismic data. Computed velocities can be used to correct for NMO so that reflections are aligned in the traces of CMP gathers before stacking.

Stacking velocity is calculated to yield the "best" stack for velocity analysis. This velocity is often called V_{NMO}.

From equation 2, Figure 5–15, the RMS velocity or stacking velocity is derived from the best-fit hyperbola over the entire spread length.

Figure 5–17 illustrates equation 2 on the $T^2(x)$ versus X^2 plane. The slope of the line is $1/V^2_{NMO}$, and the intercept value at $x = 0$ is $t^2(0)$. In practice, the least-squares fitting method is used to define line slope.

The $T^2 - X^2$ method is a reliable way to estimate stacking velocities. The accuracy is dependent upon signal-to-noise ratio, which will affect the quality of the velocity picking.

TYPES OF VELOCITY ANALYSIS

CONSTANT VELOCITY SCANS

The constant-velocity scan method of a CMP gather is one technique for velocity analysis. Figure 5–18 and Figure 5–19 display a CMP gather that is repeatedly NMO corrected, using a velocity range from 5,000 ft/sec to 13,600 ft/sec with a constant increment of 300 ft/sec. The NMO-corrected gathers are displayed side by side in the form of panels.

Two events of interest (A and B) need to be corrected for NMO. At low velocities, both events show that they are over-corrected (taking out too much moveout). Event A is flat at a velocity of 8,300 feet/sec. This

FIGURE 5–17. Best stacking velocity ($T^2 - X^2$ method)

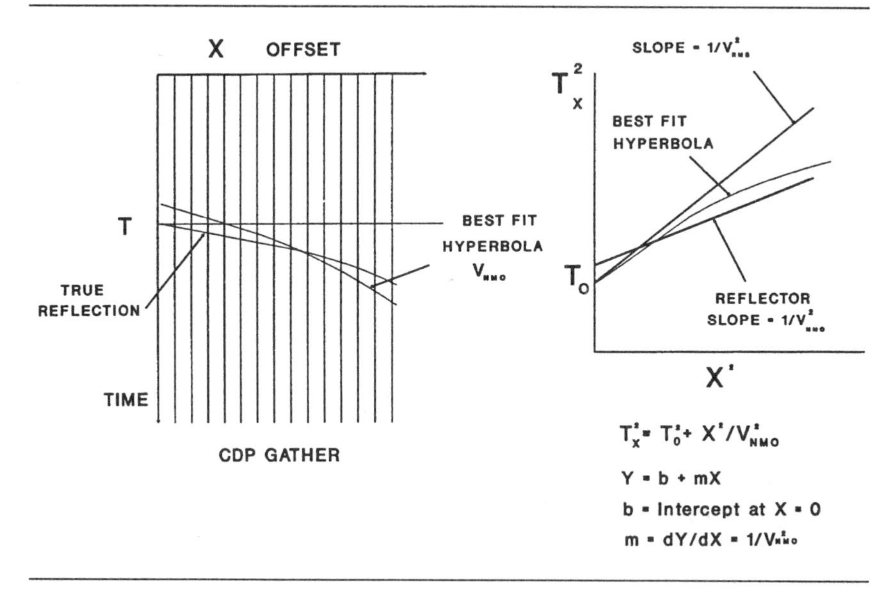

FIGURE 5–18. Constant velocity moveout corrections applied to a single CDP (velocity range 5,000–13,600 ft/sec) (Reprinted from O. Yilmaz, *Seismic Data Processing*, 1987, courtesy of the Society of Exploration Geophysicists.)

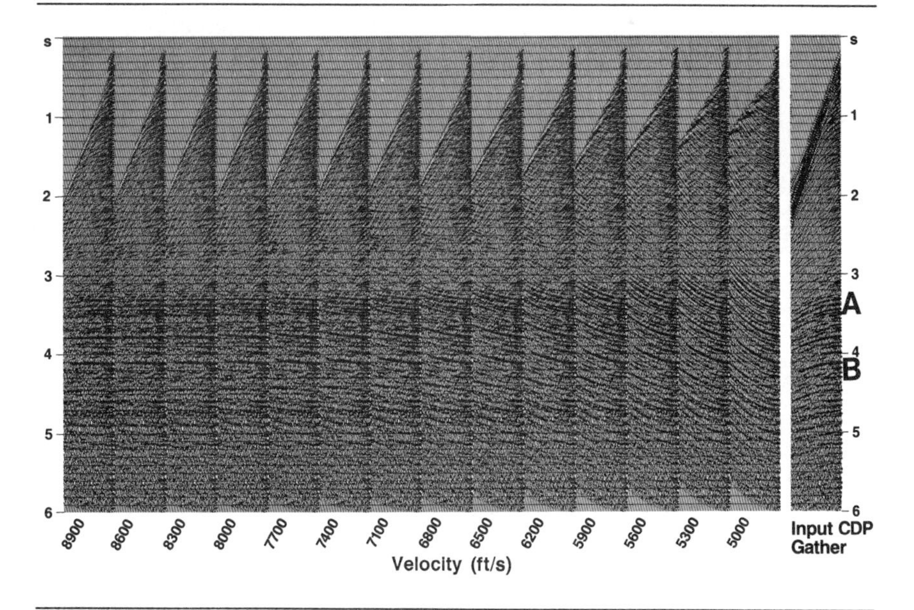

FIGURE 5–19. Constant velocity moveout corrections applied to a single CDP (velocity range 5,000–13,600 ft/sec) (Reprinted from O. Yilmaz, *Seismic Data Processing,* 1987, courtesy of the Society of Exploration Geophysicists.)

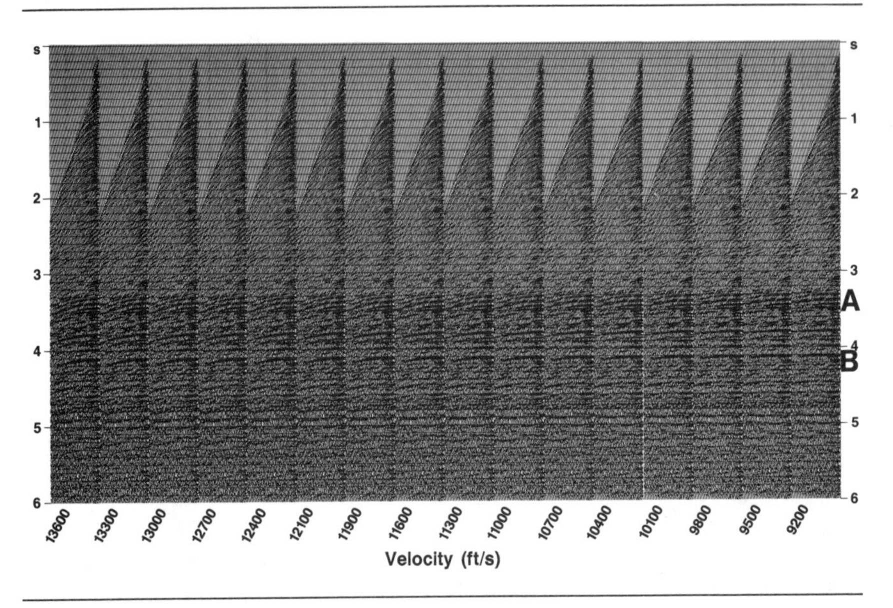

is the best stacking velocity for event A. Notice that at higher velocities event A is under-corrected (not enough moveout was taken out). Refer to the normal moveout equation.

Event B is flat at a velocity of 9,200 feet/sec. This is the best stacking velocity for this event. By proceeding this way, we can build up a velocity function that is appropriate for NMO correction for this gather.

The most important reason for obtaining reliable stacking velocities is to get the best coherency on the stack. For this reason velocities are estimated from constant velocity stack panels.

CONSTANT VELOCITY STACKS (CVS)

Stacking velocities are often estimated from gathers, stacked with a range of velocities on the basis of stacked-event amplitude and continuity.

Figure 5–20 shows this method of estimating stacking velocities. In this example, a portion of a line consisting of 24 common-depth-point gathers has been NMO corrected and stacked with velocities ranging from 5,000 ft/sec to 13,600 ft/sec. The resulting stacked 24 traces, displayed as one panel, represent one constant velocity. These panels are displayed side by side with the velocity values indicated, where velocity values increase from right to left. Stacking velocities are picked directly from these panels by selecting the velocity that yields the best coheren-

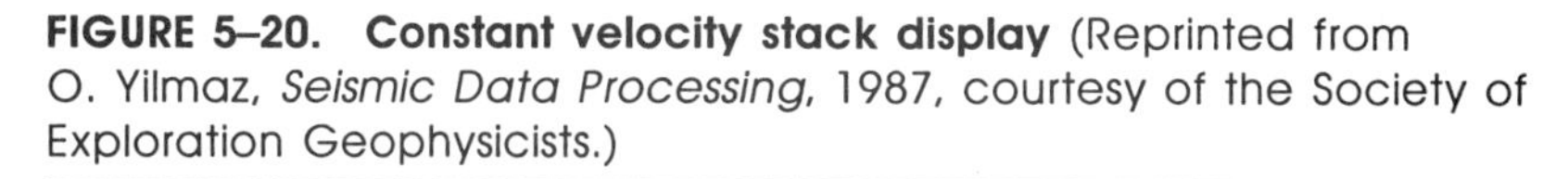

FIGURE 5–20. Constant velocity stack display (Reprinted from O. Yilmaz, *Seismic Data Processing,* 1987, courtesy of the Society of Exploration Geophysicists.)

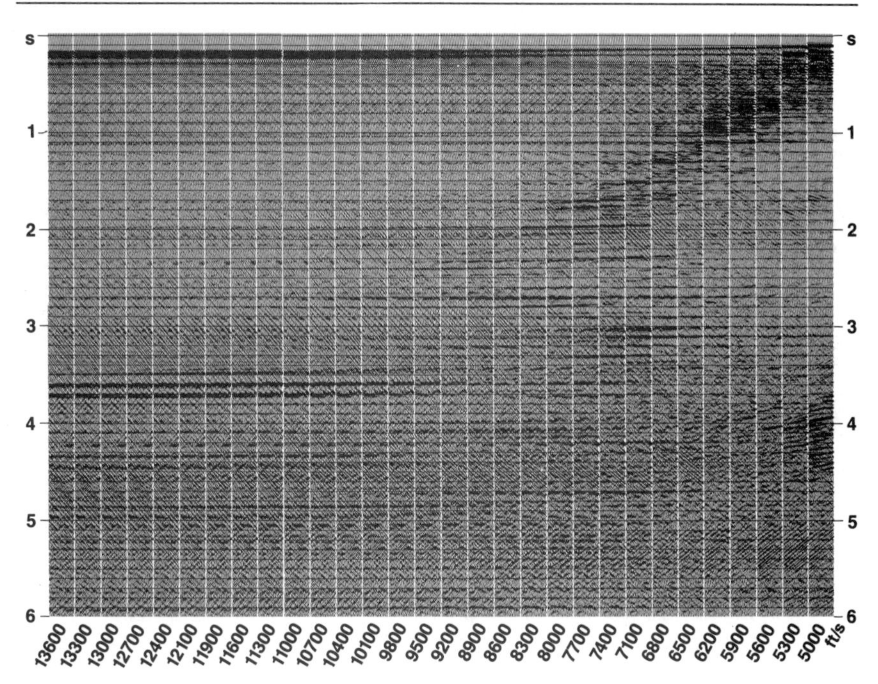

cy and the strongest amplitude for a velocity value at a certain center time.

Note that the deep event between 3.0 and 4.0 seconds seems to stack on a wide range of velocities. This represents the decrease of the resolution of the velocities with depth, due to the decrease of NMO (increase of velocity) with depth.

Figure 5–21 shows another presentation of constant velocity stacks in which the signal-to-noise ratio is low. The velocity scale is in 200 ft/sec velocity increments. The data are populated with multiple reflections.

Care must be taken in using this kind of velocity analysis to estimate the best stacking velocities. One should know the velocity range of an area, especially if there are structure changes. The velocity increment should be selected to be able to stack in all the small lateral and vertical changes. In this case, it is advisable to use a constant velocity increment.

VELOCITY SPECTRUM METHOD

The *velocity spectrum approach* is unlike the CVS method. It is based on the correlation of the traces in a CMP gather, and not on lateral continuity of the stacked events. This method, compared with the CVS method, is

FIGURE 5–21. Contant velocity stack display (courtesy of Seismograph Service Corporation)

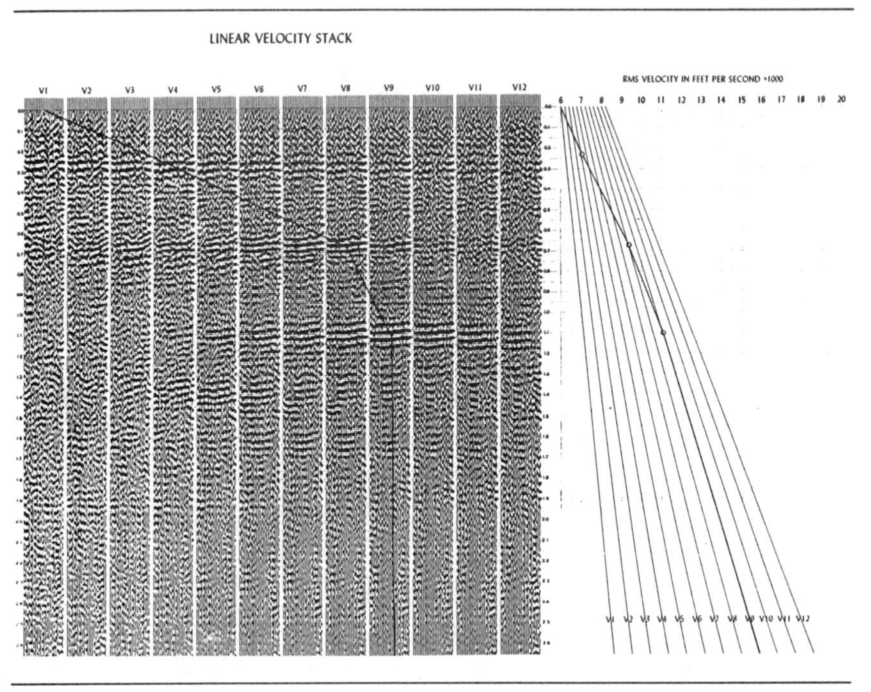

more suitable for data with multiple reflection problems, and it is less suitable for highly complex structure problems.

To demonstrate this approach to estimating the stacking velocities, we will use a model of a single reflector as demonstrated in Figure 5–22a. The input gather contains a single hyperbola from a flat reflector. Suppose we repeatedly correct the gather using constant velocity values from 2,000–4,300 m/sec, then stack the gather and display the stacked traces side by side. The result is a display of velocity versus two-way time, called a "velocity spectrum," shown in Figure 5–22b.

We have transferred the data from offset versus two-way time to stacking velocity versus two-way, zero-offset time. As you observe, the best stacking velocity for this single layer is 3,000 m/sec. If we apply the same approach to a multi-layer case (Figure 5–22a), based on the stacked amplitudes, these layers have best stacking velocities of 2,700, 2,800 and 3,000 m/sec, respectively, as shown on Figure 5–22b.

There are two commonly used ways to display the velocity spectrum: *power plot* and *contour plot.* Figure 5–23 illustrates these two kinds of displays. Figure 5–23a is the gather, 5–23b is the power spectrum presentation, and 5–23c is the coontoured version.

FIGURE 5–22. Velocity spectrum method (Reprinted from O. Yilmaz, *Seismic Data Processing*, 1987, courtesy of the Society of Exploration Geophysicists.)

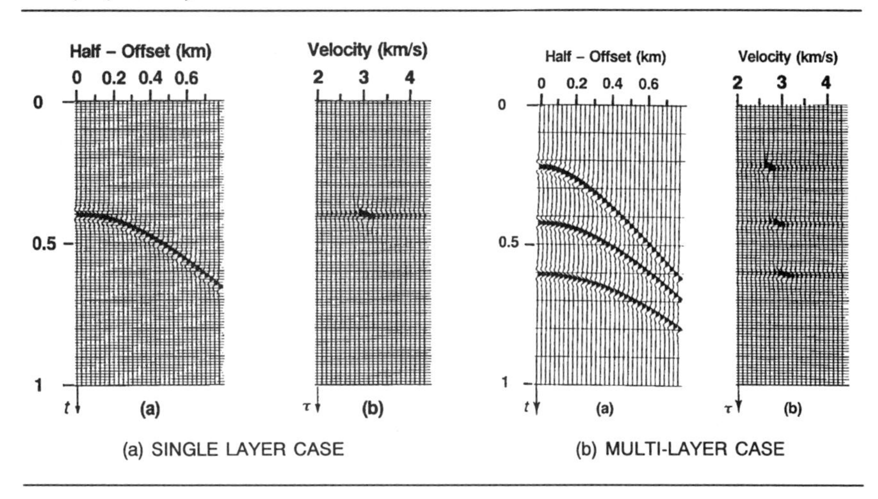

FIGURE 5–23. Velocity spectrum plots (Reprinted from O. Yilmaz, *Seismic Data Processing*, 1987, courtesy of the Society of Exploration Geophysicists.)

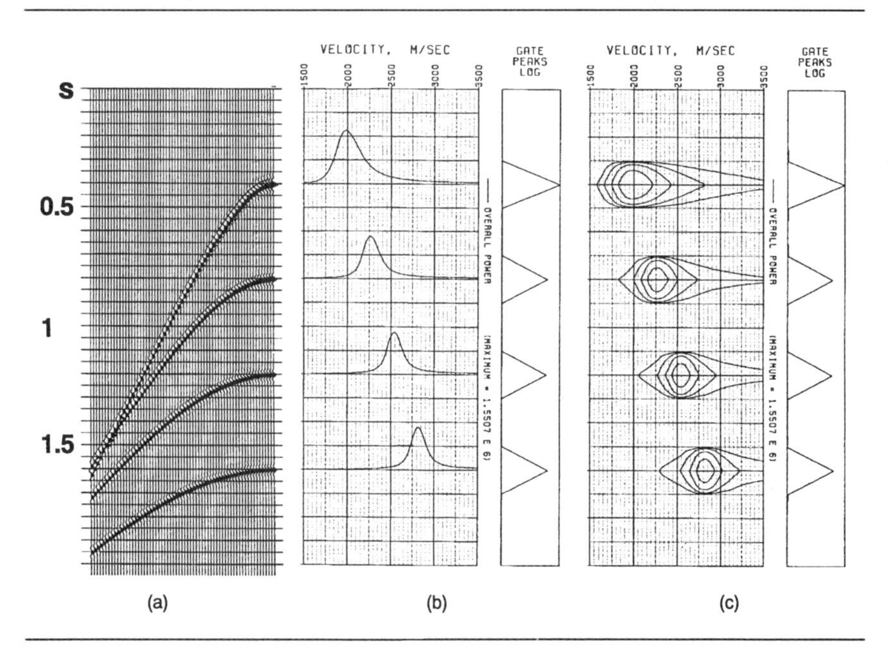

Figure 5–24 shows a suite of velocity spectra on a land line. The velocity spectra were pulled at CDPs 88, 188, 408, 498, and 578. Notice the low resolution on the velocity picks down the section; this may be due to high-noise level and velocities, as these cause lack of discrimination of the velocity values.

HORIZON VELOCITY ANALYSIS

One method to estimate velocities with enough accuracy for structural and stratigraphic applications is to analyze the velocities of a certain horizon of interest continuously. Such a detailed velocity analysis is called *Horizon Velocity Analysis.* The velocity is estimated at every CMP along the selected key horizons of interest on the stack section. The principle of estimating the velocities by this method is the same as that of the velocity spectrum. The output coherency values derived by hyperbolic time gates are displayed as a function of velocity and CMP position.

Horizon times are picked from the stacked section, digitized, and input to the horizon velocity-analysis program. The output can be displayed as shown in Figure 5–25. Whenever there are discontinuities on

FIGURE 5–24. Examples of velocity spectra (Reprinted from O. Yilmaz, *Seismic Data Processing,* 1987, courtesy of the Society of Exploration Geophysicists.)

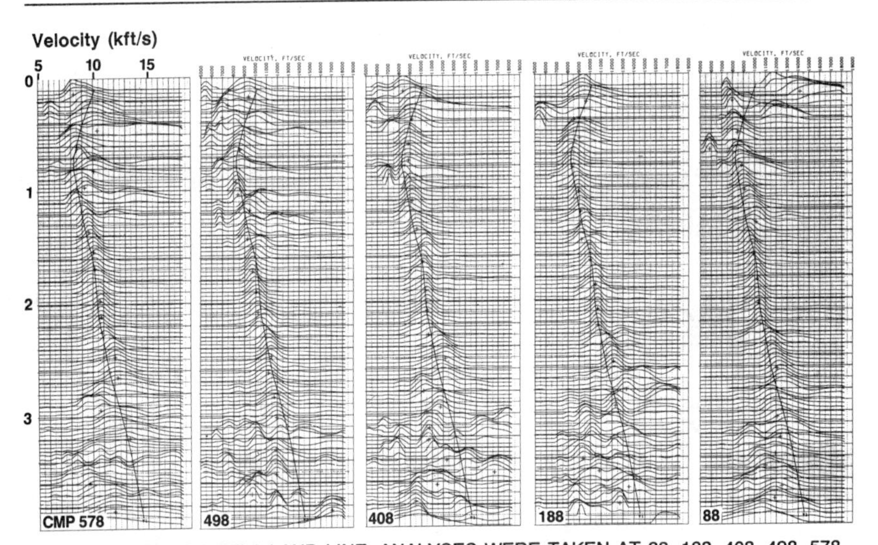

VELOCITY SPECTRA OF A LAND LINE. ANALYSES WERE TAKEN AT 88, 188, 408, 498, 578.
VELOCITY RANGES FROM 5,000–19,000 FT/SEC.
NOTE THE LACK OF RESOLUTION DOWN DEEP ON THE ANALYSES BELOW 2.5 SEC., DUE TO HIGH VELOCITY VALUES AND NOISE.

FIGURE 5–25. Horizon velocity analysis (Reprinted from O. Yilmaz, *Seismic Data Processing,* 1987, courtesy of the Society of Exploration Geophysicists.)

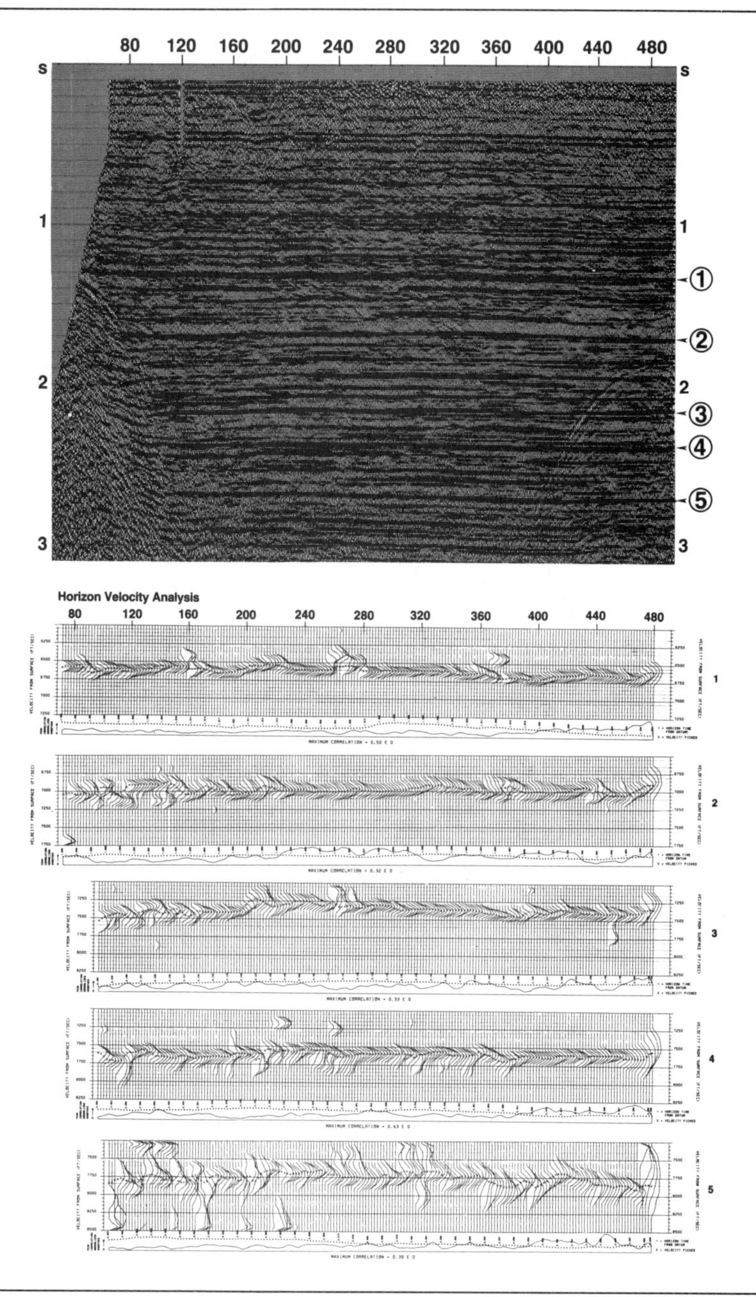

a stack section, HVA is carried out on segments of the horizon that are separated by faults.

One of the applications of horizon velocity analysis is to improve the layered velocity variations along marker horizons, especially if these velocities are used in post-stack depth migration.

PRESTACK ANALYSIS

In the case of marine data, a near-trace monitor is displayed for quality control, to design velocity analysis locations, for editing purposes, and for design of deconvolution parameters. In the case of land data, especially if it is noisy, a prestack analysis is done by running a filter test and deconvolution test. The purpose is to enhance the signal-to-noise ratio by filtering out some undesired signal patterns and to apply deconvolution to attenuate the short-period multiples by designing a short operator (Figure 5–26).

For marine data, a debubbling program may be used instead of (or in conjunction with) decon. Marine data are normally free from the random and coherent noise seen in land data, but they suffer from a high degree of multiple reflections in many offshore areas.

A common depth point tape is normally generated to construct many process steps, such as applying elevation statics, velocity analysis, residual statics, and stack.

FIGURE 5–26. Pre-stack analysis

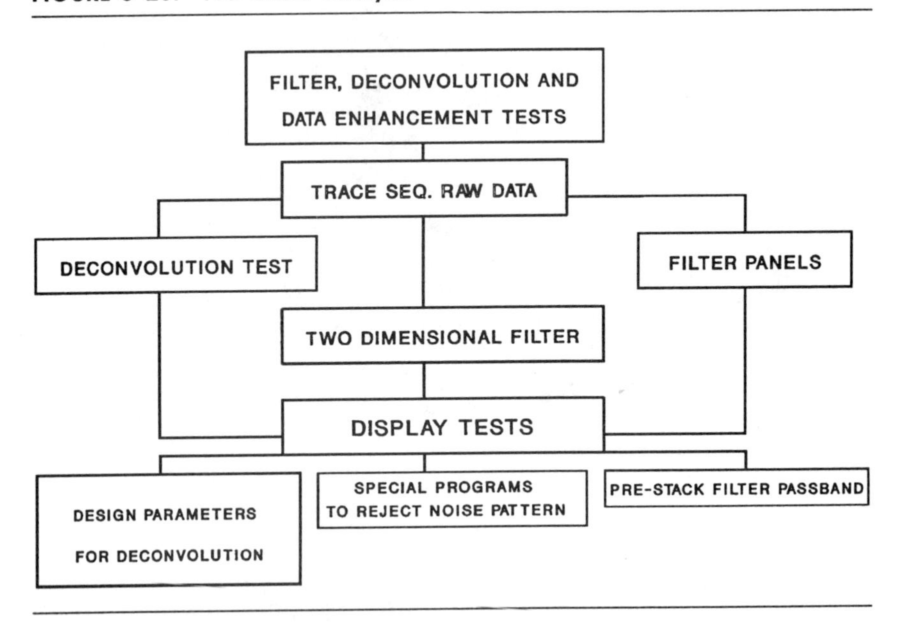

To sort the data, distances and angle changes on the seismic line will be taken into consideration to obtain the correct distance from the source to the receiver for each trace. Elevation statics are usually applied before normal moveout to derive velocity analysis from seismic data.

There is no routine processing sequence that will fit every seismic survey acquired in any part of the world, but there is an outline that must be followed. The processing sequence must be designed according to the area, record quality, and any specific problems involved.

MUTE

Mute is the process of excluding a part of the trace that contains only noise or more noise than signal. The two types of mute used are *front-end mute* and *surgical mute.*

FRONT-END MUTE

In modern seismic work, the far geophone groups are quite distant from the energy source. On the traces from these receivers, refractions may cross and interfere with reflection information from shallow reflectors. However, the nearer traces are not so affected. When the data are stacked, the far traces are muted down to a depth at which reflections are free of refractions.

Mute changes the relative contribution of the components of the stack as a function of record time. In the early part of the record, the long offset may be muted from the stack because the first arrivals are disturbed by refraction arrivals, or because their frequency content after applying normal moveout correction is lower than that of other traces (normal moveout stretch).

The transition where the long offsets begin to contribute may be either gradual or abrupt. However, an abrupt change may introduce frequencies that will distort the design of the deconvolution operator. Figure 5–27 illustrates the front-end mute process.

SURGICAL MUTE

Muting may be over a certain time interval to keep ground roll, air wave, or noise patterns out of the stack. This is especially applicable if the noise patterns are in the same frequency range as the desired signal. Convolution to filter out the noise may also attenuate the desired signal. Figure 5–28 illustrates the surgical mute approach.

FIGURE 5–27. Front-end mute (courtesy of Seismograph Service Corporation)

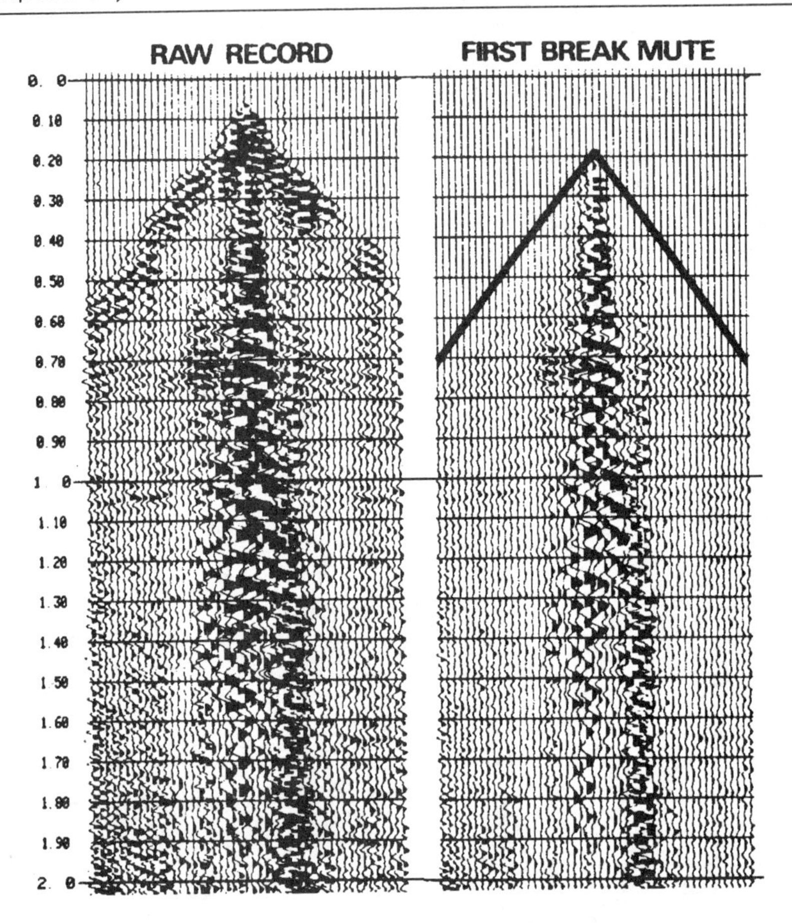

BEFORE STACKING THE DATA, A MUTE PATTERN IS DESIGNED TO CUT OFF THE INTERFERING FIRST ARRIVALS AND OTHER NOISE TRAINS. THIS CUT-OFF IS DOWN TO A TIME WHERE REFLECTIONS ARE FREE FROM THE CHANGE OF FREQUENCY DUE TO NMO STRETCH

STATIC CORRECTIONS

Two main types of correction need to be applied to reflection times on individual seismic traces in order that the resultant seismic section gives a true representation of geological structure. These are *dynamic* and *static corrections.* The static correction is so-called because it is a fixed time correction applied to the entire trace. A dynamic correction varies as a function of time. We have already discussed the dynamic corrections, which involve the application of NMO.

FIGURE 5–28. Surgical mute (courtesy of Seismograph Service Corporation)

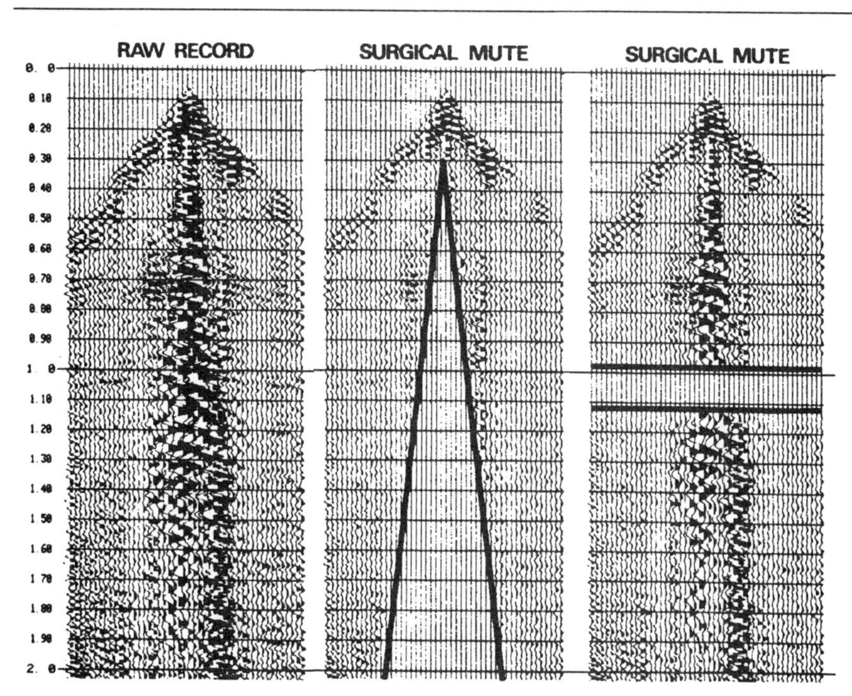

THIS IS A FIELD RECORD WITH A CRISP FIRST ARRIVALS THAT INTERFERES WITH THE SHALLOW REFLECTOR. FRONT-END MUTE MAY BE DESIGNED TO CUT OUT THE FIRST ARRIVALS FROM THE RECORD. INSIDE THE RECORD IS A CONE OF NOISE CALLED "GROUND ROLL." IN SOME CASES ITS FREQUENCY IS THE SAME AS THAT OF A REFLECTOR OF INTEREST. SURGICAL MUTE IS USED TO CUT OFF THE GROUND ROLL, SINCE CONVOLUTION MAY ATTENUATE THE SIGNAL.

ELEVATION STATICS

In order to obtain a seismic section that accurately shows the subsurface structure, the reflection must be reduced to a defined reference time. This is normally taken as a horizontal datum plane at a known elevation above mean sea level and below the base of the variable-velocity weathered layer (see Figure 5–29). The value of the total statics (ΔT) depends on the following factors:

1. The perpendicular distance from the source to the datum plane.
2. Surface topography; that is, the perpendicular distance from the geophone to the datum plane.
3. Velocity variations in the surface layer along the seismic line.
4. Irregularities in thickness of the near-surface layer.

FIGURE 5–29. Surface-consistent statics model

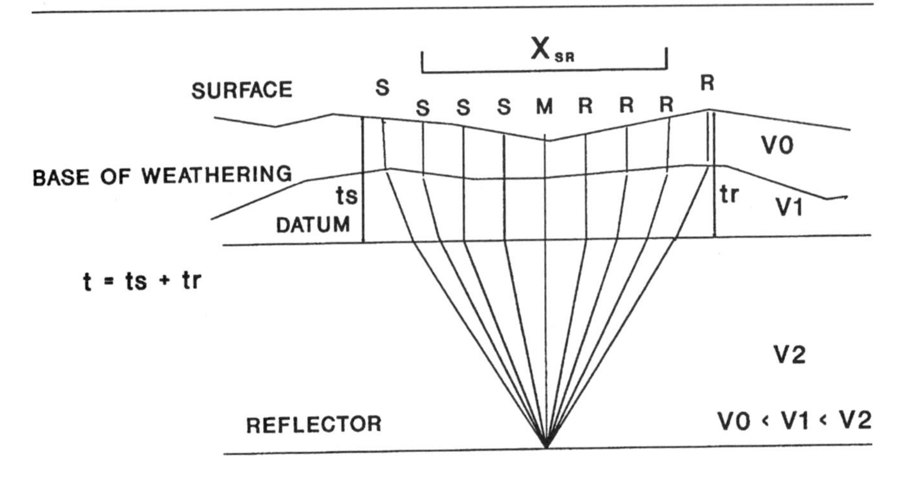

THE KEY ASSUMPTION IS THAT THE RESIDUAL STATICS ARE SURFACE CONSISTENT. THIS MEANS THAT STATICS SHIFTS ARE TIME DELAYS THAT DEPEND ON THE LOCATIONS OF SOURCE AND RECEIVER ON THE SURFACE. THIS ASSUMPTION IS VALID IF ALL RAYPATHS, REGARDLESS OF RECEIVER OFFSET, ARE VERTICAL IN THE NEAR-SURFACE LAYERS. VELOCITY IS QUITE LOW IN THE WEATHERED LAYER, AND REFRACTION IN ITS BASE TENDS TO MAKE THE TRAVEL PATH VERTICAL. THE SURFACE CONSISTENT ASSUMPTION USUALLY IS A GOOD AND VALID ONE. IN HIGH VELOCITY AREAS, AS IN PERMAFROST, IT MAY NOT BE VALID, AS HIGH VELOCITY CAUSES RAYS TO BEND AWAY FROM THE VERTICAL.

In computing ΔT, it is usually assumed that the reflection raypath in the vicinity of the surface is vertical.

The total correction is:

$$\Delta T = \Delta t_s + \Delta t_r$$

where

Δt_s = the source correction, in ms
Δt_r = the receiver correction, in ms

Figure 5–30 shows common shot gathers from a land line, where statics (due to near-surface formation irregularities) caused the departure from hyperbolic travel times on the gathers at the right side of the display.

RESIDUAL STATICS

They are statics deviation from a perfect hyperbolic traveltime after applying NMO and elevation statics corrections to traces within the CMP gather.

FIGURE 5–30. Near-surface statics problem (Reprinted from O. Yilmaz, *Seismic Data Processing*, 1987, courtesy of the Society of Exploration Geophysicists.)

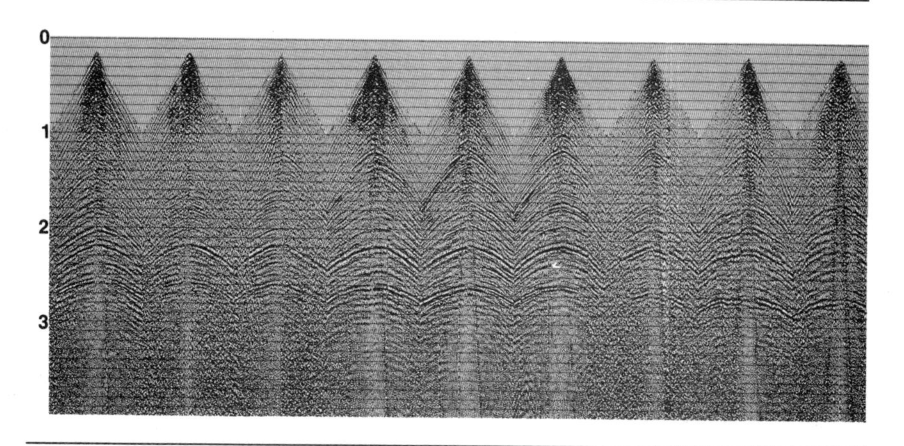

These statics cause misalignment of the seismic events across the CMP gather and generate a poor stack trace. We need to estimate the time shifts from the time of perfect alignment, then correct them using an automatic procedure.

A model is needed for the moveout-corrected traveltime from a source location to a point on the reflecting horizon, then back to a receiver location (see Figure 5–29). The key assumption is that the residual statics are surface consistent, meaning that static shifts are time delays that depend on the sources and receivers on the surface (Hileman et al., 1968 and Taner et al., 1974). This assumption is valid if all raypaths, regardless of source-receiver offset, are vertical in the near-surface layers.

Since the near-surface weathered layer has a low velocity value, and refraction in its base tends to make the travel path vertical, the surface consistent assumption usually is valid. However, this assumption may not be valid for high-velocity permafrost layers in which rays tend to bend away from the vertical.

Residual static corrections involve three stages:

1. Picking the values.
2. Decomposition of its components, source and receiver static, structural and normal moveout terms.
3. Application of derived source and receiver terms to traveltimes on the pre-NMO corrected gathers after finding the best solution of residual static corrections. These statics are applied to the deconvolved and sorted data, and the velocity analysis is re-run. A refined velocity analysis can be obtained to produce the best coherent stack section.

FIGURE 5–31. Residual statics corrections

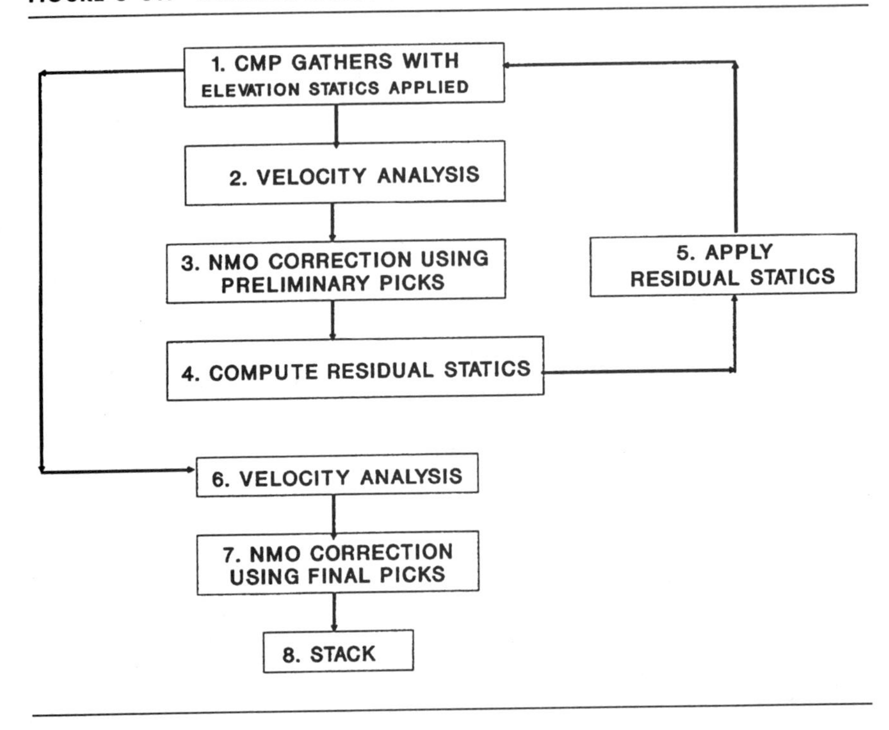

Figure 5–31 is a recommended flow chart for residual static correction and velocity analysis.

Figure 5–32 shows the effect of residual static application on velocity analysis and velocity picks.

Figure 5–33 shows a group of CMP families with normal moveout applied. Figure 5–33a is before residual static application, and 5–33b is after application of the static to each individual trace within the CMP family.

One expects better coherency on the stack after residual static is applied. Figure 5–34a shows more severe static problems, and Figure 5–34b shows the dramatic improvement after application of the surface consistent residual static.

Figure 5–35a is a CMP stack obtained by applying elevation static for elevation changes. Figure 5–35b is the stack after applying surface-consistent static. Notice the great improvement at the right side of the section.

Velocity analysis refinement can yield better velocity picks that result in a better stack. Figure 5–36 illustrates the resolving power of the

FIGURE 5–32. Effect of residual statics on velocity picks (Reprinted from O. Yilmaz, *Seismic Data Processing*, 1987, courtesy of the Society of Exploration Geophysicists.)

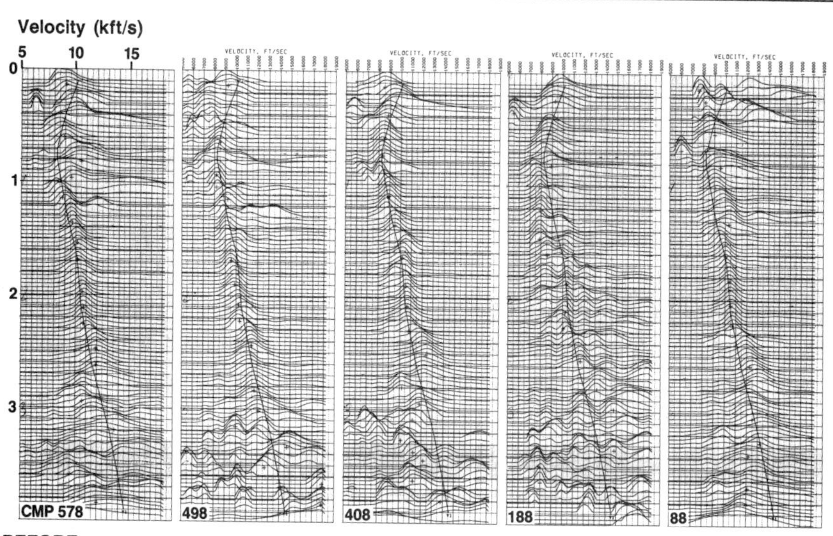

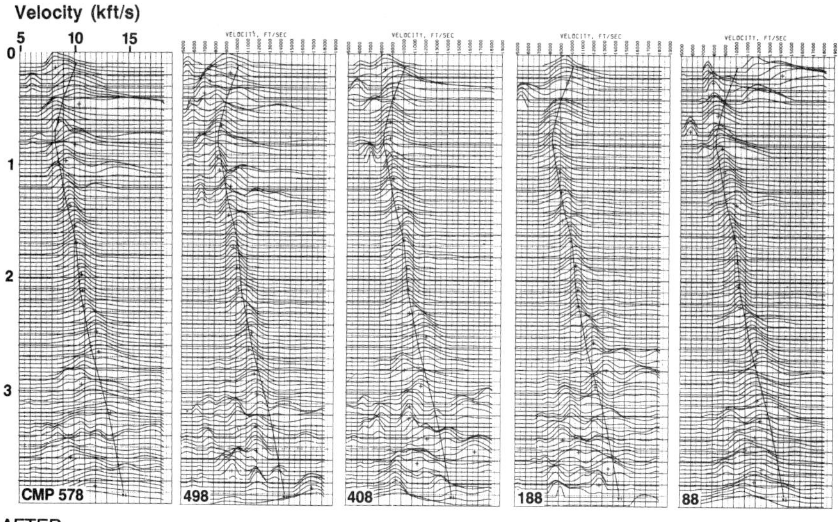

FIGURE 5–33. NMO gathers before and after residual statics (Reprinted from O. Yilmaz, *Seismic Data Processing*, 1987, courtesy of the Society of Exploration Geophysicists.)

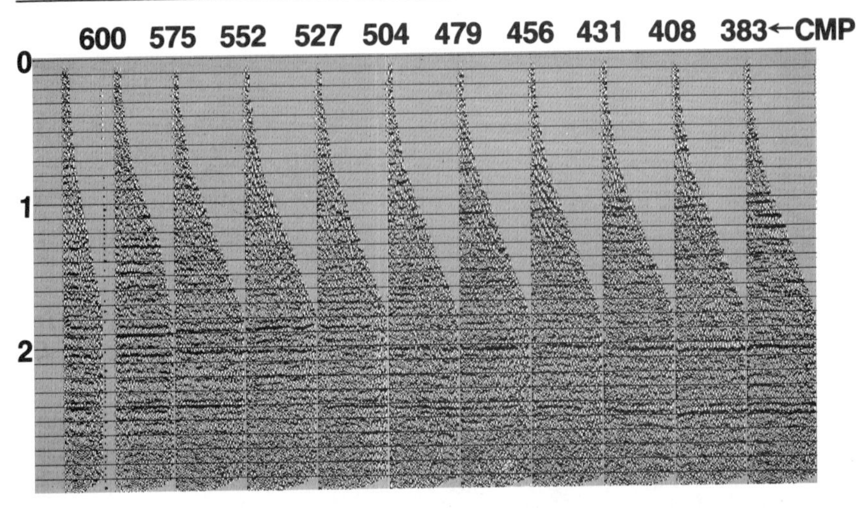

BEFORE

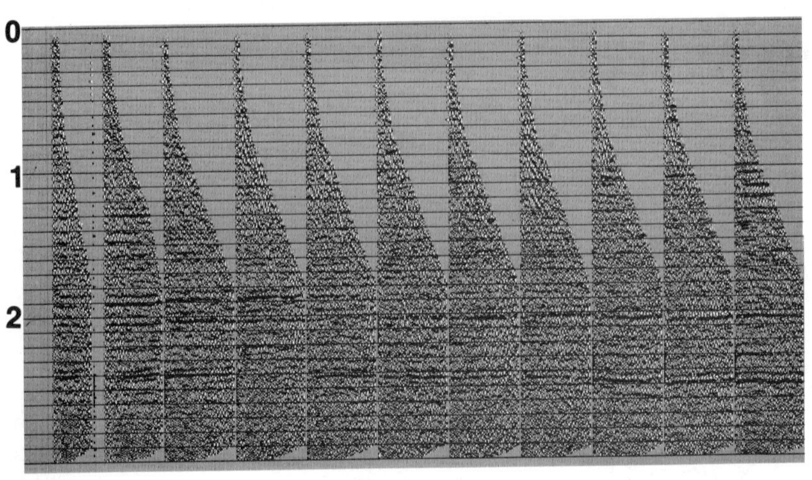

AFTER

FIGURE 5–34. NMO gathers before and after residual statics (Reprinted from O. Yilmaz, *Seismic Data Processing,* 1987, courtesy of the Society of Exploration Geophysicists.)

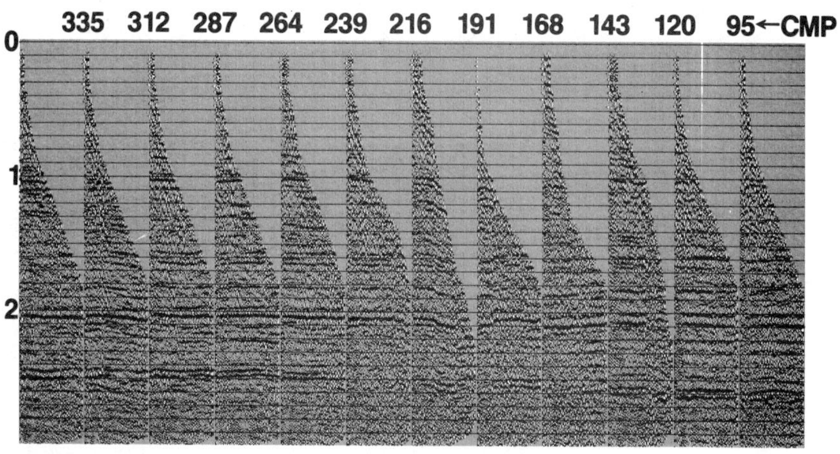

BEFORE

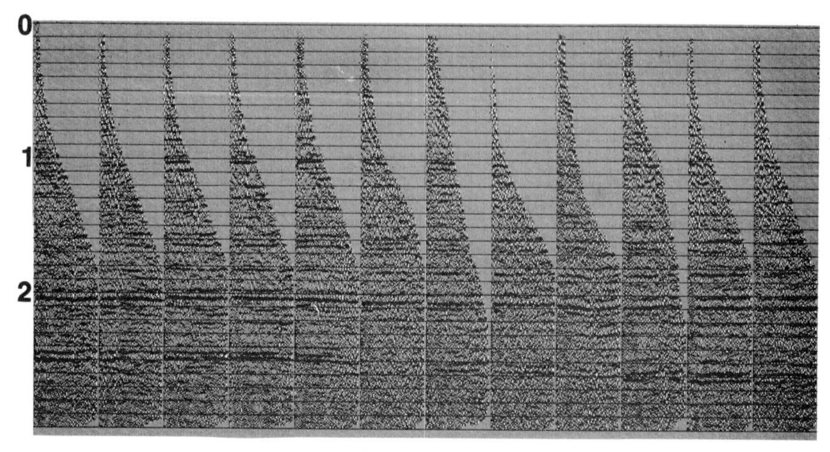

AFTER

FIGURE 5–35. Residual statics corrections (Reprinted from O. Yilmaz, *Seismic Data Processing*, 1987, courtesy of the Society of Exploration Geophysicists.)

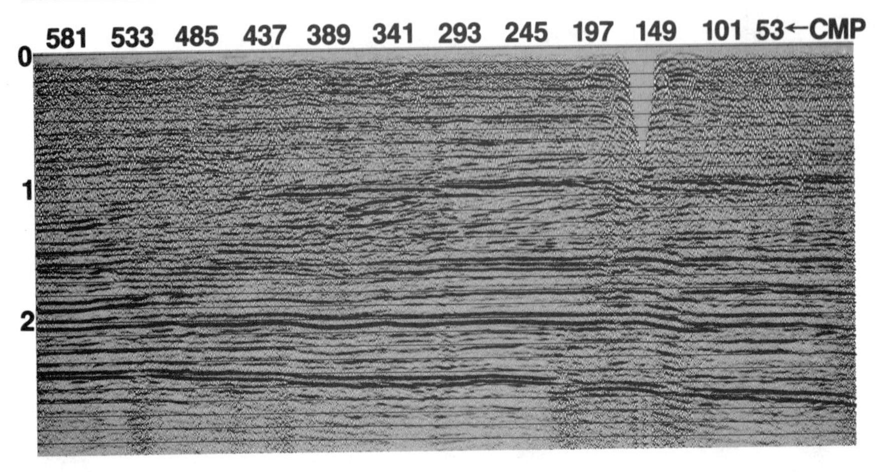

BEFORE

AFTER

FIGURE 5–36. Residual statics corrections (Reprinted from O. Yilmaz, *Seismic Data Processing,* 1987, courtesy of the Society of Exploration Geophysicists.)

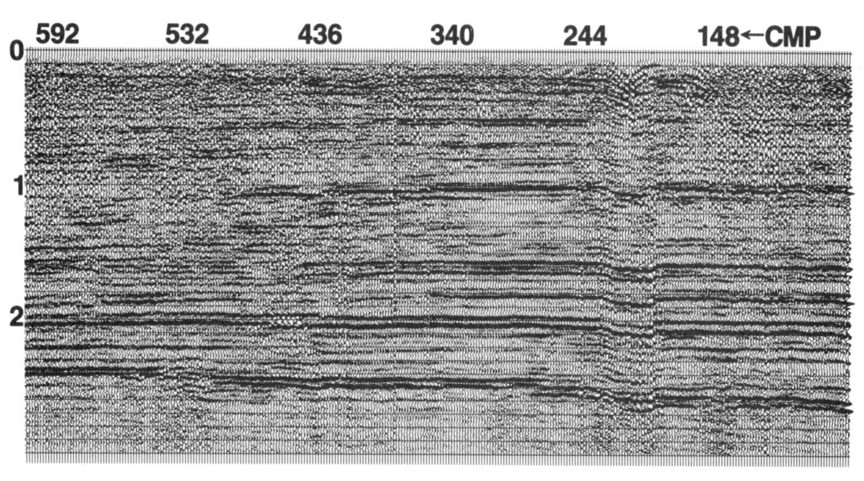

BEFORE

AFTER

surface consistent static on the near-surface weathered layer. It is obvious that the static solution did minimize the effect of the irregularities of the near-surface formation and corrected the normal moveout hyperbola to its desirable shape. By applying pre-NMO sort data and re-running the velocity analysis, new velocity picks are obtained. These generate a superior stack, as shown in the post-static version of the stack.

REFRACTION STATICS

In the example of Figure 5–36, surface-consistent residual static corrections solved the irregularities in the near-surface and corrected them. By time-shifting from trace to trace, it built better continuity and coherency on the stack section. This kind of static is called a *short-period static.*

An important question in estimating the shot and receiver static is the accuracy of the results as a function of the wavelength of static anomalies. Surface-consistent residual statics cannot solve long-period static components. Figure 5–37 shows that the residual static application yields a much improved stack response. Short-period static shifts (less than one spread length) cause traveltime distortion, which degrades the stack section. Correcting for short-period statics is not enough.

The structures at points A and B are probably created by long-period static. This is identified by tracking the shallow horizon and the elevation profile. This example suggests that the elevation statics were not applied adequately.

Residual static corrections are needed because the elevation or datum static corrections are not enough to compensate for irregularities in the near-surface formations. This is because the lateral variations of velocity in the weathered layer are unknown.

Residual static correction does a good job on the short-period static, but it does a poor job on the long-period static. The reason is that the residual static programs work on the arrival time difference between traces and not on absolute time values.

The refraction static method works on the absolute values of the first break arrival times.

Figure 5–38a illustrates selected CMP gathers from a stack section.

Figure 5–38b shows the gathers after applying linear moveout correction. A velocity value was chosen to flatten the refraction breaks.

Figure 5–38c is the stack of the section after applying linear moveout to the gathers.

Figure 5–38d is the stack after applying long-period statics. Statics were applied after picking values and derive a solution from 5–38b.

Figure 5–39 shows the dramatic difference before and after removing long-period statics by the refraction statics method.

FIGURE 5–37. Residual statics corrections—Long-period statics (Reprinted from O. Yilmaz, *Seismic Data Processing*, 1987, courtesy of the Society of Exploration Geophysicists.)

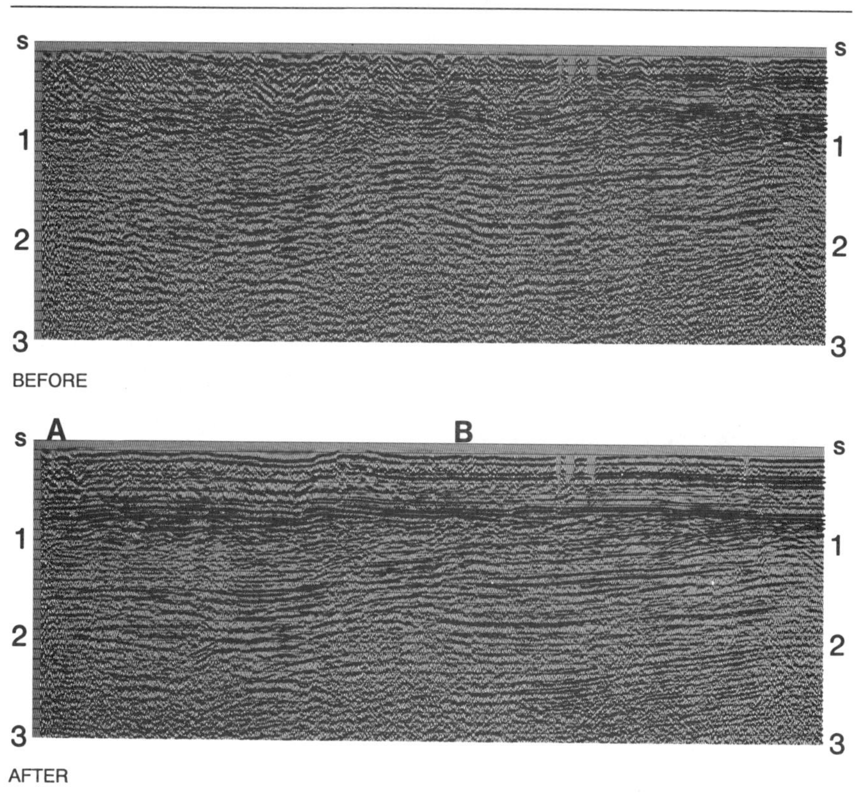

From the previous discussions, we can conclude that in case of severe weathering problems on land data statics may be applied as follows:

1. Elevation static corrections to account for elevation changes.
2. Refraction-based static corrections to remove long-period anomalies in a surface-consistent manner.
3. Surface-consistent residual static corrections to remove any remaining short-period static shifts.

STACKING

Stacking is combining two or more traces into one. This combination takes place in several ways. In digital data processing, the amplitudes of the traces are expressed as numbers, so stacking is accomplished by adding these numbers together.

FIGURE 5–38. Refraction statics method (Reprinted from O. Yilmaz, *Seismic Data Processing*, 1987, courtesy of the Society of Exploration Geophysicists.)

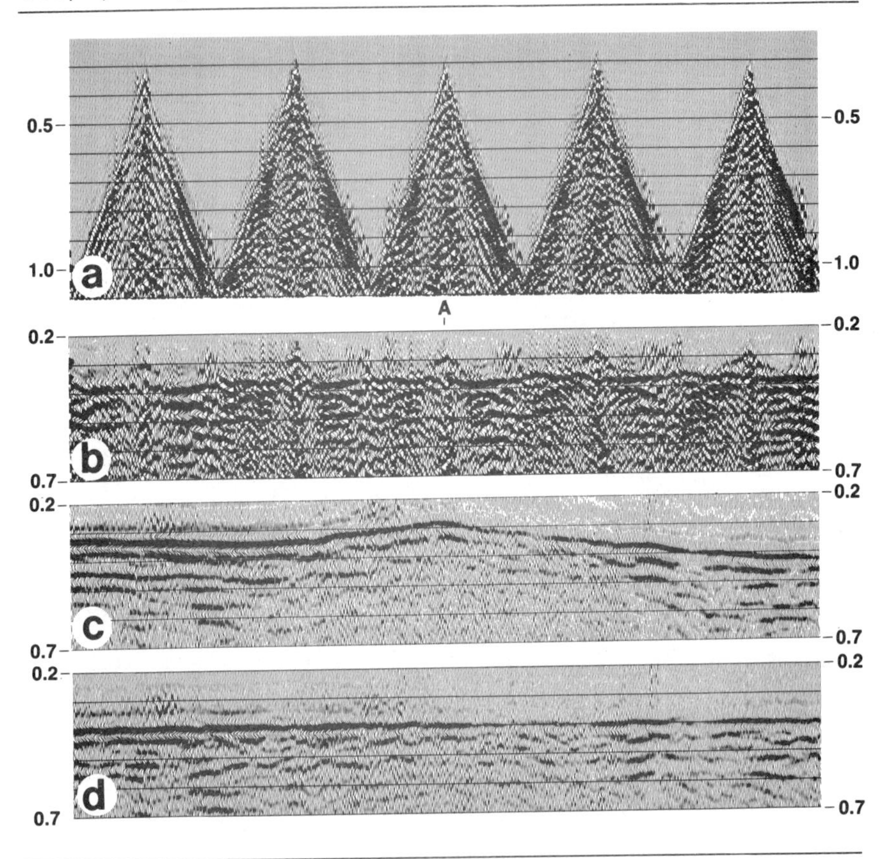

Peaks appearing at the same time on each of two traces combine to make a peak as high as the two added together. The same is true of two troughs. A peak and a trough of the same amplitude at the same time cancel each other, and the stack trace shows no energy arrival at that time. If the two peaks are at different times, the combined trace will have two separate peaks of the same sizes as the original ones. After stacking, the traces are "normalized" to reduce the amplitude so that the largest peaks can be plotted.

Figure 5–40 illustrates the principle of stacking.

Applications of stacking include testing normal moveout, determining velocities, and attenuating noise to improve signal-to-noise ratio.

Six traces in a CMP family is called (after stack) a 600% stack. Similarly, 24 traces in a CMP family is called a 2,400% stack.

FIGURE 5–39. Refraction statics (Reprinted from O. Yilmaz, *Seismic Data Processing*, 1987, courtesy of the Society of Exploration Geophysicists.)

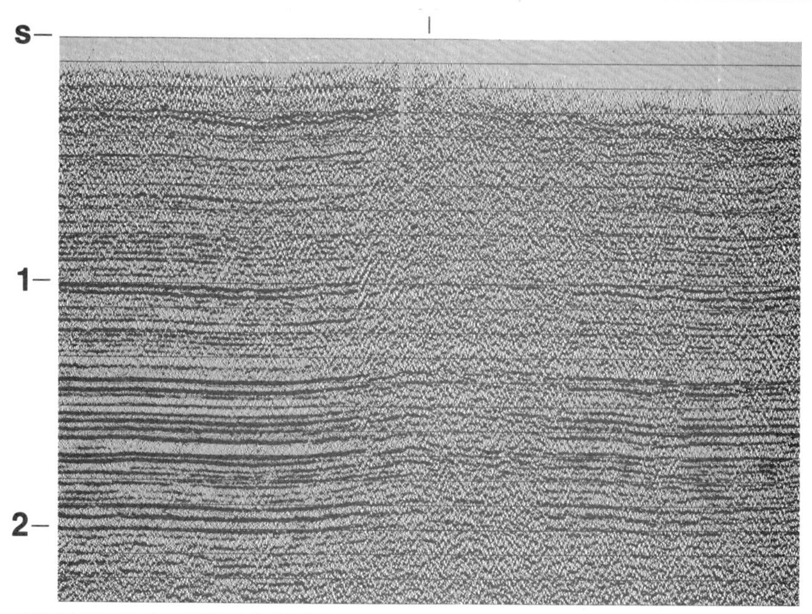

STACK WITH ELEVATION STATICS

STACK AFTER LONG-PERIOD STATICS REMOVED USING REFRACTION ARRIVALS

FIGURE 5–40. The stacking process

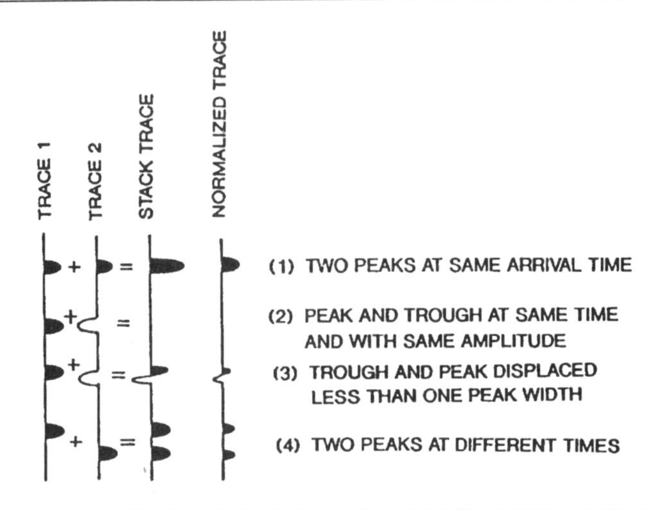

DATA PROCESSING OBJECTIVES

One of the most important objectives of seismic data processing is to produce readily interpretable sections. These sections must be geologically sound and represent the subsurface of the earth. They should be good enough to image the subsurface traps, where other seismic techniques may be used for qualitative analysis of some petrophysical properties of the reservoir rocks. Seismic sections are used for special applications such as direct hydrocarbon indicators, porosity presence, and some other applications that will be reviewed later in the text.

FACTORS AFFECTING SEISMIC DATA PROCESSING OBJECTIVES

Seismic data processing objectives are strongly affected by field data-acquisition parameters. They are also affected by the quality of the field work and the condition of the equipment. There is an impression that field problems can be solved in the data processing stages, using the magic of the computer software and the muscle of the hardware. The bad news is that some (but not all) field problems can be solved in the data processing center.

Marine conditions strongly affect the quality of the acquired data. Rough seas can cause streamer feathering, and a hard bottom will generate a variety of multiple reflections.

Seismic data are usually collected under conditions that are less than ideal. Although we may not have the most desired information, data processing techniques may help to obtain the best information available in the seismic data.

Like any other exploration tool, seismic method has its limitations. These start in the way the data was acquired in the field (i.e., the data resolution) which is determined by the frequency band recorded. We can improve the distribution of frequencies within the band limit, but we cannot generate genuine frequencies outside of this band. If we try to artificially create frequencies outside the resolution limit, we are stretching the seismic tool beyond its capability and creating pitfalls in its applications.

Near-surface formations play a critical role in the integrity of the subsurface sections; many dry holes have been drilled on apparent anticlinal features caused by near-surface anomalies.

Data processing is a part of the interpretation phase; better and more reliable seismic sections are generated if the processing geophysicist understands the structure and stratigraphy of the area. Good communication between the geophysicist, the geologist, and the engineer is necessary to get the best possible subsurface image.

OVERVIEW OF DATA PROCESSING FLOW

One of the most challenging jobs in the exploration and exploitation phases of the industry is to get a geologically sound seismic section that is reliable enough to map the subsurface, describe the lithologic facies, to detect a hydrocarbon accumulation, and to define its lateral extent.

Every basin has its geological settings, rock properties, configuration, water depth, water bottom conditions, near-surface formations, outcrops, tectonics, and sedimentary environment. Target horizons vary from shallow to deep. Traps vary from structural to stratigraphic or combinations of the two. Sizes of these traps vary from small to large. Pay zones vary from thin to thick.

Accordingly, field data-acquisition techniques should cover all these factors in order that the recorded data will reveal the desired information. In turn, the seismic data-processing sequence will be designed to manipulate the recorded information to extract the correct image of the subsurface.

So, as one can see, each area must be treated as a unique case. There is no such thing as a flow chart of data processing techniques that may be applied to every seismic line acquired in any part of the world, or a cookbook method to follow that will produce a valid seismic section.

Extensive testing should be conducted after the data demultiplexing stage in order to design the proper parameters to apply to the seismic data in hand. Prestack analyses such as filter and decon tests should be conducted to study the frequency spectra. Then one can design the optimum parameters to obtain better distribution of the frequencies within the usable bandwidth and attenuate noise to enhance signal-to-noise ratio.

In the case of marine seismic data with strong multiples, tests are conducted to attenuate the multiples, revealing and enhancing the desired signal. Special problem tests are done, such as velocity filter and other two-dimensional filters, to help solve the problems before starting the routine flow of processing.

One can test the surface-consistent static problem. This will solve the short-period static, which affects the continuity of the seismic marker. On the other hand, there is the problem of long-period statics (i.e., static with a wavelength longer than a cable length), which can cause false structural phenomena.

A refraction statics program may solve this kind of problem. In deriving the velocities from seismic data, one should know the velocity range of the area. This depends on the geological setting, type of rocks, and depth of the target.

Near-surface statics can affect the velocities derived from surface seismic data. A static solution should be reached, residual statics should be applied to the sorted data, and velocities should be run and repicked, as discussed previously.

Selecting the proper front-end mute can be very important in revealing shallow events, especially in areas where geological markers outcrop. Space-variant mute may be used to follow the variation in the subsurface structure in the shallow part of the section.

Gain is one of the most important parameters, particularly if the data is to be processed for the direct detection of hydrocarbons. The type of gain applied is important in every stage of the data-processing flow chart. Selecting the gain to display a process stage can either help or hinder the interpretation of the optimum parameters.

After we reach stack stage, post-stack filter and decon tests must be done routinely in the test line to see if filter and/or decon will help the data after stack. If it does not help, there is no reason to tamper with a good stack, and the filters are not applied.

Then, it is apparent that there is no universal rule for processing a seismic line. A typical land data processing sequence may include:

Stage 1. Field tapes
Data reduction—demultiplex and cross-correlation
Line geometry
Prepare prestack analysis
Decon, filter analysis, special problems tests

Stage 2. Common depth sort (gather)
Prestack filter and/or decon applications if needed
Special enhancement to improve signal-to-noise, if needed
Application of elevation statics

Stage 3. Velocity analysis
Velocity spectra—constant-velocity stack
Design of front-end mute
Stage 4. Velocity applications
Normal moveout check
Stage 5. Correlation statics
Surface-consistent residual statics
Application of statics to sort
Re-run velocity analysis with refined static
Refined surface-consistent residual statics
Correlation statics
Stage 6. Final stack
Post-stack filter and decon tests
Post-stack enhancement programs
Migration
Seismic inversion
Special post-stack applications

SUMMARY AND DISCUSSION

There is a belief that the powerful computers and the sophisticated software applications can solve field problems. The bad news is that the computer software will not solve all the problems. If the problem of the data is resolution and insufficient data acquired in the field, or errors in the field parameters or malfunctions of the field equipment, computer software cannot retrieve such information because it is not present in the field seismic data.

Data processing is a very powerful tool in manipulating the field information. It converts the raw data with all information, including noise and distorted information, to meaningful seismic cross-sections that represent vertical slices through the subsurface. These slices represent geological information and potential hydrocarbon traps.

We have learned a generalized seismic data-processing flow. Our intention is to explain the physical meaning of some "buzz" words you hear, such as *demultiplexing*, which is rearranging the samples in the traces and across the records boundaries to make the data in trace sequential form for further software application. *Cross-correlation* is used to extract information from seismic data acquired by the Vibroseis method; it is used to solve statics problems and for many more seismic data-processing applications. *Gain* is applied to compensate for spherical divergence and to preserve the relative true amplitude, which can be used as hydrocarbon indicator.

Convolution is used to filter out some undesirable frequencies while *deconvolution* is a process to attenuate short-period multiple reflections and to enhance the vertical resolution. *Normal moveout* is applied to correct for the field geometry and produce a zero-offset stack section as if the source and receiver were in the same location.

Elevation static are applied to minimize the effect of the variation of the surface elevations. *Refraction static* are applied to solve the irregularities in the near-surface formation called "weathering layer" or "layers." It may vary in thickness and velocities, both laterally and vertically. These near-surface layers distort the deep structural and stratigraphic features that are obtained from seismic sections.

Data processing is both an art and science. The processing steps should be handled in logic order. Every step should be checked carefully before proceeding to the next step.

Remember that the goal of processing the seismic data is to obtain geologically sound sections and not just a pretty one with man-made information. The map is only as reliable as your seismic data. Many dry holes have drilled on what appeared to be structural highs due to processing problems.

As we discussed in Chapter 4, it is difficult to calculate the cost to acquire a mile of seismic data; you will also find that it is not easy to give a price range for processing a mile of seismic data.

The cost per mile depends upon the type of data, whether it is marine or land, the signal-to-noise ratio of each type, the field problems involved, the field configuration such as the number of channels, fold, and sampling interval; and the processing sequence needed including special software programs to perform certain tasks.

If you elect to use a contractor to process newly acquired seismic data or you would like to evaluate a prospect and a reprocessing job is needed, it is advisable to consult with several data processing contractors. Software applications for basic processing are very similar from one contractor to another. The knowledge and the experience of the processing geophysicist is the major factor in the quality and the validity of the processed data.

The interaction of you or your staff in the data processing sequence is very important; your geological input and any information may help to solve a problem is a very rewarding experience. It will help you to make a sound judgment on a drilling location.

Seismic work stations may be used for post-stack enhancement programs such as modelling, synthetic seismograms, 3-D interpretation, and other applications. The cost of having one of these stations varies according to the hardware configuration required to apply the various software packages. It varies from $10,000 for a small system to $200,000 for a powerful hardware configuration.

KEY WORDS

Absorption
Ambient noise
Autocorrelation
Convolution
Correlation
Cross-correlation
Deconvolution
Demultiplex
Dynamic correction
Gain
Gate
Migration
Multiplex
Mute
Residual statics
RMS (velocity)
Stacking
Static corrections

EXERCISES

1. Given the wavelets *A.* (–1, 3, 2) and *B.* (1, –1),
   a. Find the autocorrelation of each wavelet.
   b. If the energy is defined as the square of the amplitude, find the cumulative energy of each wavelet.
2. Find the cross-correlation of the following wavelets:
   A. (3, –4, 2, 1) *B.* (1, 0, –6, 2).
3. Find the convolution of *A.* (4, –2, 1, 3) and *B.* (–1, 0, 1).
4. List three important geologic factors that influence the velocity of seismic waves in rocks, and discuss their effects on velocity.
5. The time-distance curve for the single flat-layer model is determined by applying the Pythagorean theorem to the sides of a right triangle. The Pythagorean equation is:

$$(VT)^2 = X^2 + (VT_0)^2$$

or

$$T^2 = T_0^2 + X^2/V^2$$

where

VT is the slanted ray length.
VT_0 is the vertical distance
X is the full offset
T is the two-way slant time and

T_0 is the two-way vertical time.

The NMO equation for the single-layer model is:

$$\Delta T = T - T_0 \left(T_0 = 2dZ/V \right)$$

$$\Delta T = \left(T_0^2 + X^2/V^2 \right)^{0.5} - T_0$$

A. For the case of one layer of thickness ΔZ = 4,000 ft, interval velocity V_1 = 8,000 ft/sec, and offset distance X = 6,000 ft, calculate:
   1. The two-way traveltime T_0
   2. The two-way slant time T
   3. The normal moveout ΔT

B. For the case of two layers with interval velocities V_1 = 8,000 ft/sec, and V_2 = 12,000 ft/sec, thicknesses ΔZ_1 = 2,000 ft and ΔZ_2 = 4,000 ft. Calculate T_0, T, and ΔT.
   Note: $T_0 = T_1 + T_2$ and $V_A = 2\left(\Delta Z_1 + \Delta Z_2 \right) / T_0$

6. Define the following terms:
   a. Stacking velocity
   b. Velocity spectrum
   c. Horizon velocity analysis
   d. Migration velocity
7. Compute and plot the time-distance curve for the direct arrival and the reflection for depth = 1 km and velocity = 1.5 km/sec. Let X range from 0 to 1 km, in increments of 200 m.

   Keep in mind that for direct arrival time, $T = X/V$, and for reflected time $T = (X^2/V^2 + 4h^2/V^2)^{0.5}$
8. A velocity survey was conducted in a well gives the following one-way travel times:

Depth (feet)	One-way time (sec)
2,600	0.40
8,500	1.00
12,400	1.30
15,700	1.50

   a. Plot four small graphs to show the variations of both interval velocity and average velocity with both depth and one-way time. Draw straight-line segment graphs as appropriate.
   b. Comment on the errors involved in using these representations by comparing the depths computed for a two-way travel time of 2.8 sec, from both the interval velocity and average velocity graphs.
9. An earth model consists of two horizontal layers. The first layer is 3,300 ft thick and has an interval velocity of 1.25 miles/sec. The second layer is 4,950 ft thick and has an interval velocity of 1.875 miles/sec. Consider three rays, with Snell's parameter P equal to 0, 0.333, and .167, respectively.

a. Calculate the angles θ_1 and θ_2 for each ray.
Use the equation:

$$P = (\sin\theta_1)/V_1 = (\sin\theta_2)/V_2$$

b. Calculate the half-offsets for each of the three rays.
Use the equation:

$$h = \Delta h_1 + \Delta h_2 = \Delta Z_1 \tan\theta_1 + \Delta Z_2 \tan\theta_2$$
$$\Delta h_k = \Delta Z_k \tan\theta_k$$

c. Calculate the one-way traveltimes for each ray. Remember that: $\Delta t_n = \Delta Z / V_n$, where n is the layer number.

$$T = \Delta t_1 + \Delta t_2 = \Delta Z_1 / V_1 + \Delta Z_2 / V_2$$

BIBLIOGRAPHY

Morley, L. and J. Claerbout. "Predictive Deconvolution in Shot-Receiver Space." *Geophysics* 48 (1983): 515–531.

Al-Chalabi, M. "Series Approximations in Velocity and Traveltime Computations." *Geophysics Prospect* 21 (1973): 783–795.

Al-Chalabi, M. "An Analysis of Stacking, RMS, Average, and Interval Velocities over a Horizontally Layered Ground." *Geophysics Prospect* 22 (1974): 458–475.

Al-Sadi, H. N. *Seismic Exploration.* Boston: Birkhouser Boston, Inc., 1980.

Anstey, N. A. "Signal Characteristics and Instrument Specification." *Seismic Prospecting Instruments,* 1 (1970).

Anstey, N. A. "Seismic Interpretation." *Physical Aspects.* Boston: IHRDC, 1977.

Anstey, N. A. *Simple Seismics.* Boston: IHRDC, 1982.

Bakus, M. M. "Water Reverbrations, Their Nature and Elimination." *Geophysics* 24 (1959): 233–261.

Berkhout, A. J. "Least Squares Inverse Filtering and Wavelet Deconvolution." *Geophysics* 42 (1977): 1369–1383.

Dix, C. H. "Seismic Velocities from Surface Measurements." *Geophysics* 20 (1955): 68–86.

Dix, C. H. *Seismic Prospecting for Oil.* Boston: IHRDC, 1981.

Dobrin, M. B. *Geophysical Prospecting.* New York: McGraw-Hill, 1976.

Dobrin, M. B. *Introduction to Geophysical Prospecting.* New York: McGraw-Hill, 1960.

Duncan, J. W. and F. K. Levin. "Effect of Normal Moveout on a Seismic Pulse." *Geophysics* 38 (1973): 635–642.

Grant, F. S. and G. F. West. *Interpretation Theory in Applied Geophysics.* New York: McGraw-Hill, 1965.

Hampson, D. and B. Russell. "First-Break Interpretation using Generalized Inversion."*J. Can. Soc. Explor. Geophyics* 20 (1984): 40–54.

Hileman, J. A., P. Embree, and J. C. Pfleuger. "Automated Static Corrections." *Geophysics Prospect* 16 (1968): 328–358.

Hilterman, F. J. "Three-Dimensional Seismic Modeling." *Geophysics* 35 (1970): 1020–1037.
Hilterman, F. J. "Amplitudes of Seismic Waves. A Quick Look." *Geophysics* 40 (1975): 745–762.
Hubral, P. and T. Krey. "Interval Velocities from Seismic Reflection Time Measurements." *Soc. Explor. Geophys. Monograph.* (1980).
Levin, F. K., "Apparent Velocity from Dipping Interface Reflections." *Geophysics* 36 (1971): 510–516.
Lindsey, J. P. "Elimination of Seismic Ghost Reflections by Means of a Linear Filter." *Geophysics* 25 (1960): 130–140.
Mayne, W. H. "Common Reflection Point Horizontal Stacking Techniques." *Geophysics* 27 (1962): 927–938.
Neidell, N. S. and M. T. Taner. "Semblance and Other Coherency Measures for Multichannel Data." *Geophysics* 36 (1971): 482–497.
Newman, P. "Divergence Effects in a Layered Earth." *Geophysics* 38 (1973): 481–488.
Osman M. O. Discrimination between intrinsic and apparent attenuation in layered media: M.S. thesis, The Univ. of Tulsa, Tulsa, OK, 1988.
Palmer, D. "The Generalized Reciprocal Method of Refraction Seismic Interpretation." *Geophysics* 46 (1981): 1508–1518.
Robinson, E. A. "Dynamic Predictive Deconvolution." *Geophys. Prospect.* 23 (1975): 779–797.
Robinson, E. A., and S. Treitel. *Geophysical Signal Analysis.* Englewood, N. J.: Prentice-Hall, 1980.
Robinson, E. A. *Seismic Velocity Analysis and the Convolutional Model.* Boston: IHRDC, 1983.
Schneider, W. A. "Developments in Seismic Data Processing and Analysis (1968–1970)" *Geophysics* 36 (1971): 1043–1073.
Schneider, W. A. and S. Kuo. "Refraction Modeling for Static Corrections." *55th Ann. Int. Soc. Explor. Geophys. Mtg.* (1985).
Sheriff, R. E. "Encyclopedic Dictionary of Exploration Geophysics." *Soc. Explor. Geophys.* Tulsa, OK, 1973.
Sheriff, R. E. *A First Course in Geophysical Exploration and Interpretation.* Boston: IHRDC, 1978.
Sherwood, J. W. C. and P. H. Poe. "Constant Velocity Stack and Seismic Wavelet Processing." *Geophysics* 37 (1972): 769–787.
Taner, M. T. and F. Koehler. "Velocity Spectra." *Geophysics* 34 (1969): 859–881.
Taner, M. T., F. Koehler, and K. A. Alhilali. "Estimation and Correction of Near-Surface Time Anomalies." *Geophysics* 39 (1974): 441–463.
Wiggens, R. A., K. L. Larner, and R. D. Wisecup. "Residual Static Analysis as a General Linear Inverse Problem." *Geophysics* 41 (1976): 992–938.
Taner, M. T. and F. Koehler. "Velocity Spectra—Digital Computer Derivation and Applications of Velocity Functions." *Geophysics* 39 (1969): 859–881.
Tatham, R. H., and P. Stoffa, "A Potential Hydrocarbon Indicator." *Geophysics* 41 (1976): 837–849.
Telford, W. M., L. P. Geldart, R. E. Sheriff, and D. A. Keys. *Applied Geophysics.* Cambridge, England: Cambridge University Press, 1976.
Waters, K. H., *Reflection Seismology.* New York: John Wiley, 1978.
Yilmaz, O. "Seismic Data Processing." Tulsa: OK: *Soc. Explor. Geophys.* 1987.

CHAPTER 6

MIGRATION

INTRODUCTION

A seismic section is assumed to represent a cross-section of the earth. The assumption works best when layers are flat, and fairly well when they have gentle dips. With steeper dip the assumption breaks down; the reflections are in the wrong places and have the wrong dips.

In estimating the hydrocarbons in place, one of the variables is the areal extent of the trap. Whether the trap is structural or stratigraphic, the seismic section should represent the earth model.

Dip migration, or simply *migration*, is the process of moving the reflections to their proper places with their correct amount of dips. This results in a section that more accurately represents a cross-section of the earth, delineating subsurface details such as fault planes. Migration also collapses diffractions.

THE NORMAL INCIDENCE

Figure 6–1a shows the simple model of a flat reflector. Energy from the source goes straight down to the reflecting horizon and back up to a geophone at the shot point. If the horizon dips, the energy goes to and from it by the most direct path, which is along a normal to the reflector. Energy that strikes the reflector at other angles goes off in another direction, as shown in Figure 6–1b.

The normal incidence principle is the basic idea behind dip migration. All migration techniques follow this principle. Structure and velocities cause the raypath to follow a nonstraight path down to the horizon and back up, but right at the reflecting surface the energy path is normal to it. The reflection point is not directly under the shot point but offset from it; and after applying normal move out corrections, the source and

FIGURE 6–1. Normal-incidence reflection

(a)

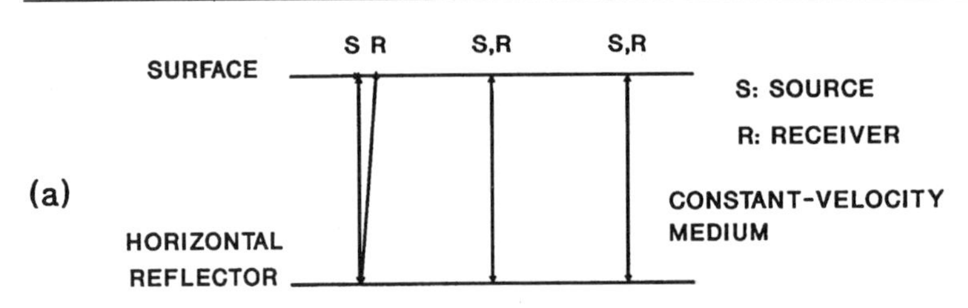

(b)

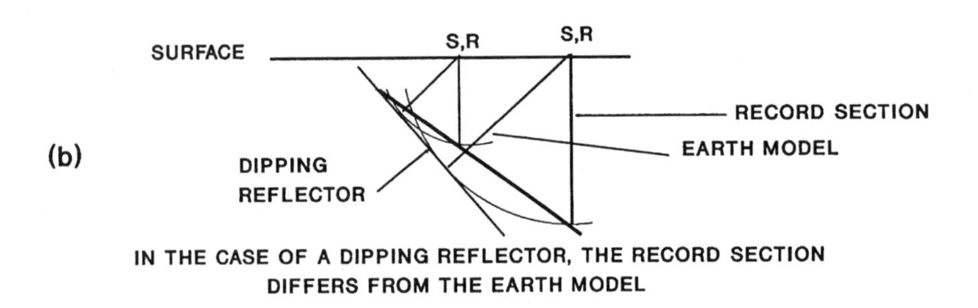

the receiver are in the same location. This section is called "zero offset section," and the ray path is perpendicular to the dipping reflecting horizons.

This is what happens to the reflected energy, but how does it appear on the record section? We display the traces vertically and parallel, because they are only measurements of the times for a sequence of energy that strike the geophones. At this point, the time during which the ray traveled to and from a dipping reflector appears on the section as though the path had been straight down (Figure 6–2a). Figure 6–2b shows the normal incidence principle applied to the record section to convert to earth model; note that events moved updip after migration.

Using the normal incidence principle, we can discuss some subsurface features and how they look when converted from the record section to the earth model. Then some rules can be formed for how the features on the record section change when migrated back to the correct configuration.

EFFECTS OF MIGRATION ON REFLECTIONS

For simplicity, we shall assume that the acoustic velocity is constant throughout the geologic section, and that the sections are shot along the direction of dip so that there are no reflections from one side or the other of the line.

FIGURE 6–2. Migration of dipping events

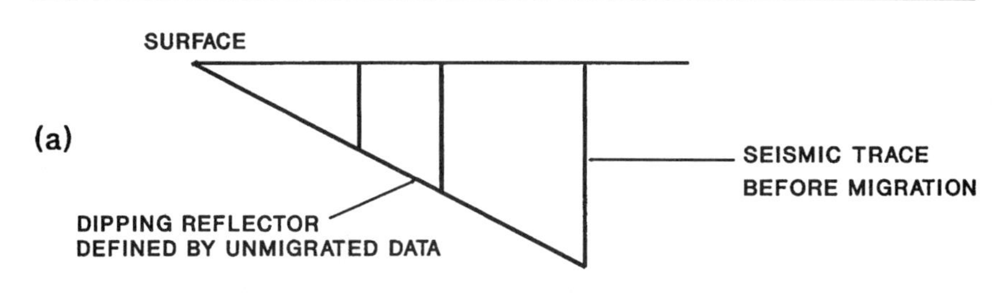

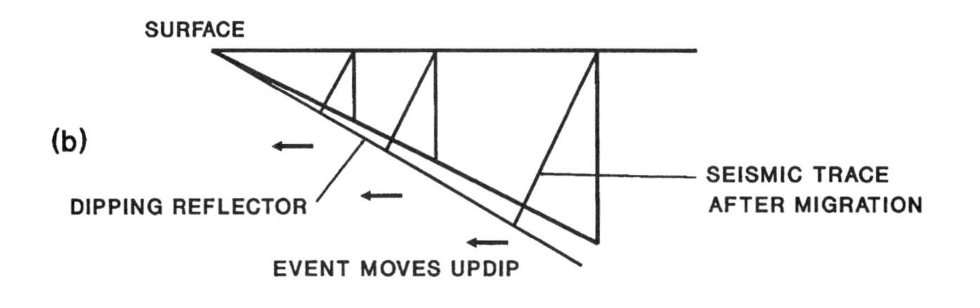

Although the travel paths are normal to the dipping reflectors, the traces are displayed on the record section as if the travel paths were vertical. The purpose of migration is to move them back up-dip (see Figure 6–2b). Note that the lateral extent of the dipping reflector is shortened and the dip angle is steeper after migration.

ANTICLINE

In Figure 6–3a the anticline is defined by seismic traces displayed vertically and parallel to each other and the feature appears spread out. Figure 6–3b shows the effect of the migration by applying the normal incidence principle. Events on the flanks of the anticline moved updip cause the lateral extent of the feature to become smaller. Since the crest of the anticline is horizontal, migration has no effect on it. The closure (i.e., the maximum height from the crest to the lowest closed contour) will be the same or a little less.

SYNCLINE

In Figure 6–4a the syncline is displayed in the record section with seismic traces displayed vertically and parallel to each other. After migration (see Figure 6–4b) the feature becomes broader as the ray paths have to reach out to be reflected normally from the reflector. The trough did not

FIGURE 6–3. Migration of anticline (adapted from Coffeen, 1986, courtesy PennWell Books)

BEFORE MIGRATION

CREST

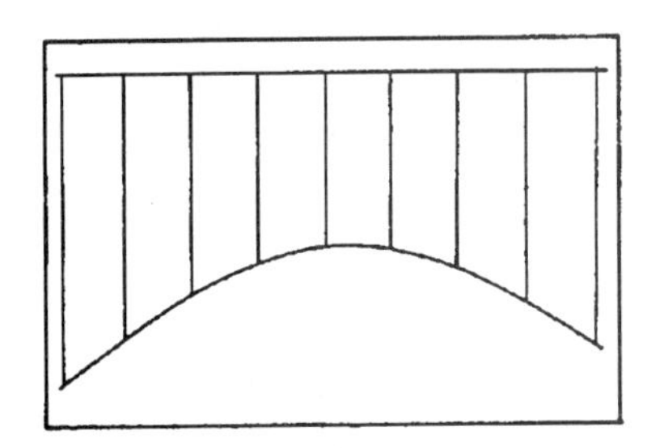

RECORD SECTION

(a)

AFTER MIGRATION

CREST

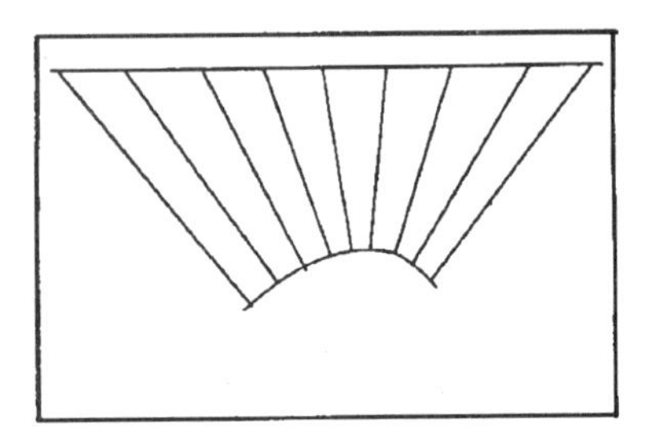

EARTH MODEL

(b)

MIGRATION OF ANTICLINE NARROWS THE LATERAL EXTENT AND REDUCES DIP OF ISOLATED ANTICLINES, EQUAL OR LESS CLOSURE. THE CREST OF THE ANTICLINE DOES NOT MOVE.

FIGURE 6–4. Migration of a syncline (adapted from Coffeen, 1986, courtesy PennWell Books)

RECORD SECTION

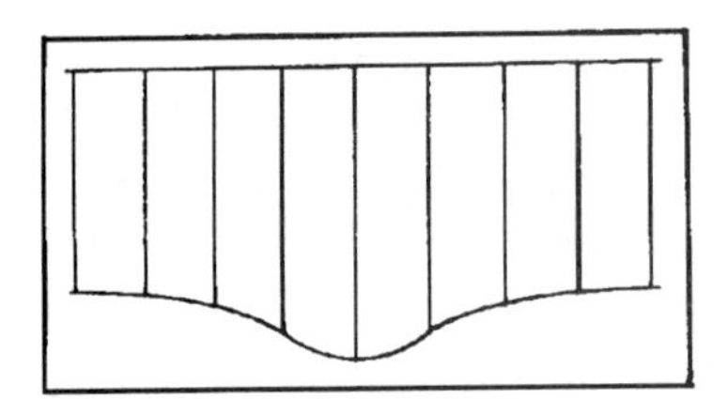

TROUGH

BEFORE MIGRATION

(a)

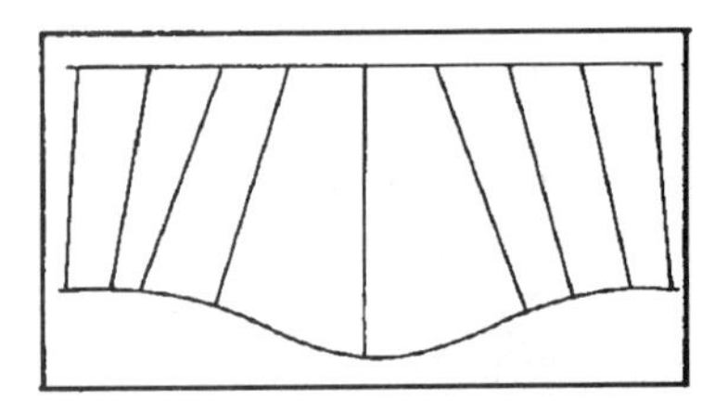

TROUGH

AFTER MIGRATION

(b)

AFTER MIGRATION A SYNCLINE IS BROADER, THE LOWEST POINT IS FLAT AND DOES NOT MOVE, AND CLOSURE IS THE SAME OR LARGER

move after migration because it is horizontal and the closure is the same or larger.

The amount of relief of the syncline is not very critical, as this structure never has any hydrocarbon accumulation.

CROSSING REFLECTIONS FROM BURIED FOCUS (BOW-TIE EFFECT)

Synclines can behave in another way on the record section, if they are relatively narrow or are deep in the section. Deeper or narrower synclines have raypaths that cross on the way down, with one trace being in a position to receive information from two or even three parts of the syncline (Fermat's principal).

The case of two crossing lineups of energy, with perhaps an apparent anticline visible beneath them, is illustrated in Figure 6–5b. This is called the *bow-tie effect*, or *buried focus*. Accordingly, a sharp syncline may be revealed by migration of crossing reflections as shown in Figure 6–5a.

FAULT

When a fault breaks off reflections sharply, or if for some other reason there is a point (or edge) in the subsurface, that point returns energy to

FIGURE 6–5. Migration of a buried focus (adapted from Coffeen, 1986, courtesy PennWell Books)

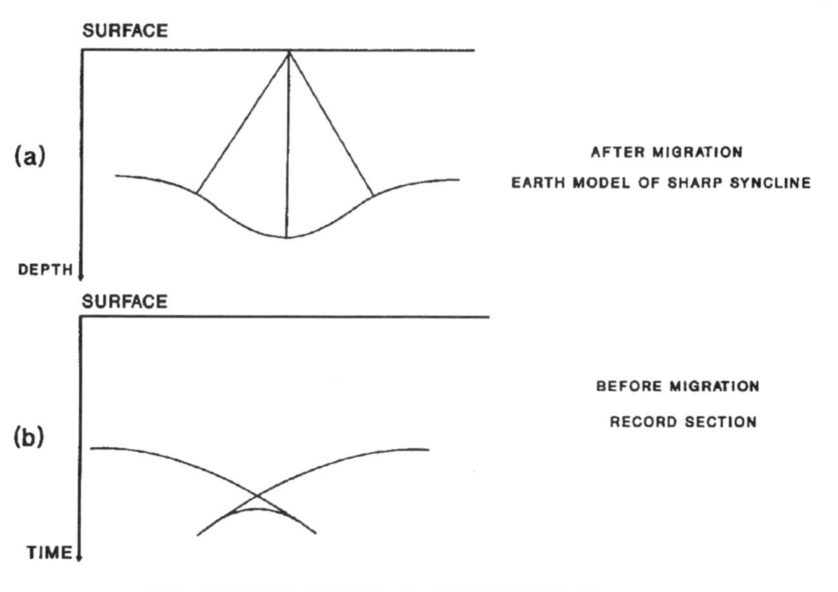

FIGURE 6–6. Point source (adapted from Coffeen, 1986, courtesy PennWell Books)

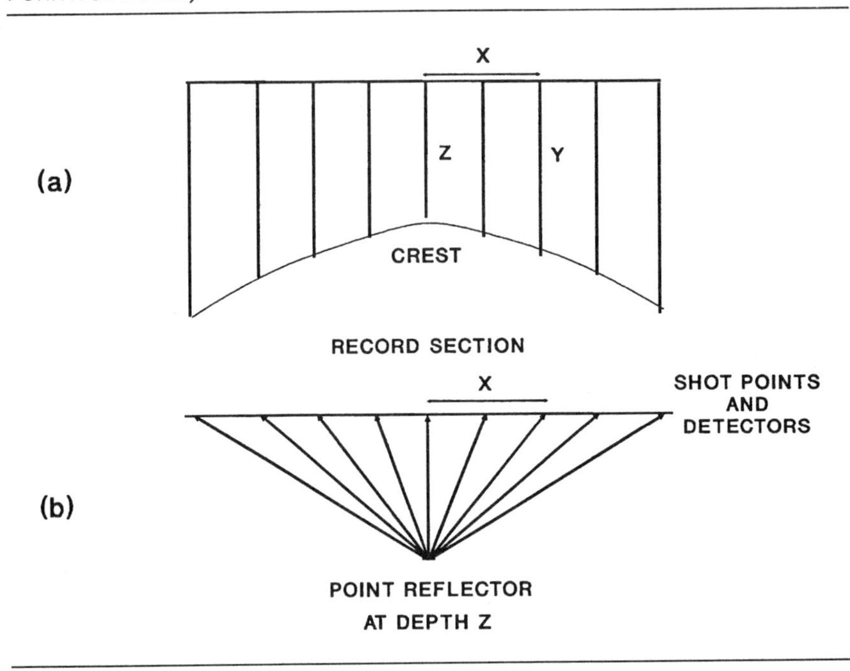

any source within the range. That is, it behaves as a new source of energy.

Energy is returned to a number of receivers at different distances from the point source as shown in figure 6–6b. In cross-section, with reflected energy vertically below the shot point, it is an apparent anticline. The form is actually a hyperbola as shown in Figure 6–6a. This is a *diffraction*, and it is recognized easily by its regular shape. Sometimes half of it is visible, so the broken-off formation appears to continue in a smooth curve downward.

Although it is not a normal reflection, a diffraction pattern is created by the seismic traces displayed vertically in the record section, so the same process of migration applies.

A diffraction pattern is collapsed to a point after migration.

MIGRATION PRINCIPLES

Figure 6–7 uses the harbor example to illustrate the physical principle of migration, Claerbout (1985).

Assume that a storm barrier exists at some distance Z_3 from the beach, and there is a gap in the barrier. The gap in the barrier acts as a Huygens secondary source, causing circular wavefronts that approach the beach line (see Figure 6–7a). The gap in the barrier is referred to

FIGURE 6–7. Migration principles (Reprinted from O. Yilmaz, *Seismic Data Processing*, 1987, courtesy of the Society of Exploration Geophysicists.)

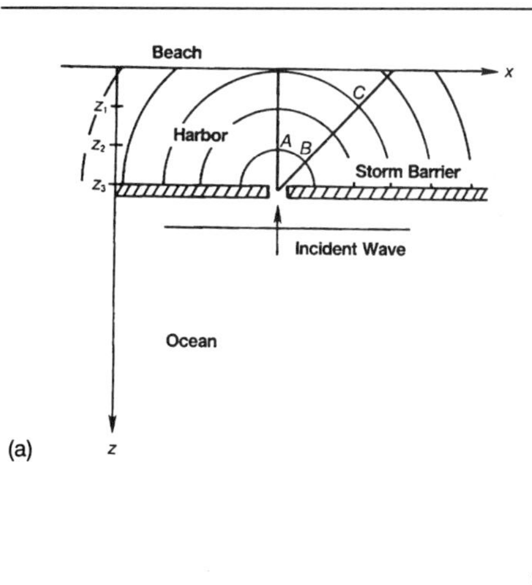

(a)

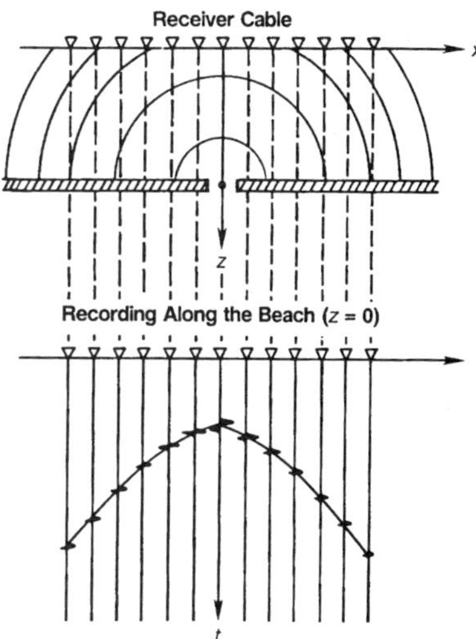

(b)

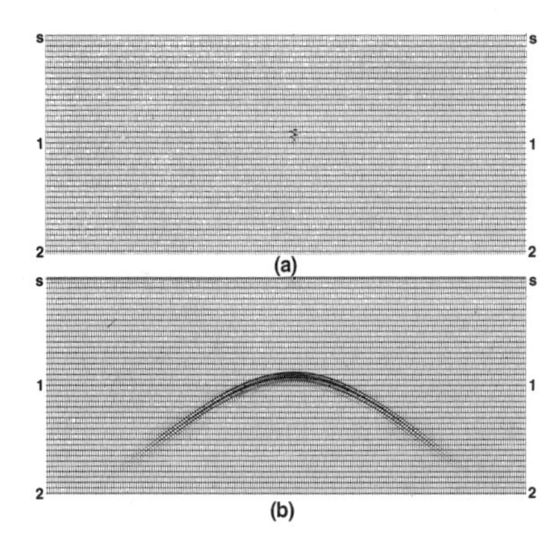

(c)

as a *point aperture.* It is similar to a point source in the subsurface, since both generate circular wavefronts.

From this experiment, we find that a Huygens secondary source responds to a plane incident wave and generate:

a. A semicircular wavefront in the (X,Z) plane.
b. A hyperbolic diffraction in the (X,T) plane, as in Figure 6–7b.

Figure 6–7c represents a Huygens source in depth section. The upper section (a) maps into a point on the zero-offset time section. The vertical axis in the lower section (b) is two-way time.

Refer to Figure 6–8A, and imagine that the subsurface consists of points along each reflecting horizon. This model behaves much as the

FIGURE 6–8. Migration principles (Reprinted from O. Yilmaz, *Seismic Data Processing,* 1987, courtesy of the Society of Exploration Geophysicists.)

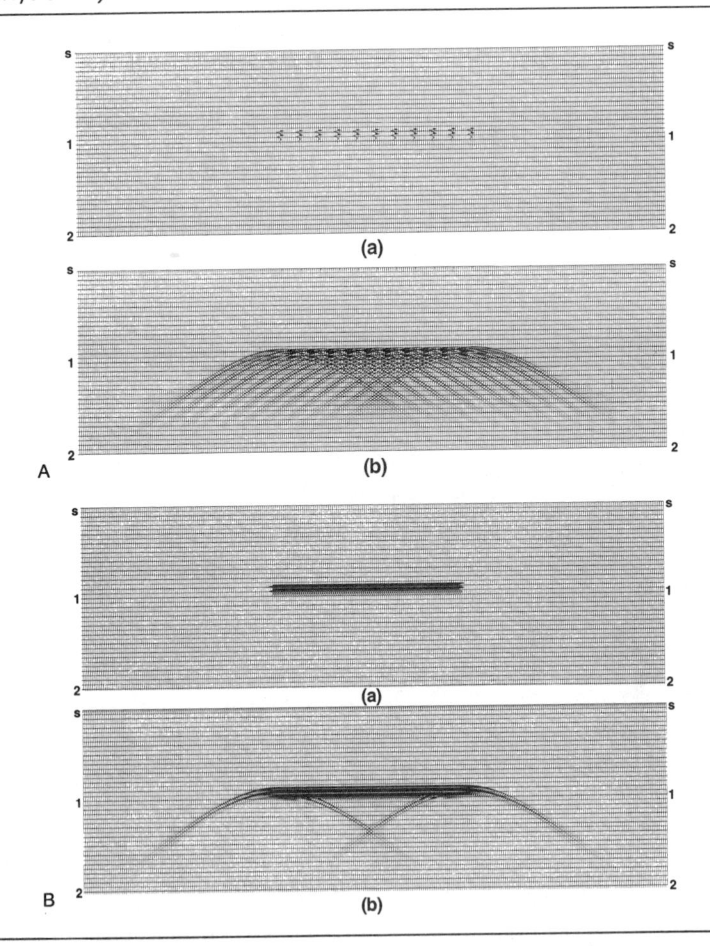

gap in the barrier. So each of these points acts as a Huygens secondary source and produces hyperbolas in the (X,T) plane.

As the sources get closer to each other, superposition of the hyperbolas produces the response of the actual reflecting interface, as shown in Figure 6–8B. These hyperbolas are comparable to the diffractions seen at fault boundaries on the stacked sections.

In summary:

- Reflectors in the subsurface can be visualized as being composed of many points that act as Huygens' Secondary Sources.
- The stacked section (zero offset) consists of the superposition of many hyperbolic traveltime responses.
- When there are discontinuities (faults) along the reflector, diffraction hyperbolas stand out.
- A Huygen's secondary source signature is semicircular in the (X,Z) plane and hyperbolic in the (X,T) plane.

Figure 6–9 shows the principles of migration based on diffraction summation. Figure 6–9a is a zero offset seismic section. (The trace interval is 25 meters and constant velocity of 2,500 m/sec.) Figure 6–9b is the migrated version; the amplitude at point B on the hyperbola is mapped onto the apex A along the hyperbolic traveltime equation.

FIGURE 6–9. Migration principles (Reprinted from O. Yilmaz, *Seismic Data Processing*, 1987, courtesy of the Society of Exploration Geophysicists.)

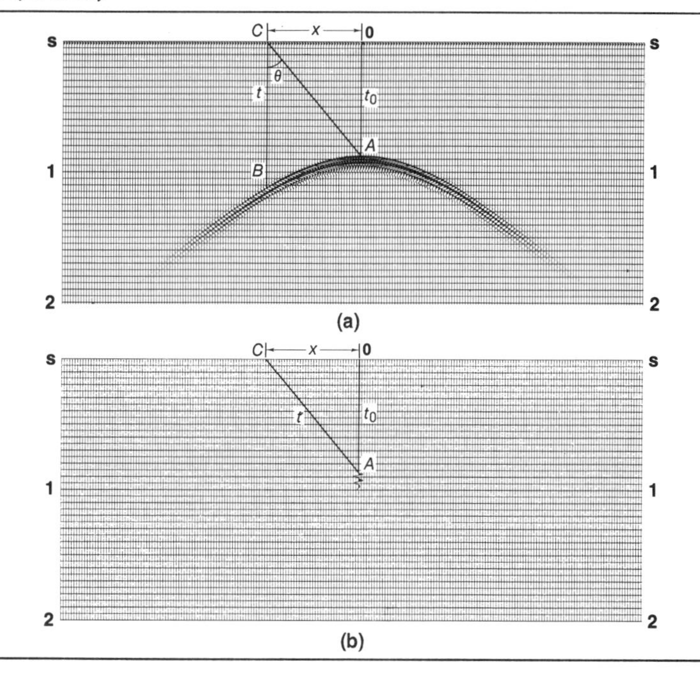

METHODS OF MIGRATION

KIRCHHOFF MIGRATION

Diffraction migration or *Kirchhoff's migration* is a statistical approach technique. It is based on the observation that the zero-offset section consists of a single diffraction hyperbola that migrates to a single point. Migration involves summation of amplitudes along a hyperbolic path. The advantage of this method is its good performance in case of steep-dip structures. The method performs poorly under low signal-to-noise ratio.

FINITE-DIFFERENCE MIGRATION

Finite difference migration is a deterministic approach; it is modeled by an approximation of the wave equation that is suitable for use with computers. One advantage of the finite difference method is its ability to perform well under low signal-to-noise ratio condition. Its disadvantages include long computing time and difficulties in handling steep dips.

FREQUENCY DOMAIN OR F-K MIGRATION

It is a deterministic approach via the wave equation instead of using the finite difference approximation. The two-dimensional Fourier transform is the main technique used in this method. Some of the advantages of *F-K method* are fast computing time, good performance under low signal-to-noise ratio, and excellent handling of steep dips. Disadvantages of this method include difficulties with widely varying velocities. Those who are interested in learning more about migration methods should refer to appendix A.

EXAMPLES OF MIGRATION

Figure 6–10 is a stack section. Note that diffractions mask out the true subsurface structure due to the bow-tie effect.

Figure 6–11 is a migrated version of the same stack in time domain using the diffraction collapse method, or Kirchhoff migration. One can see that migration unties the bow-ties and turns the structure to a series of synclines.

Figure 6–12 is another stack section. Strong diffractions are apparent in the center of the section, presumably from a fault. Note the bow-tie effect and buried focus phenomena down deep on the section.

Figure 6–13 shows the diffraction-collapse migrated section. The structure looks like a thrust fault or perhaps a sharp syncline.

FIGURE 6–10. The "bow tie" effect (Reprinted from O. Yilmaz, *Seismic Data Processing*, 1987, courtesy of the Society of Exploration Geophysicists.)

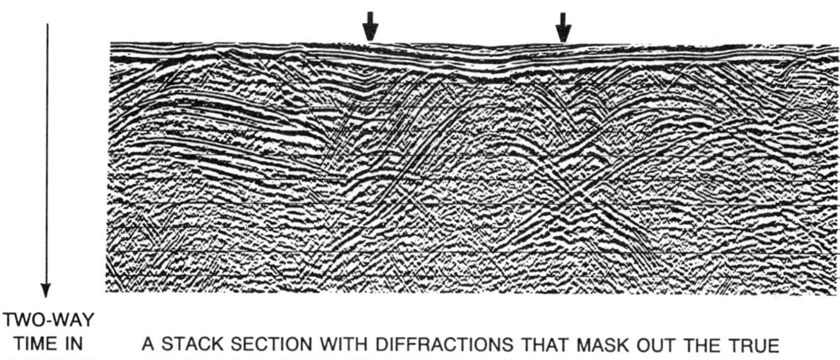

A STACK SECTION WITH DIFFRACTIONS THAT MASK OUT THE TRUE SUBSURFACE STRUCTURE. THESE PATTERNS ARE CALLED "BOW TIES."

FIGURE 6–11. After migration (Reprinted from O. Yilmaz, *Seismic Data Processing*, 1987, courtesy of the Society of Exploration Geophysicists.)

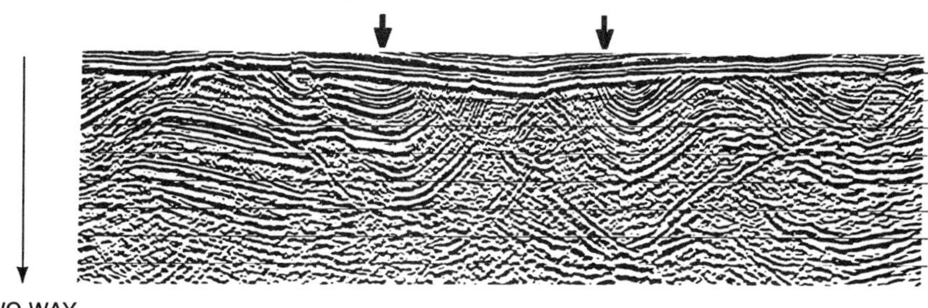

THE SAME SECTION AS IN FIGURE 6–10 AFTER MIGRATION. THE BOW TIES HAVE BEEN "UNTIED" AND ARE REVEALED AS SYNCLINES.

Figure 6–14 is a comparison between a time stack section and the migrated section. (a) Shows a lot of diffractions, a highly faulted structure, and fault orientations are not clear. Finite-difference time migration (b) collapses the diffractions, delineates fault planes, and gives a more interpretable section.

Figure 6–15 compares a stack section and a time migration section. This example from the Gulf Coast Basin illustrates a growth fault as seen in (a). Time migration (b) reveals fault planes. Note the graben and horst faulting. Migration clearly shows the structural complexity on the

FIGURE 6–12. Time migration (courtesy GX Technology)

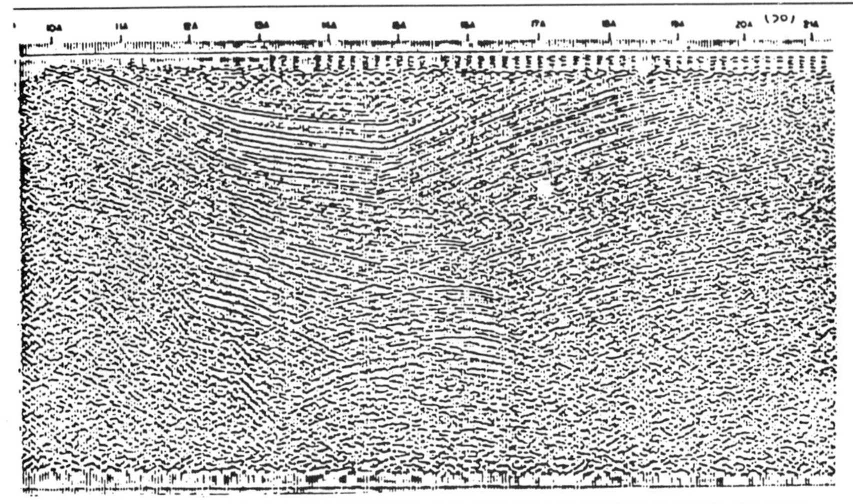

A STACK SECTION. STRONG DIFFRACTIONS ARE APPARENT IN THE CENTER OF THE SECTION, PRESUMABLY FROM A FAULT.

FIGURE 6–13. Time migration (courtesy GX Technology)

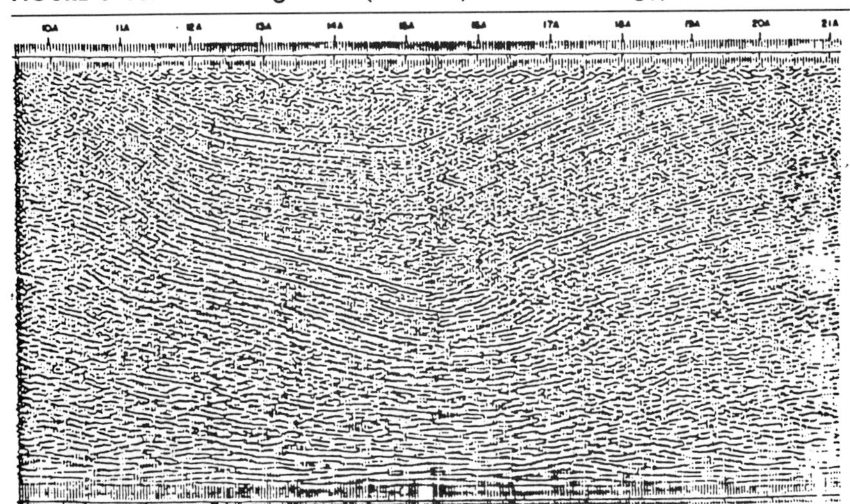

THIS IS THE SAME SECTION AS IN FIGURE 6–12, AFTER MIGRATION. THE DIFFRACTIONS ARE COLLAPSED. THE STRUCTURE NOW LOOKS LIKE A THRUST FAULT OR PERHAPS A SHARP SYNCLINE.

FIGURE 6–14. Time migration (Reprinted from O. Yilmaz, *Seismic Data Processing*, 1987, courtesy of the Society of Exploration Geophysicists.)

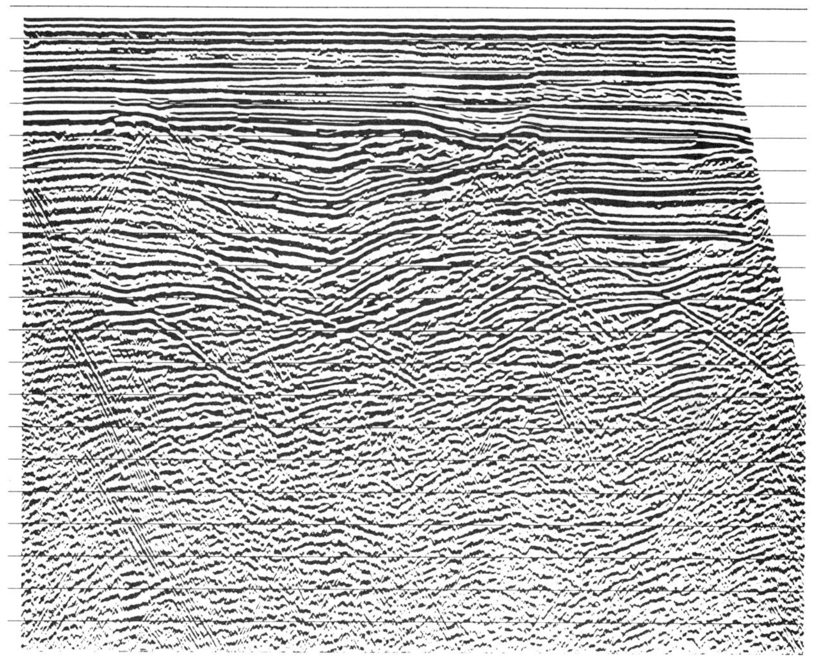

(a) TIME STACK SECTION IN A HIGHLY FAULTED STRUCTURE. THERE ARE MANY DIFFRACTIONS, AND THE PATTERNS ARE UNCLEAR.

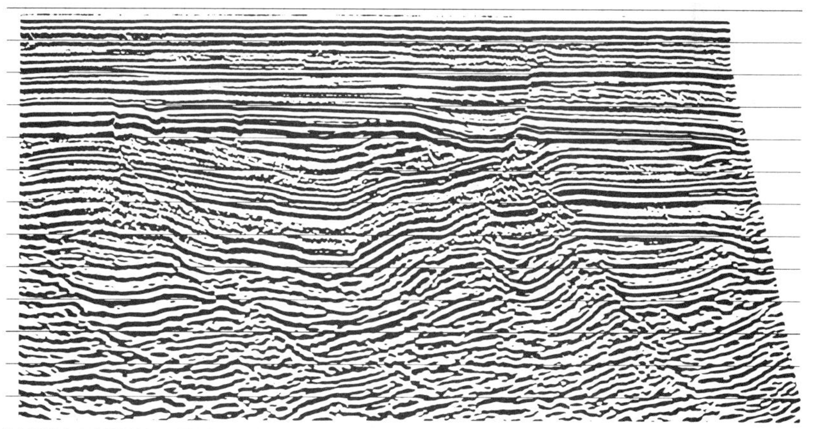

(b) THE SECTION AFTER FINITE-DIFFERENCE TIME MIGRATION. DIFFRACTIONS ARE COLLAPSED AND FAULT PLANES ARE DELINEATED.

FIGURE 6–15. Time migration (courtesy GX Technology)

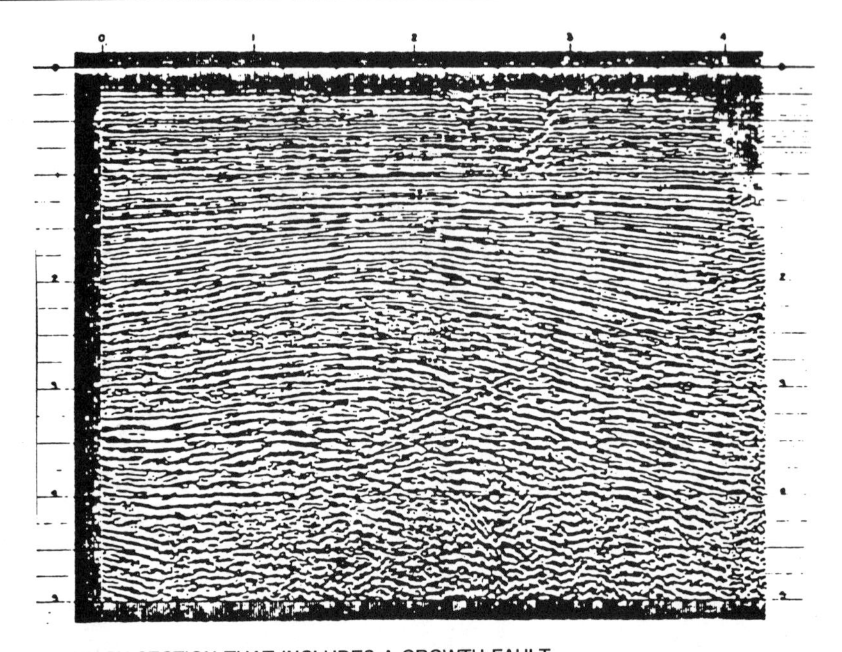

(a) STACK SECTION THAT INCLUDES A GROWTH FAULT.

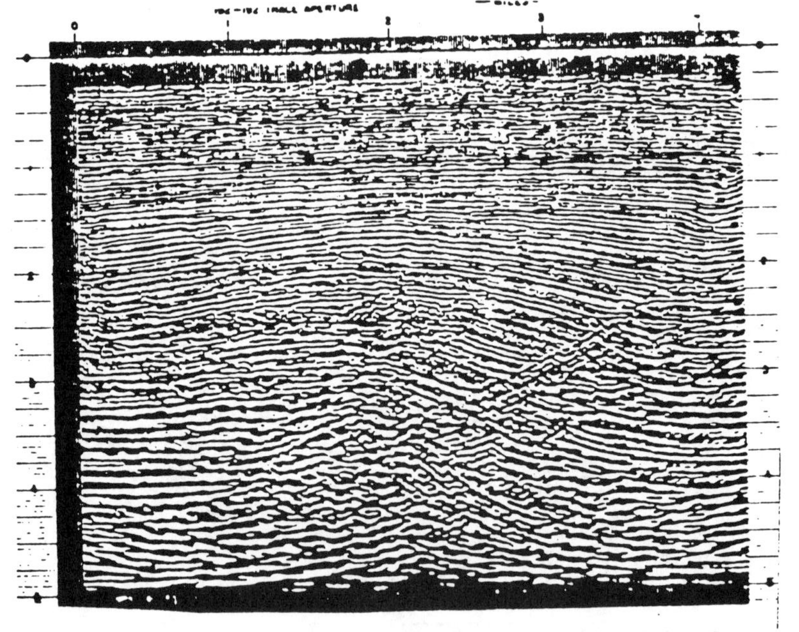

(b) THE MIGRATED STACK SECTION REVEALS THE FAULT PLANE. NOTE THE GRABEN AND HORST FAULTING. STRUCTURAL COMPLEXITY ON THE TOP OF CLOSED FEATURE IS REVEALED BY COLLAPSING MINOR DIFFRACTIONS.

top of the closed feature by collapsing minor diffractions. Wave-equation time migration was used to migrate the stacked section.

Figure 6–16a shows complex folding and thrust faulting. Figure 6–16b, the migrated version, is more interpretable. It has better fault definition and the subthrust fold structure is better defined.

The stack section in Figure 6–17 has poor signal-to-noise ratio; the F-K migrated section shows better fault detail due to the collapse of diffractions.

Figure 6–18 is another comparison between a stack section and the F-K migrated version. The migrated section shows more details, diffraction collapse, and excellent fault plane definition.

DEPTH MIGRATION

Time migration is appropriate as long as lateral velocity variations are moderate. When these variations are substantial, depth migration is needed to obtain a true picture of the subsurface. In particular, the geologist prefers to have a depth-migrated section of the subsurface. Unfortunately, due to the lateral change of the velocity and the structure complexity, it is very difficult to get a reliable section.

3-D MIGRATION

In time domain, 3-D migration is needed when the stack section contains events out of the profile. This is the common type of 3-D migration. This technique will be covered in Chapter 10, 3-D Surveys.

PRESTACK PARTIAL MIGRATION (DIP MOVEOUT—DMO)

Post-stack migration is acceptable when the stacked data are zero-offset. If there are conflicting dips with varying velocities or a large lateral velocity gradient, a prestack partial migration is used to attenuate these conflicting dips.

By applying this technique before stack, it will provide a better stack section that can be migrated after stack.

Prestack partial migration only solves the problem of conflicting dips with different stacking velocities.

The application of prestack partial migration versus post-stack migration can be summarized as follows:

- Post-stack migration is acceptable when the stacked data is zero-offset. This is not the case for conflicting dips with varying velocity or large lateral velocity variations.
- Prestack partial migration (PSPM) or dip moveout (DMO) provides a better stack, which can be migrated after stack.
- PSPM solves only conflicting dips with different stacking velocities.

FIGURE 6–16. Time migration (courtesy GX Technology)

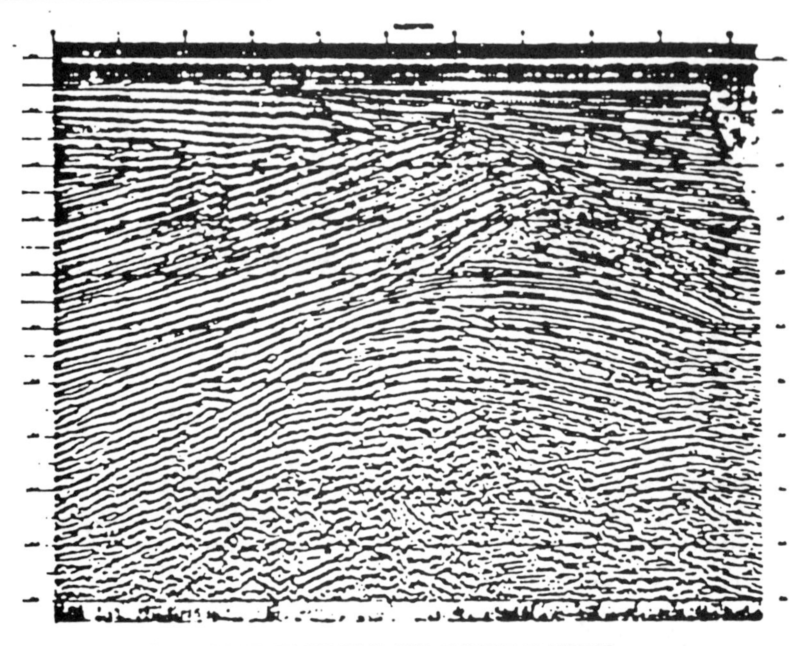

(a) STACK SECTION COMPLEX FOLDING AND THRUST FAULTING.

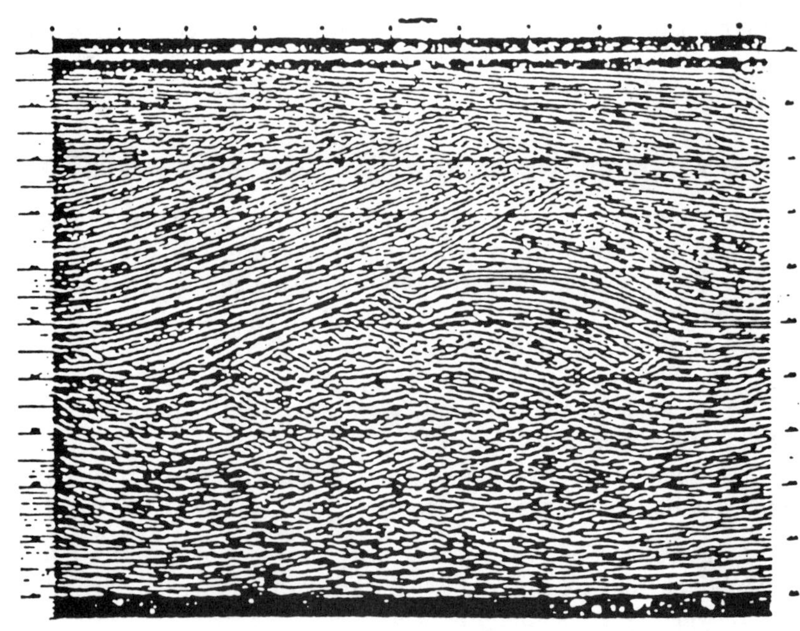

(b) MIGRATED STACK MORE INTERPRETABLE SECTION WITH BETTER FAULT DEFINITION. THE FOLD STRUCTURE IS CLEARER.

FIGURE 6–17. F-K migration (courtesy of Seismograph Service Corporation

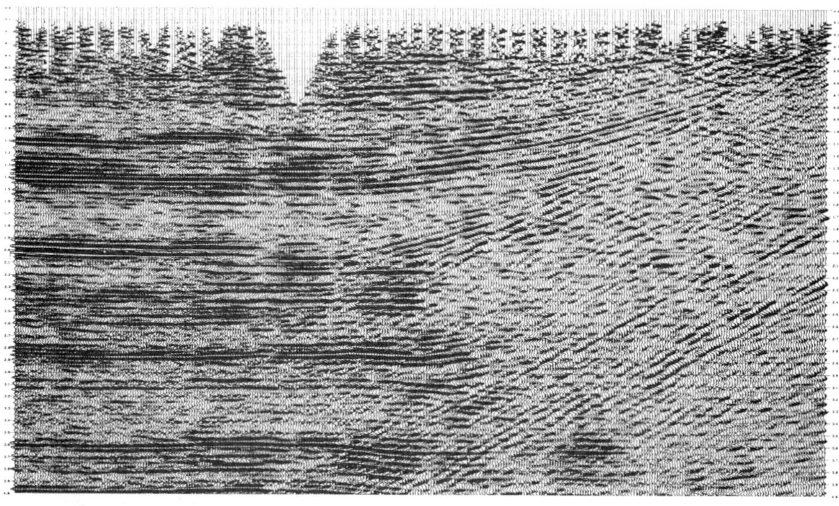

(a) STACKED SECTION

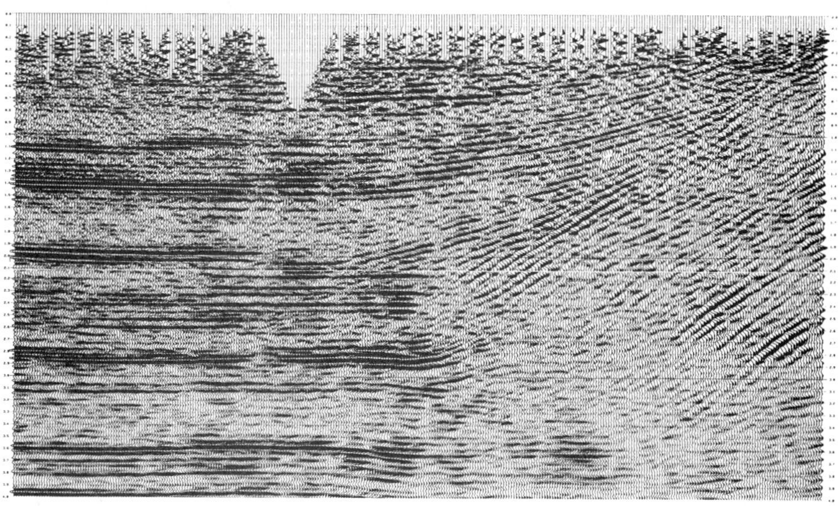

(b) AFTER F-K MIGRATION

FIGURE 6–18. F-K migration (courtesy of Seismograph Service Corporation)

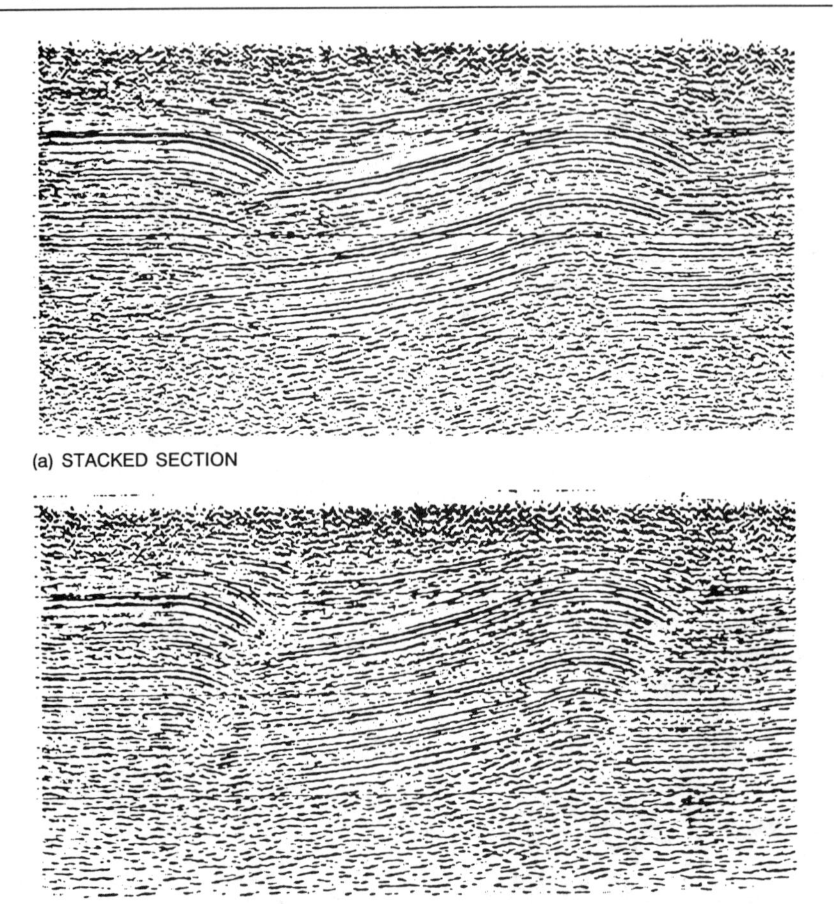

SUMMARY AND DISCUSSION

Migration is a process that moves seismic events to their proper subsurface positions due to the way record sections are displayed. After applying normal moveout, we assume that the seismic section is a series of arrival time with the normal incidence ray path at the reflectors. This section is called zero-offset because the NMO process move the source and the receiver at the same position.

In case of flat or a fairly flat horizons, the record section (seismic data display) and the earth model are the same.

In case of dipping reflectors, they are not the same. The normals to the reflectors are displayed vertically and parallel to each other; their true subsurface positions are rotated down from their original position.

Migration process is preformed in order to move these reflectors to their proper positions so that the record section matches the earth model.

The process of migration moves events updip and makes the ray travel path normal to the reflectors. It shortens the lateral extent of the event, and the angle of dip will be steeper after migration. By applying this principle, one can observe that an anticline will have a narrower areal extent after migration while a syncline will have a broader areal extent. The crest of the anticline and the trough of the syncline do not move because their position is horizontal, while the closure in both cases may not change.

Diffractions from the top of the fault plane are observed as hyperbolas. They collapse to a point after migration. This will help the interpreter to more accurately delineate and orient fault patterns.

In case of a deep sharp syncline (buried focus), migration unties the bow-tie feature on the record section and reveals the true synclinal features.

Migration improves the horizontal resolution and gives more accurate subsurface picture of the geologic structures, which give more realistic size of the hydrocarbon accumulations in place.

There are different methods of migration techniques. Each of which is designed to handle a different geologic setting of the target structure.

KEY WORDS

Buried focus
Closure
Diffractions
Dip move-out
F-K
Migration
Trap
Wave equation
Zero-offset section

BIBLIOGRAPHY

Baysal, E., D. D. Kosloff and J. W. C. Sherwood. "Reverse Time Migration." *Geophysics*, 48 (1983): 1514–1524.

Berkhout, A. J. *Seismic Migration—Imaging of Acoustic Energy by Wave Field Extrapolation.* Amsterdam, Netherlands: Elsevier Science Publ. Co., Inc., 1980.

Black, J. L., I. T. McMahon, H. Meinardus and I. Henderson. "Applications of Prestack Migration and Dip Moveout." *Paper presented at the 55th Ann Int. Soc. Explor. Geophys. Mtg.*, 1985.

Chun, J. H. and C. Jacewitz. "Fundamentals of Frequency-Domain Migration." *Geophysics*, 46 (1981): 717–732.

Claerbout, J. F. *Fundamentals of Frequency-Domain Migration.* New York: McGraw-Hill, 1976.

Claerbout, J. F. "Imaging the Earth's Interior." *Blackwell Scientific Publications.* (1985).

Claerbout, J. F. and S. M. Doherty. "Downward Continuation of Moveout-Corrected Seismograms." *Geophysics* 37 (1972): 741–768.

Fowler, P. "Velocity-Independent Imaging of Seismic Reflectors." *Presented at the 54th Ann. Int. Soc. Explor. Geophys. Mtg.*, Atlanta, December, 1984.

Gardner, G. H. F., W. S. French and T. Matzuk. "Elements of Migration and Velocity Analysis." *Geophysics* 39 (1974): 811–825.

Gadzag, J. "Wave-Equation Migration by Phase Shift." *Geophysics* 43 (1978): 1342–1351.

Gadzag, J. and P. Squazzero. "Migration of Seismic Data by Phase Shift Plus Interpolation." *Geophysics* 49 (1984): 124–131.

Hubral, P. and T. Krey. "Interval Velocities from Seismic Time Measurements." *Soc. Expl. Geophys. Monograph.* (1980).

Lee, M. W. and S. H. Suh. "Optimization of One-Way Wave Equations." *Geophysics* 50 (1985): 1634–1637.

Levin, F. K. "Apparent Velocity from Dipping Interface Reflections." *Geophysics* 36 (1971): 510–516.

Robinson, E. A. *Migration of Geophysical Data.* Boston: IHRDC, 1983.

Rothman, D., S. Levin and F. Rocca. "Residual Migration: Applications and Limitations." *Geophysics,* 50 (1985): 110–126.

Schneider, W. "Integral Formulation for Migration in Two and Three Dimensions." *Geophysics* 43 (1978): 49–76.

Stolt, R. H. "Migration by Fourier Transform." *Geophysics* 43 (1978): 23–48.

Taner, M. T. and F. Koehler. "Velocity Spectra—Digital Computer Derivation and Applications of Velocity Functions." *Geophysics* 32 (1969): 859–881.

Yilmaz, O. and R. Chambers. "Migration Velocity Analysis by Wave Field Extrapolation." *Geophysics* 49 (1984): 1664–1674.

Yilmaz, O. and J. F. Claerbout. "Prestack Partial Migration." *Geophysics* 45 (1980): 1753–1777.

Yilmaz, O. "Seismic Data Processing." *Soc. Explor. Geophys.*, Tulsa, OK, 1987.

CHAPTER 7

MODELING

INTRODUCTION

Although modern well-logging programs provide an abundance of information about the formations penetrated by a well, there is a discontinuity of information between wells. A seismic section can be used to fill this information gap, showing changes in facies between the wells.

The main method of calibrating a seismic record and identifying lithology for stratigraphic interpretation is by creating a synthetic seismogram from borehole measurements. The logs used for this are the sonic (acoustic velocity) and the density logs. Some other logs are useful for establishing specific lithologies with depth.

The synthetic seismogram is a simple form of one-dimensional seismic modeling. Generating a synthetic seismogram consists of calculating the seismic record, which should be observed for a given sequence of rock units.

USE OF SYNTHETIC SEISMOGRAMS

Synthetic seismograms provide a mean of linking borehole logs with actual seismic records. Their principal use is in identifying the event on a seismic record that relates to a particular interface or sequence of interfaces. By comparing actual field records with synthetic seismograms made for primary reflections and for primaries-plus-multiples, it is possible to determine which events are primary reflections (see Figure 7–1).

Another very important use of synthetic seismograms is to see the effect of a change in a geological section. One can vary the input data to

FIGURE 7–1. Using a synthetic seismogram to identify lithology (courtesy GX Technology)

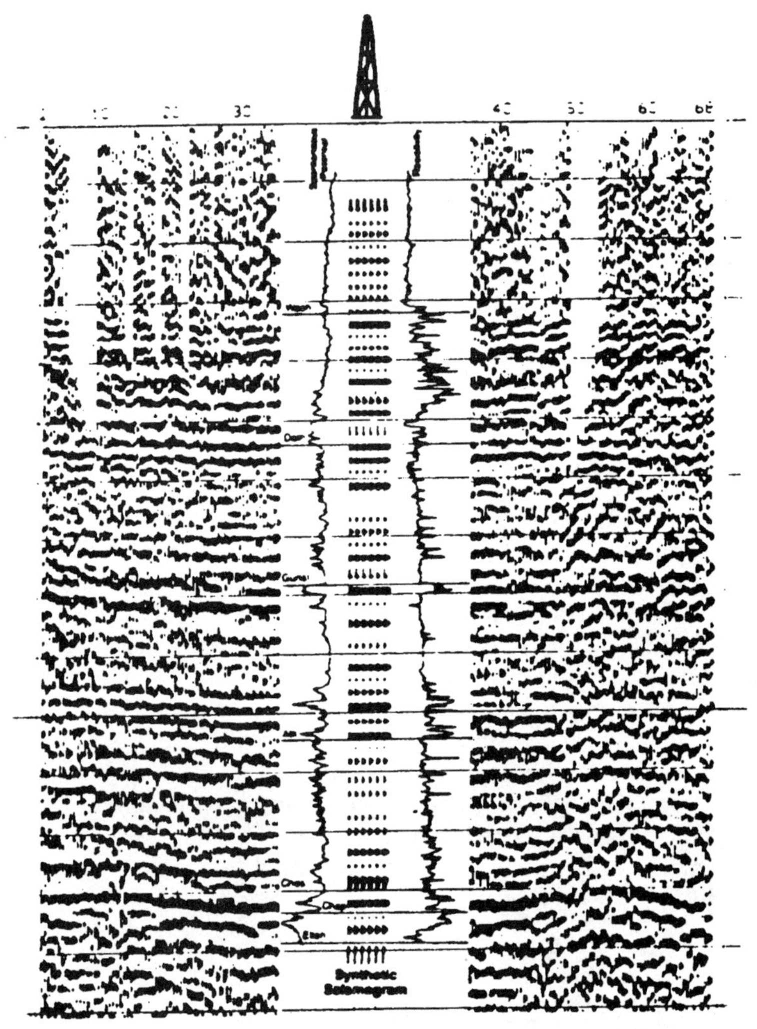

COMPOSITE OF DISPLAYS OF SEISMIC CDP DATA, DOWNHOLE LOGS AND SYNTHETIC SEISMOGRAM.

simulate changes in the thickness of a unit, disappearance of a unit, or a change of lithology. Figure 7–2 shows how a synthetic seismogram can be a guide to what to look for on a seismic section as evidence of changes to be expected, such as pinch-outs, channel sand development, or other facies changes.

FIGURE 7–2. Modeling (courtesy of Seismograph Service Corporation)

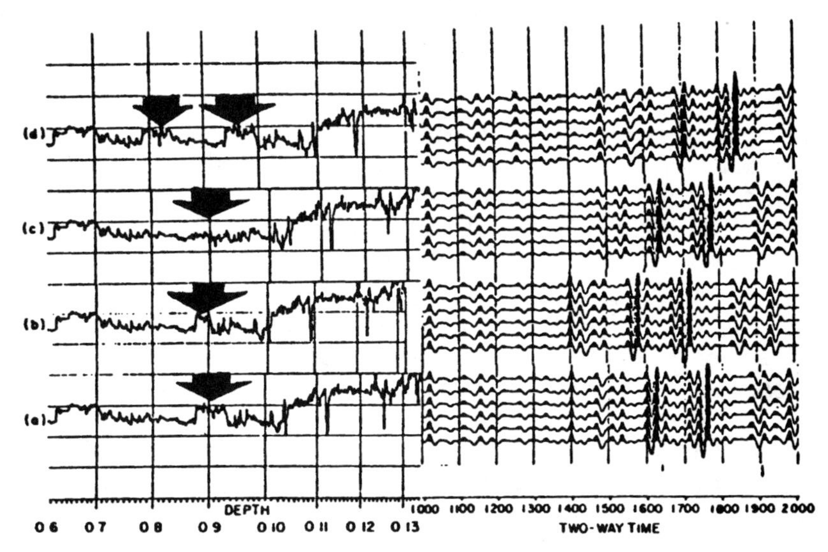

(a) THE VELOCITY LOG SHOWS A THICK SAND INTERVAL. TO THE RIGHT IS THE CORRESPONDING PORTION OF A SYNTHETIC SEISMOGRAM.
(b) TO MAKE THE SAND BODY THINNER, AND THE SYNTHETIC SEISMOGRAM INDICATES HOW SUCH A THIN SAND SHOULD APPEAR ON THE SEISMIC SECTION.
(c) THE LOG HAD BEEN MODIFIED BY REPLACING THE SAND WITH SHALE.
(d) THE SINGLE SAND BODY HAS BEEN REPLACED BY TWO SEPARATE SANDS.

ACOUSTIC IMPEDANCE

The *acoustic impedance* of a rock is defined as the product of its density and velocity, ρV, where ρ is the density of the medium in gm/cm^3 and V is its interval velocity in unit distance/time. A change in acoustic impedance gives rise to reflections.

If a seismic ray strikes an interface perpendicular to it, the acoustic impedance is expressed simply in terms of the product of the density and velocity on each side of the interface.

The reflection coefficient R is the ratio of the amplitude A_1 of the reflected ray to the amplitude A_0 of the incident ray. For a normally incident ray, the coefficient is written in terms of acoustic impedance:

$$R = \left(\rho_2 V_2 - \rho_1 V_1\right) / \left(\rho_2 V_2 + \rho_1 V_1\right) \tag{1}$$

where $\rho_1 V_1$ and $\rho_2 V_2$ are the acoustic impedances of the first and second layers, respectively.

The transmission coefficient is the ratio of the amplitude transmitted to the incident amplitude:

$$T = A_2 / A_0 \tag{2}$$

GENERATING A SYNTHETIC SEISMOGRAM

The synthetic seismogram used by the interpreter for calibrating a seismic record in terms of lithology is developed along the following procedure:

1. Digitizing the sonic log and density log at uniform depth intervals, normally every 6 inches or 1 foot.
2. Decimation in depth by averaging to avoid the introduction of aliasing. Data sampled at 6-inch intervals are averaged over 2-foot intervals. The four values in the 2-foot interval are averaged by filtering to permit a valid resampling to the larger sample interval.
3. Conversion of depth values of velocity and density to time values. For example, the sonic log travel time values at 2-ft intervals are integrated (summed) to accumulate an assigned time interval. This is typically one millisecond, if the final synthetic seismogram is to be sampled at 2-ms intervals. Depth intervals established by this process for the desired uniform time intervals are used to sample the density log.
4. Combining transit time and density samples to produce acoustic impedance. The acoustic impedance is derived as a function of one-way time by multiplying the velocity and density values from step 3.
5. Computing the reflectivity series as a function of time. The successive reflection coefficients are calculated from the layered acoustic impedances from step 4 by taking the ratio of the difference in acoustic impedances to the sum of acoustic impedances for adjacent layers. Using the equation in Figure 7–3, an increase in acoustic impedance with depth gives positive reflection coefficient and a decrease in acoustic impedance with depth gives negative reflection.
6. Filtering, or convolution, of the reflectivity series. The reflectivity series computed in step 5 is filtered to the same bandwidth of the seismic section with which the synthetic seismogram will be compared. It is preferable to use the wavelet from the seismic data, if it is known, for better correlation and match between the seismic data and the synthetic seismogram. These initial steps complete the calculations of the primary reflections only. The seismogram should be refined to the time-depth relationship by using a check shot survey or a Vertical Seismic Profile (VSP) survey (see Chapter 8). A check shot survey is used to calibrate the transit time integration.

Figure 7–3 illustrates the process.

REFINING THE TIME-DEPTH RELATIONSHIP

1. Correlation of synthetic seismogram events with a check shot survey.
2. Time remap of the reflectivity series. This is normally done by linear interpolation between events. This step causes stretch or squeeze of intervals to conform with the correlation process from the above step.

FIGURE 7–3. Acoustic impedance log to seismic trace

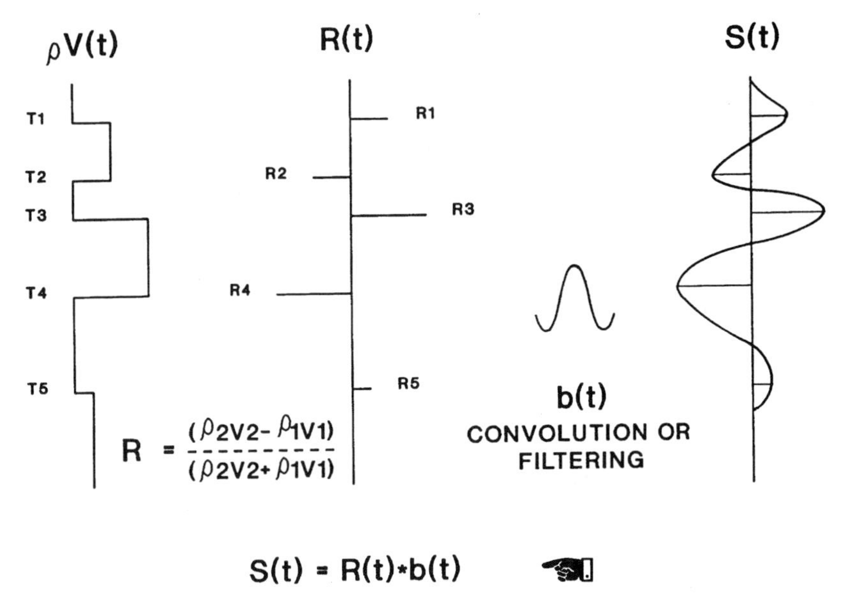

SEISMIC TRACE = REFLECTIVITY SERIES CONVOLVED WITH A WAVELET

3. Reconvolution of the modified reflectively series before correlating it with the seismic data.

These steps can be repeated until a good correlation between the synthetic seismogram and the seismic section is established.

Basic Assumptions in Generating a Synthetic Seismogram

1. Time-depth relationship is known to a detail level.
2. Most events on the seismic data are from valid subsurface markers.
3. The wavelet on the synthetic seismogram can be estimated in order to get a good correlation with the seismic data.

Note that the synthetic seismogram provides a small lateral area of extent with high vertical resolution, while the seismic section provides a large lateral area of extent with less vertical resolution.

Figure 7–4 presents an example of the correlation of synthetic seismograms with stacked seismic data.

Two-Dimensional Modeling

The one-dimensional modeling techniques are useful for studies of a few points on a subsurface to resolve fairly detailed changes in a localized

FIGURE 7–4. Correlation with seismic data (courtesy of Seismograph Service Corporation)

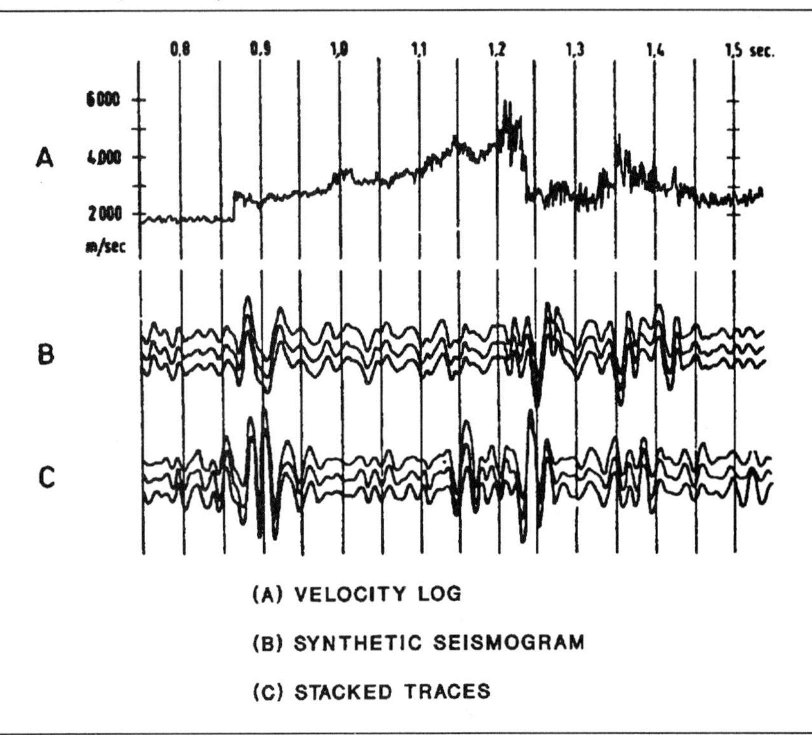

area. They do not, however, aid in defining lateral variations between points, focusing, shadow zones, or diffractions. Two-dimensional modeling is used to fill the above needs. These models can range from very simple to extremely complex, and they will be discussed in this chapter.

The explorationist, whether a geologist, an engineer, or a geophysicist, seeks to learn the nature of the subsurface. The main purposes of using a two-dimensional modeling package are to:

- Enable the interpreter to synthesize a geologic interpretation into a seismic cross-section for validity testing.
- Aid the interpreter in his evaluation of the structural, stratigraphic, and reservoir conditions.
- Evaluate amplitude variations, bright spots, dim spots, and focusing.
- Solve special interpretation problems such as near-surface formation problems, velocity pull-up, and other problems that cause ambiguity in the structural and stratigraphic interpretations.
- Help in designing the field and processing parameters by studying the tuning effect.

FIGURE 7–5. Seismic distortion—anticline (courtesy of Seismograph Service Corporation)

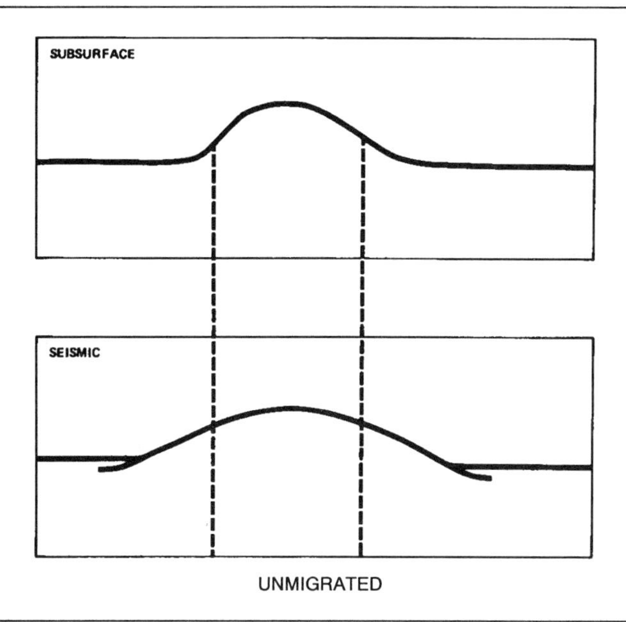

FOCUSING

Since the earth's layering is represented by reflected energy, only flat, parallel bedding yields undistorted subsurface reflections. In case of an anticline, it appears to be distorted, as shown in Figure 7–5. The seismic section shows larger lateral extent compared to the earth or subsurface model.

SHADOW ZONES

In Figure 7–6 some traces show no energy return (dead zone). These are referred to as *shadow zones*, and they are common near faults and other discontinuous areas in the subsurface. This zone is a portion of the subsurface from which reflections are not present because the ray paths do not reach the geophones on the surface.

DIFFRACTIONS

Diffractions are events observed on the seismic data that occur at discontinuities in the subsurface, such as faults or at velocity changes, such as bright spots.

Figure 7–7 shows a seismic line across a large horst block. Notice that at the left side of the section diffractions mask out the fault plane.

FIGURE 7–6. Shadow zone—seismic model (courtesy of Seismograph Service Corporation)

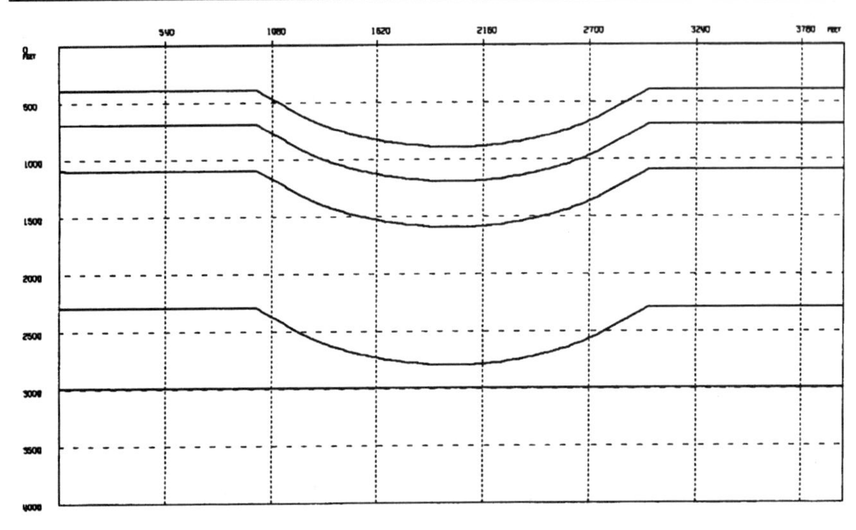

(a) EARTH MODEL

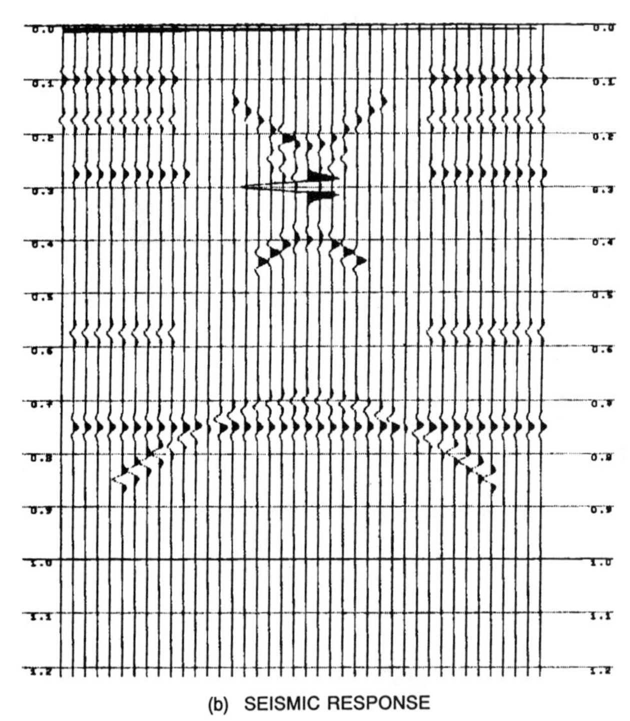

(b) SEISMIC RESPONSE

IN (b) THE NORMAL INCIDENCE RAY TRACING, NOTE THE DEAD AREA IN THE SHALLOW PART OF THE SECTION—SHADOW ZONES.

FIGURE 7–7. Diffraction model and seismic data of horst block (courtesy of Seismograph Service Corporation)

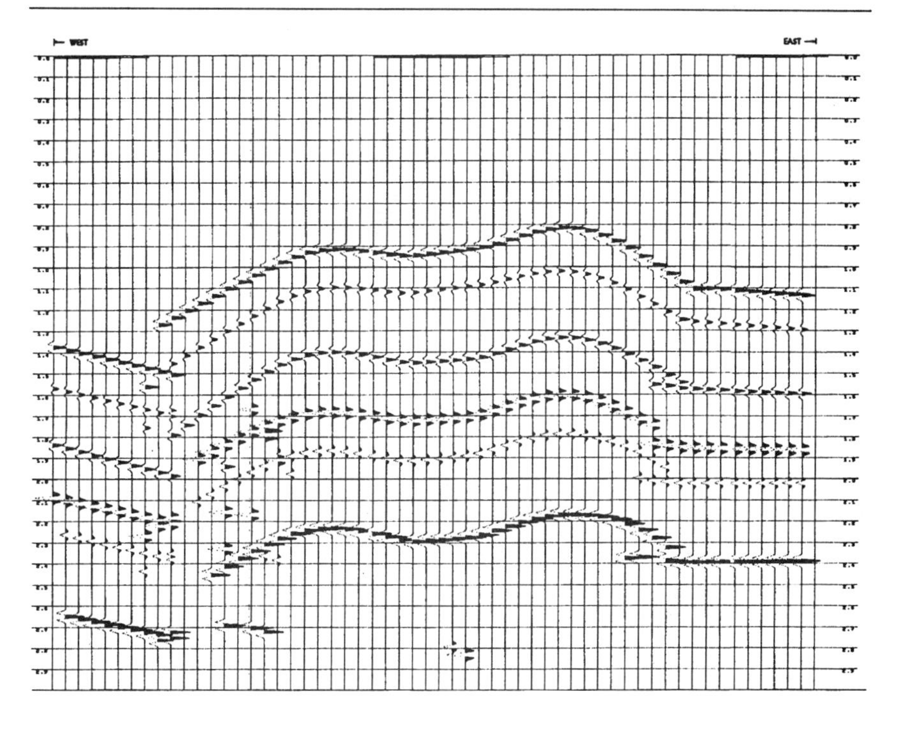

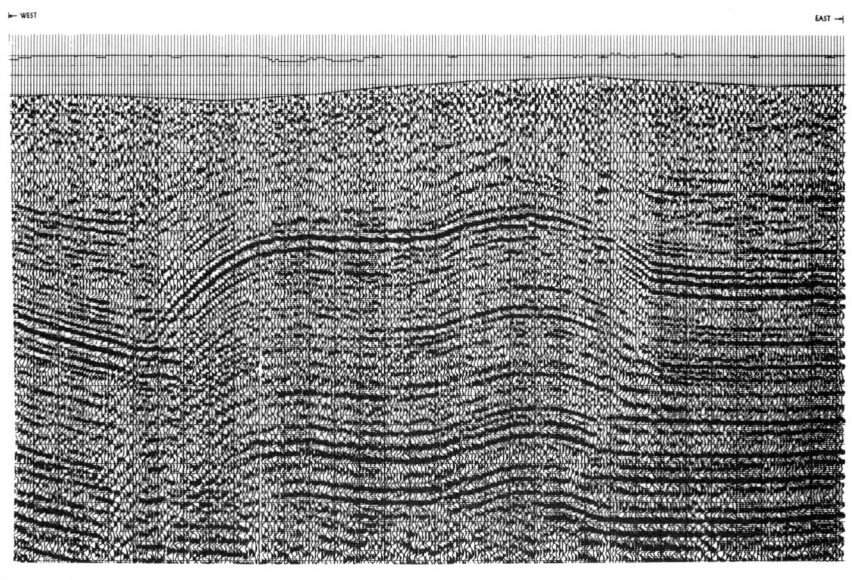

A normal incident raypath model of this feature without diffractions shows the fault more clearly, and the shadow zone can be seen in the vicinity of the fault.

Figure 7–8 shows a model of a fault (to the left) displayed separately. Notice diffraction emitted from various points on the fault plane.

FIGURE 7–8. Diffraction model (courtesy of Seismograph Service Corporation)

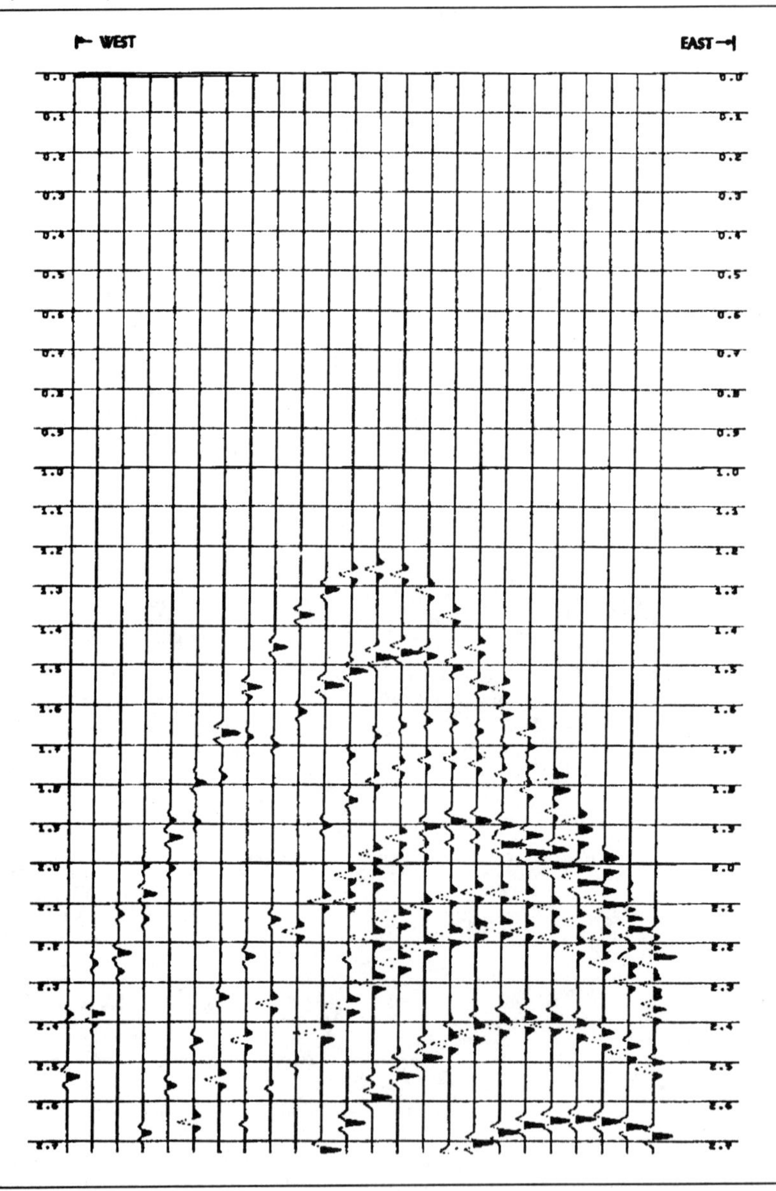

MODELING FOR DATA ACQUISITION AND PROCESSING

VERTICAL RESOLUTION AND TUNING EFFECT

Because the earth attenuates high frequencies much more than lower frequencies, the seismic reflection method uses the low end of the spectrum. Usable frequencies are normally within the range of 5 to 100 Hz. Ground roll and air blast (wind noise) may cover the lower and higher frequencies in the 5 to 100 Hz range; the useful "signal" frequency band may be in the range of 16 to 65 Hz. The limitation of the bandwidth causes a limitation in the vertical resolution of the seismic data. Therefore, whenever there is two or more closely spaced reflectors, a constructive or destructive interference occurs. The resulted composite wavelet shape depends on the time interval between the reflectors and their reflection coefficient.

With very thin layering, reflections become so close together and interference is so severe that there is no reflection present, or else one strong reflection is present.

Figure 7–9 illustrates how the variation in thin bed thickness affects the seismic response.

Figure 7–10 shows two wedge models, one with unusual velocity in the wedge, the other one with a step velocity profile. Both models were

FIGURE 7–9. Thin bed response (courtesy of Seismograph Service Corporation)

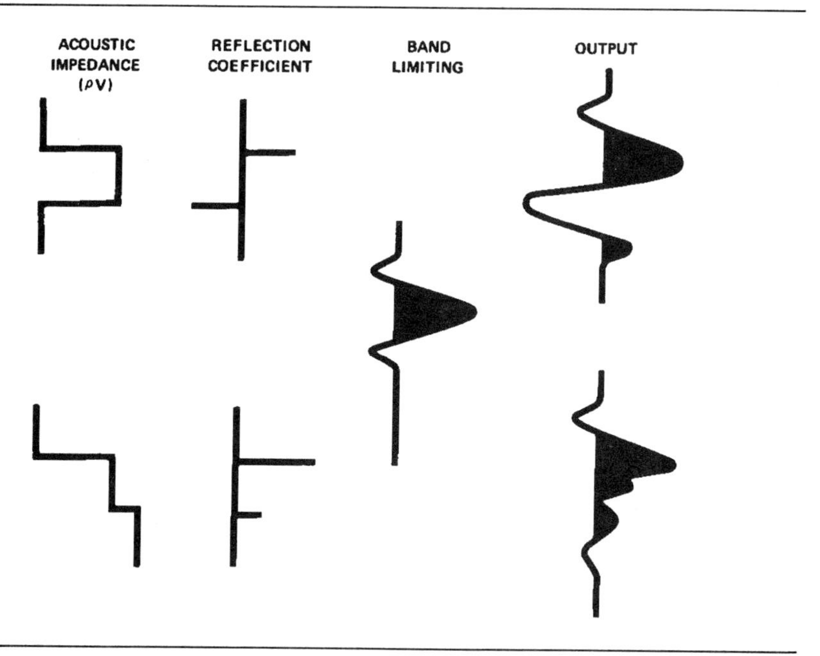

FIGURE 7–10. Wedge models (courtesy of Seismograph Service Corporation)

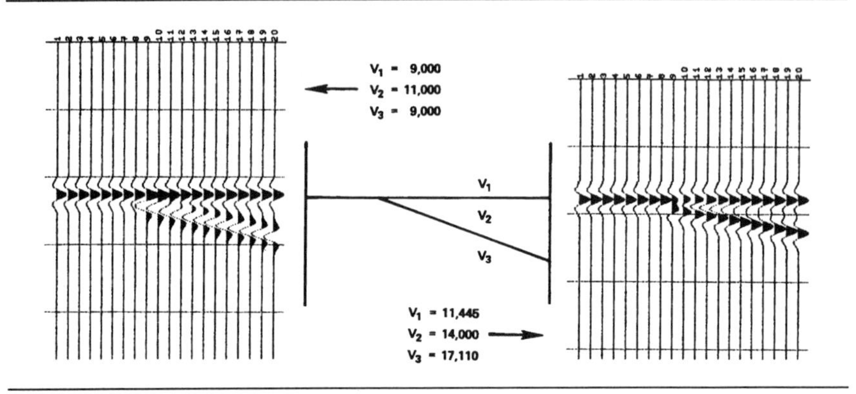

convolved with a 30 Hz wavelet. To the right, where the wedge is thick, the wavelets are separate, but as the wedge thins they merge into a complex wavelet with changing amplitude and phase.

Figure 7–11 shows the relationship between bed thickness and seismic response as a percentage. The tuning effect for the cases of anomalous velocity and step velocity are shown for the two wedge models illustrated in Figure 7–10. The tuning effect is dependent on the velocity, bed thickness, frequency, and the reflection coefficient.

Figure 7–12 is a filter test of a synthetic seismogram. It shows how one reflection (A) responds to high frequencies and another one (B) responds to low frequencies.

NOISE

Noise is defined as any signal on the seismic record other than the desired signals from geological markers. Even with carefully selected field data acquisition parameters and an optimum processing sequence, residual noise will still mask the subsurface details to a certain extent. Types of noise are random noise, residual coherent noise, out-of-plane reflections, multiples, and ghosts.

These types of noise can cause apparent changes in reflection times (spurious faults), frequencies (false stratification), and amplitudes.

TOPOGRAPHIC OR SURFACE ANOMALIES

Variations in the surface elevations and irregularities are principal causes of seismic distortion. The surface formation material causes changes

FIGURE 7–11. Thin-bed tuning effect (courtesy of Seismograph Service Corporation)

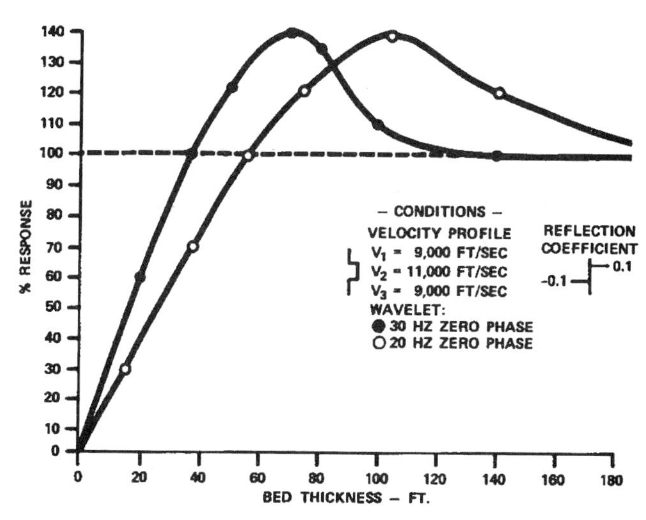

THIN BED TUNING EFFECT—ANOMALOUS VELOCITY

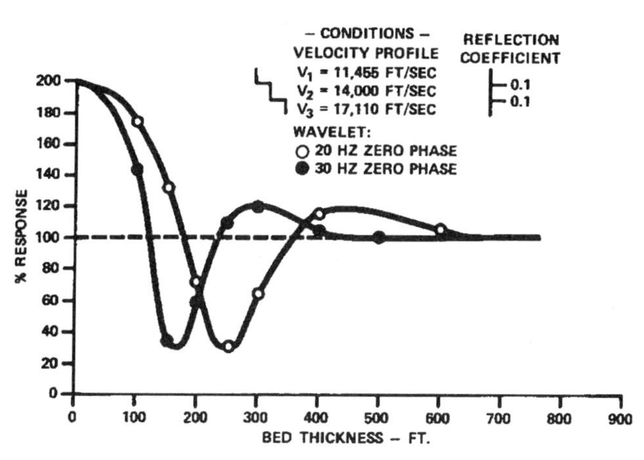

THIN BED TUNING EFFECT—STEP VELOCITY

in the reflection quality due to source/receiver response. Lateral change in the near-surface formation and sudden variation in the velocity are illustrated in Figure 7–13. Note, in this model that the deep reflector is a flat calibration horizon and shows the seismic distortion directly. One can see that there is consistent time shift and similar pattern on the three reflectors on the seismic model.

FIGURE 7–12. Frequency analysis of synthetic seismogram (courtesy of Seismograph Service Corporation)

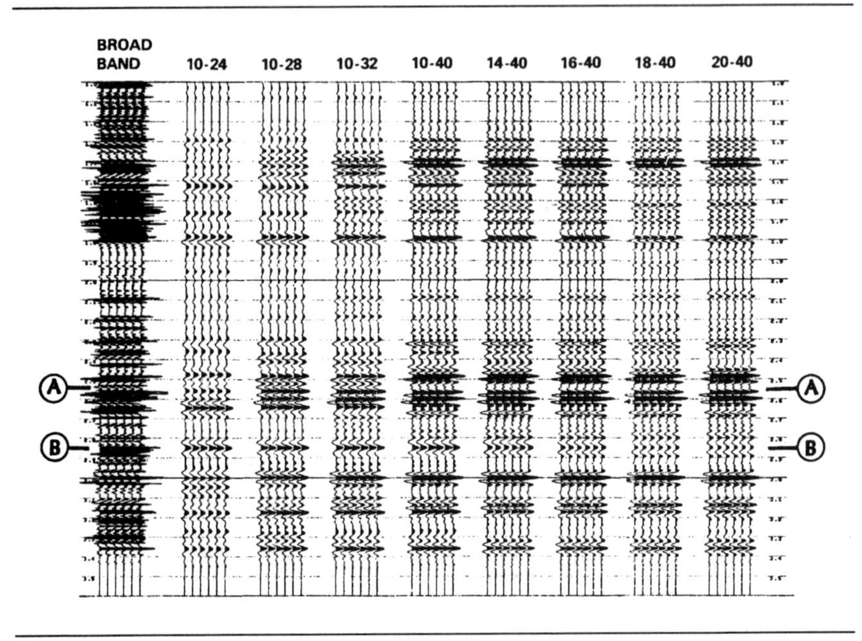

Figure 7–14 is a dramatic example of how the change in the surface elevation and near-surface velocity variations can cause an apparent turnover into a fault which really does not exist.

VELOCITY PULL-UP—APPARENT STRUCTURE CHANGE

Velocity variations in the subsurface can cause many distortions that give a false picture of the subsurface. Examples of such cases are:

1. Apparent structure due to velocity pull-up for shallower beds (see Figure 7–15).
2. Apparent thickness change, Figures 7–16 a and b. In these examples the subsurface section is apparently thinning basinward. However, as one expects, bed thickness actually *thickens* basinward.
3. Distortion of bedding due to intrusion, such as overpressured shale sections or salt plugs. Figures 7–17 a and b show the subsurface section and seismic model of an overpressured shale.

These are only a few of the several distortion mechanisms that must be dealt with in the modeling process. A geophysicist can discuss these problems intelligently with other earth scientists and take the necessary steps to avoid potential problems.

FIGURE 7–13. Distortion caused by near-surface (courtesy of Seismograph Service Corporation)

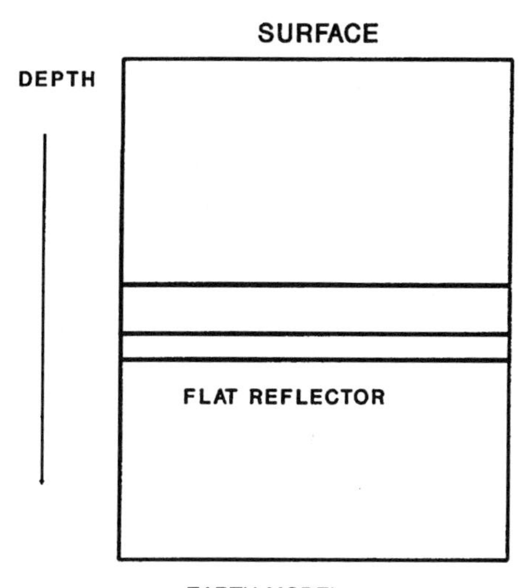

EARTH MODEL

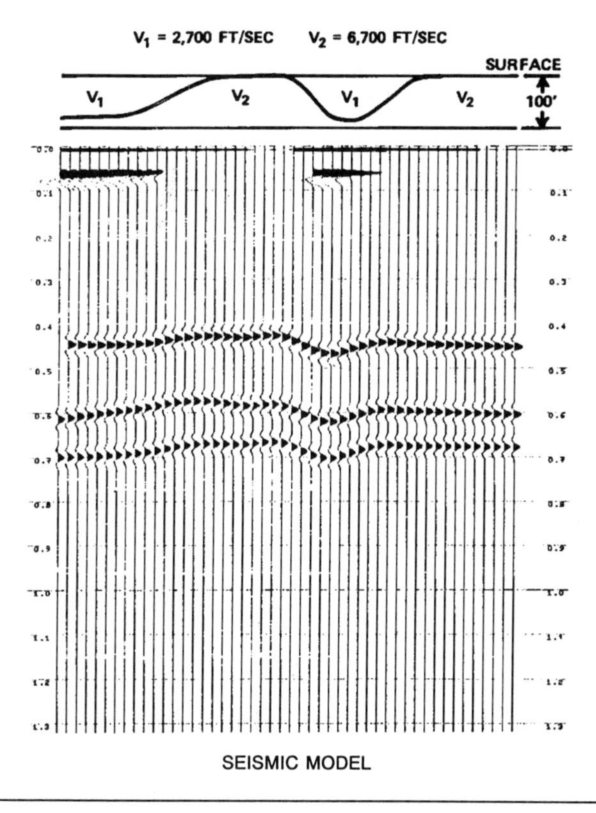

SEISMIC MODEL

FIGURE 7–14. Near-surface velocity problem (courtesy of Seismograph Service Corporation)

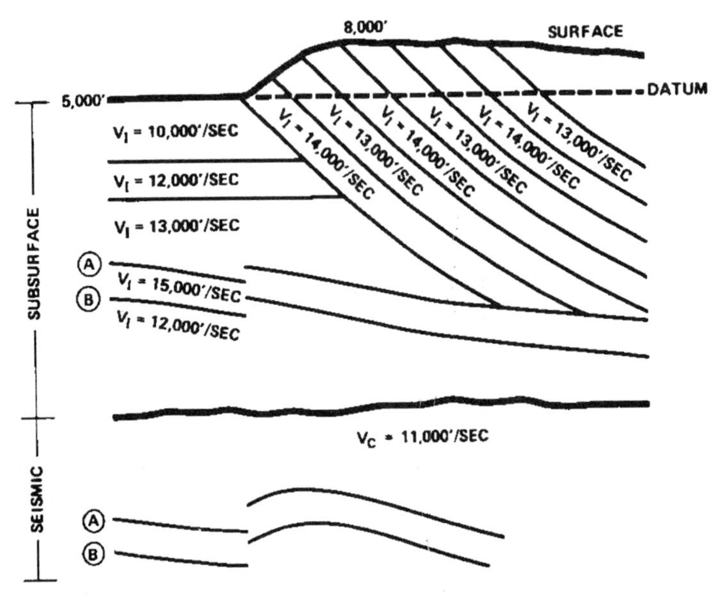

ABOVE: A GEOLOGICAL MODEL REPRESENTS CHANGES IN THE SURFACE ELEVATION AND NEAR SURFACE VELOCITIES, NOTE REFLECTORS A & B.
BELOW: SEISMIC RESPONSE OF REFLECTORS A & B, NOTE THE APPARENT TURNOVER INTO THE FAULT AS A RESULT OF THE NEAR SURFACE EFFECTS.

FIGURE 7–15. Velocity pull-up (courtesy of Seismograph Service Corporation)

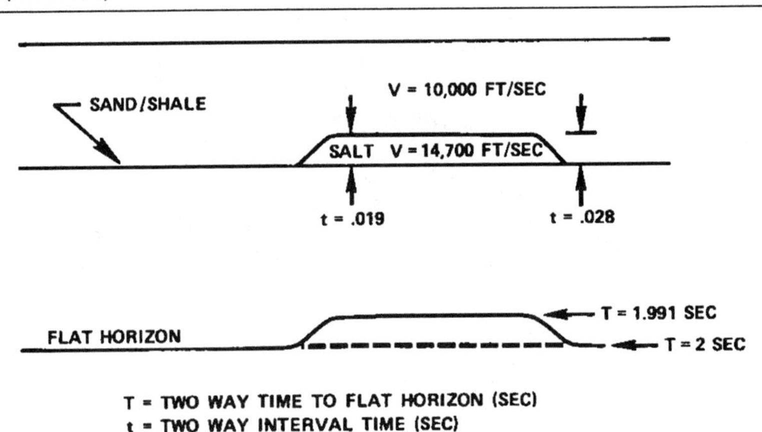

FIGURE 7–16. Subsurface section—basinward thinning (courtesy of Seismograph Service Corporation)

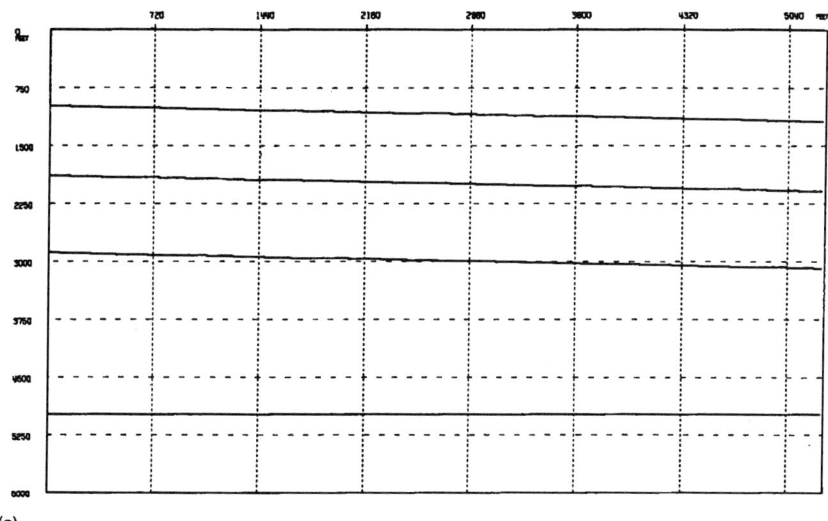

(a)

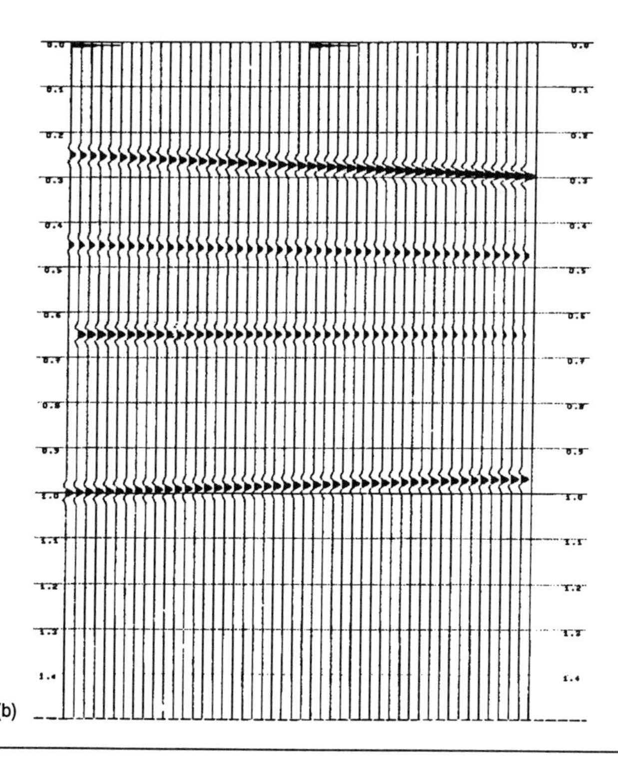

(b)

FIGURE 7–17. Subsurface section—overpressured shale (courtesy of Seismograph Service Corporation)

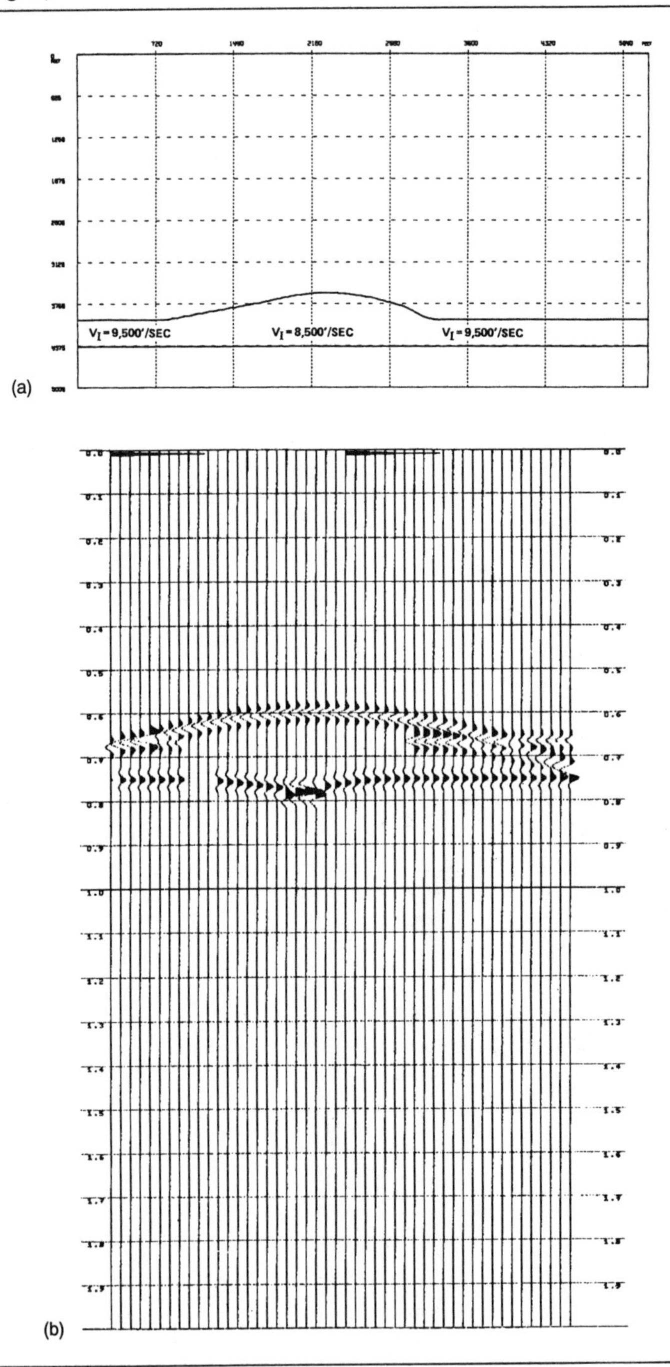

TYPES OF TWO-DIMENSIONAL MODELS

Table 7–1 lists the various types of two-dimensional modeling, with brief descriptions, and their limitations.

Since the interpreter uses two-dimensional modeling to test an assumed geological model against seismic data to validate his assumptions, and since the wave-equation modeling approach as well as CDP ray tracing are so expensive, the most widely used two-dimensional modeling package is normal-incidence ray tracing. This technique gives the desired accuracy for most problems at a reasonable cost.

To illustrate ray tracing modeling, let us follow the modeling sequence step by step:

1. Develop a scaled geologic model of the subsurface setting under investigation. This will include:
   - Geometry of bedding.
   - Interval velocities.
   - Formation densities (if available).
2. Input this data to the computer using a scanner or digitizer.
3. Perform ray tracing.
4. Calculate a spike seismic section using the equations:

$$R = (\rho_2 V_2 - \rho_1 V_1)/(\rho_2 V_2 + \rho_1 V_1) \quad \text{(reflection coefficient)}$$
$$T = A_2/A_0 \quad \text{(transmission coefficient)}$$

5. Convolve with the desired wavelet.
6. Add random noise (if desired).

TABLE 7–1. Types of two-dimensional modeling

Type	Description	Restrictions
1. Normal incidence, straight path.	Uses average velocity for times and position.	Not very accurate in modeling diffractions, and shadow zones.
2. Vertical path.	Interpolated one-dimensional.	Unable to model shadow zones and diffractions.
3. Normal incidence ray tracing.	Snell's law.	Poor results in modeling tightly folded structures.
4. Wave equation.	Finite differences, Huygens principle.	Most complete solution. Long time to run.
5. CMP ray tracing.	Offset rays by Snell's law.	Useful in modeling field and processing parameters. Expensive.

The geologic model is then interactively altered to a seismic section, while keeping the parameters confined to local physical units.

CONCLUSIONS

The seismic method, like any other exploration method, is an attempt to learn the facts about the subsurface geologic structures and lithology changes. The explorationist relies on acoustic models. One of these is the two-dimensional model, which is tested against physical constraints to match the real data.

The advantages of modeling are:

1. It provides a closer look at the seismic distortion mechanisms, which normally alter the real subsurface image of the earth.
2. The cost of modeling is much less than relying on the drilling bit to test a cursory interpretation.
3. It is the best tool to educate the explorationist in developing the insights required in the interpretation of data.

SUMMARY AND DISCUSSION

One-dimensional synthetic seismograms are used to calibrate seismic records, to identify lithology, and to assess stratigraphic interpretations. The process of generating a synthetic seismogram is a type of forward modeling. It is a process of constructing seismic traces from observed petrophysical properties of rock strata such as velocity and density.

An important use of synthetic seismograms is to see the effect of a change in a geological section by varying thickness, replacing, or eliminating part of the geological units.

Two-dimensional modeling is used to help the interpreter in analyzing geologic interpretations, in evaluating bright spots, dim spots, solve interpretational problems such as near-surface formation irregularities and velocity pull-ups. It also helps in designing the field parameters and data processing sequences.

The major advantage of modeling is to give a closer look at the seismic distortions that may alter the real subsurface image. It is the best tool to educate the explorationist in developing information that will help in upgrading the data interpretation.

Generating synthetic seismograms or two-dimensional models can be done using a desktop computer.

With the advancement in the computer hardware and software, these very important interpretational tools can be obtained and manipulated in a very short time.

The cost of a modeling package varies from a few thousand to a few hundred thousand dollars, depending on the hardware configuration and software sophistication.

KEY WORDS

Bright spot
Calibration
Horst block (faulting)
Reflectivity series
Shadow zone
Synthetic seismogram
Transit time
Tuning effect
Vertical resolution
Velocity pull-up
Well logging

EXERCISES

1. Given the following stratigraphic sequence from top to bottom:

Layer	Wave Velocity (ft/sec)	Density (gm/cm^3)
Weathered	1,600	1.5
Sandstone-1	6,500	2.0
Hard sandstone	7,500	2.4
Limestone	10,000	2.4
Sandstone-2	13,000	2.5
Basement	16,500	2.8

   a. Find the reflection coefficients for each interface in the case of a seismic incident wave.
   b. Find the transmission coefficients for the same case.
2. A certain sandstone layer 80 ft. thick has a velocity of 10,000 ft/sec. A seismic wave strikes the top of the sandstone and is reflected with reflection coefficient 0.2. The part that is transmitted into the sandstone is reflected by the bottom of the sandstone layer with reflection coefficient –0.2. Neglect amplitude loss in transmission through the top of the sandstone, and neglect intrabed multiples.
   a. If the basic wavelet has amplitudes at 4-ms sampling given by (8, 7, –7, –5, 0, 4, 2), find the composite wavelet shape that is reflected by the sandstone layer. Explain why we can say in this case that the top and base of the sandstone are "resolved."
   b. What is the dominant frequency of the composite wavelet?
3. A shale bed is incased in the middle of the sandstone layer with 40 feet of sand above the shale and 40 feet below. The thickness of the shale is such that the traveltime through it is the same as through 40 ft. of sand. Let the reflection coefficient be –0.2 at the top of the shale and 0.2 at the bottom of the shale. For the wavelet in (a) above.
   a. Find the composite reflected wavelet.
   b. Calculate the dominant frequency of the composite wavelet and explain the result.

BIBLIOGRAPHY

Anstey, N. A. "Attacking the Problems of the Synthetic Seismogram." *Geophys. Prosp.* 8 (1960): 242–260.

Arya, V. K. and H. D. Holden. "A Geophysical Application: Deconvolution of Seismic Data." *N. Hollywood, CA,* Digital Signal Processing. *Western Periodicals,* (1979): 324–338.

Baranov, V. "Film Synthetique Avec Reflexions Multiples—Theorie Et Calcul Practique." *Geophys. Prosp.* 8 (1960): 315–325.

Collins, F. and C. C. Lee. "Seismic Wave Attenuation Characteristics from Pulse Experiments." *Geophysics* 21 (1950): 16–40.

Delaplanhce, J., R. F. Hagemann and P. G. C. Bollard. "An Example of the Use of Synthetic Seismograms." *Geophysics* 28 (1963): 842–854.

Dennison, A. T. "An Introduction to Synthetic Seismogram Techniques." *Geophys. Prosp.* 8 (1960): 231–241.

Durschner, H. "Synthetic Seismograms from Continuous Velocity Logs." *Geophys. Prosp.* 6 (1958): 272–284.

Faust, L. Y. "A Velocity Function Including Lithologic Variation." *Geophysics* 18 (1953): 271–288.

Futterman, W. I. "Dispersive Body Waves." *J. Geophys. Res.* 67 (1962): 5279–5291.

Gardner, G. H. F., L. W. Gardner and A. R. Gregory. "Formation Velocity and Density—The Diagnostic Basics for Stratigraphic Traps." *Geophysics* 39 (1974): 770–780.

Gerritsma, P. H. A. "Time to Depth Conversion in the Presence of Structure." *Geophysics* 42 (1977): 760–772.

Goupillaud, P. L. "An Approach to Inverse Filtering of Near-Surface Layer Effects From Seismic Records." *Geophysics* 26 (1961): 754–760.

Hilterman, F. J. "Three-Dimensional Seismic Modeling." *Geophysics* 35 (1970): 1020–1037.

Hilterman, F. J. "Amplitudes of Seismic Waves—A Quick Look." *Geophysics* 40 (1975): 745–762.

Kelly, K. R., R. W. Ward, S. Treitel and R. M. Alford. "Synthetic Seismograms: A Finite Difference Approach." *Geophysics* 41 (1976): 2–27.

Lavergne, M. and C. William. "Inversion of Seismograms and Pseudo Velocity Logs." *Geophysics* 25 (1977): 231–250.

Peterson, R. A., W. R. Fillipone and F. B. Coker. "The Synthesis of Seismograms From Well Log Data." *Geophysics* 20 (1955): 516–538.

Sengbush, R. L., P. L. Laurence and F. J. McDonal. "Interpretation of Synthetic Seismograms." *Geophysics* 26 (1961): 138–157.

Sheriff, R. E. "Encyclopedic Dictionary of Exploration Geophysics." Tulsa, OK: *Soc. Expl. Geophys.* 1973.

Sheriff, R. E. "Factors; Affecting Amplitudes—A Review of Physical Principles, in Lithology and Direct Detection of Hydrocarbons Using Geophysical Methods." *Geophys. Prosp.* 25 (1973): 125–138.

Treitel, S. "Seismic Wave Propagation in Layered Media in Terms of Communication Theory." *Geophysics* 31 (1966): 17–32.
Trorey, A. W. "Theoretical Seismograms With Frequency and Depth Dependent Absorption." *Geophysics* 27 (1962): 766–785.
Wuenschel, P. C. "Seismogram Synthesis Including Multiples and Transmission Coefficients." *Geophysics* 25 (1960): 106–219.

CHAPTER 8

VERTICAL SEISMIC PROFILING

HISTORICAL REVIEW

The development of the exploration seismology was started by Fessenden's patent in 1917. It was the first documented seismic application involving surface sources and geophones in order to investigate a subsurface structure. Most of the modern day, seismic surveys are still largely restricted to surface geophones and energy sources.

Barton (1929) referred to Fessenden's earlier work and described the possible use of seismic borehole measurements. Barton's work is considered the beginning of what has become later the vertical seismic profile (VSP). In 1931, McCollum and Larue strongly advocated the use of existing wells in collecting data. They proposed that local geological structure could be determined by measuring travel times from surface energy sources to geophones located in wells.

Borehole seismology was largely ignored by the geophysical community. For many years the major use of boreholes for seismic purposes was limited primarily to the measurement of propagation velocities, Dix (1939). These measurements led to the development of check shot surveys, which are now commonly used in the oil industry.

The rigorous work of Jolly (1953), Riggs (1955), and Levin and Lynn (1958) showed the potential use of the borehole measurements, which led to more development of VSP technology. Although the value and applications of VSP were known early and were applied in the USSR and elsewhere, geophysicists in the Western Hemisphere continued to use boreholes for velocity measurements only.

It is, therefore, imperative that geophysicists share with explorationists and operation and production engineers the current facts about VSP. The data recorded in a vertical seismic profile can, in concept, give insight into some fundamental properties of propagating seismic wavelets, and aid in the understanding of transmission and reflection processes in the earth. This information should improve the structural, stratigraphic, and lithological interpretation of surface seismic recordings.

THE CONCEPT OF VERTICAL SEISMIC PROFILING

VSP is a technique in which a seismic signal is induced at the surface and recorded by a borehole geophone placed at various depths in the well.

In a horizontal seismic profile, the source and the receiver are on the surface, while in a VSP survey the direction of the geophone is perpendicular to the source. This difference between these two types of techniques is illustrated in Figure 8–1. A geophone positioned deep in the subsurface responds to both upgoing and downgoing seismic events, whereas only reflected events can be recorded by a geophone planted at the surface. The VSP is similar to a velocity survey, since the source and receiver geometry is the same for both techniques. However, the VSP and velocity surveys differ in two aspects:

1. The distance between geophone recording depths is much less for the VSP (every 15–40 meters) while in the velocity survey shot levels are separated by few hundreds of meters.
2. First break times are the critical information collected in a velocity survey, but first breaks *in addition to* upgoing and downgoing events are recorded in a VSP survey.

FIGURE 8–1. VSP concept

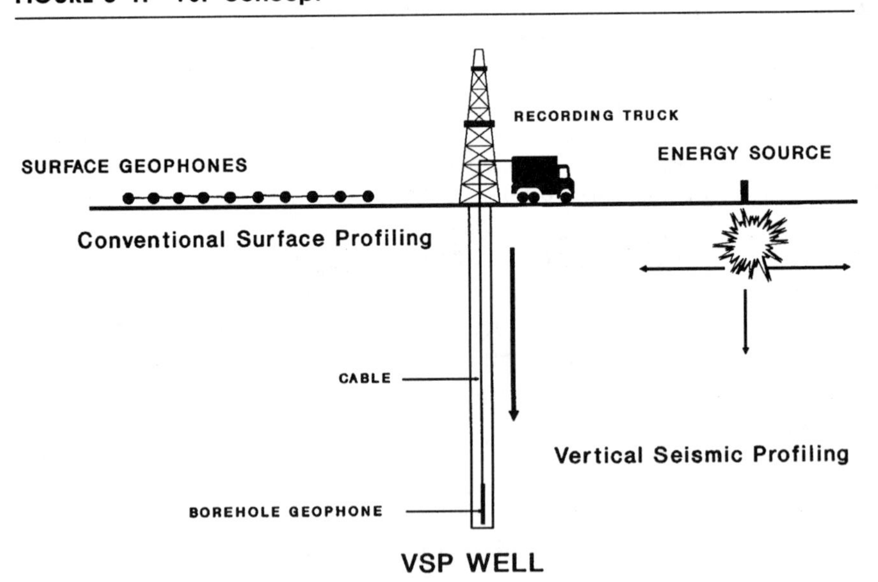

FIELD EQUIPMENT AND PHYSICAL ENVIRONMENT NEEDED

The basic components required for a VSP survey are (1) a borehole, (2) an energy source, (3) a downhole geophone, and (4) a recording system.

Other equipment and physical factors that can be involved in VSP data acquisition will be discussed later. Design of the equipment will not be discussed because of the space limitation. Readers who are interested in learning more about equipment design are referred to the bibliography at the end of this chapter.

BOREHOLES

An appropriate borehole must exist to run a VSP. Several factors should be considered in selecting a hole.

HOLE DEVIATION

Data collection is more economical and the interpretation is easier if the VSP survey is conducted in a vertical hole. In a deviated hole, there is uncertainty about the position of the downhole geophone relative to the energy source, complicated by the fact that the source may be moved to several different locations during the data acquisition phase. Therefore, an accurate deviation survey should be run in deviated holes.

Problems of interpretation are more common and more severe on VSP's in offshore wells drilled from a platform, because they are usually highly deviated. However, there are some advantages to recording a VSP in a deviated well, since this type of well allows the subsurface beneath the borehole to be imaged laterally with great resolution. But if the intent is to identify the depth and one-way time of primary reflectors, a vertical hole is the better choice, as a VSP can be recorded quicker and easier, and the results are more accurate.

CASING & CEMENTING

A good VSP survey can be recorded only when seismic body waves in the earth are transmitted across the borehole/formation interface to the downhole geophone with minimum waveform character distortion.

It is preferable to record the VSP in a cased hole, where the geophone is protected from hole caving and differential pressure problems. The survey time is not limited in a cased hole, whereas an uncased hole might have to be re-entered periodically for conditioning. The casing should be cemented, because there must be a medium between the casing and the borehole that is a good transmitter of seismic energy, and the best medium is cement.

The four common borehole environments, in order of preference for VSP data recording, are:

1. Single casing, cemented.
2. Uncased.
3. Single casing, uncemented, old enough for mud and cuttings in annulus to have solidified.
4. Recently cased, uncemented.

BOREHOLE DIAMETER

If the proposed hole is uncased, roughness of the borehole wall will affect the clamping of the geophone to the formation. In particular, clamping may be impossible in large washouts because the geophone locking arm is too short to reach the borehole wall. A caliper log (measures the hole diameter) of the uncased well should be used to select recording depths.

Blair (1982) concluded that the seismic detector can be installed at any point on the circumference of a cylindrical borehole and still record the same particle motion, as long as the wavelengths of interest in the wavelet propagating past the borehole are greater than 10 times the circumference of the hole.

For smaller wavelengths, the detector should be coupled to the *opposite* side of the hole from that which is first displaced by the propagating signal in order to minimize amplitude distortions and phase shifts created by the presence of the borehole, Hardage (1982). This preferred position of the detector is called the *wave shadow* side of the borehole. However, when recording VSP data in a standard oil or gas well it should make no difference where the geophone is clamped around the circumference of the borehole, since the borehole is small compared with the seismic wavelength.

BOREHOLE OBSTRUCTIONS

A cased hole may contain packers or trace rings that can prevent the tool from reaching those depth intervals. This may be critical to the survey. The presence or absence of obstructions must be confirmed before starting a VSP survey. A cheap tool of the same diameter as the geophone may be run into the well to check for possible obstructions before the VSP survey starts.

BOREHOLE INFORMATION

A complete interpretation of VSP data can be done only if a suite of independent data, which specifies the physical properties of the formations around the borehole, has been recorded in the proposed VSP survey well.

A well in which a suite of logs, such as caliper, sonic, density, resistivity, and radioactive logs were recorded, in addition to which cores were taken and drill cuttings are preserved, would be much preferred over a borehole for which no data is available. A cement bond log and measurements of the depths of all casing strings are required to be sure of the nature of the acoustic coupling between the VSP geophone and the formation.

VSP ENERGY SOURCE

An important use of VSP data is to clarify the interpretation of surface-recorded seismic data.

It is advisable that both the VSP and surface-recorded seismic data have the same wavelet and high-frequency content for better correlation and tie. In many instances this cannot be achieved and the match, if it can be accomplished, must be done in the data processing stage. One must observe the following points:

1. The VSP energy source must generate a consistent and repeatable shot wavelet. Otherwise, it will not be easy to correlate equivalent characters of upgoing and downgoing wavelets throughout the vertical section over which data is gathered. Figure 8–2 illustrates the preferred energy source wavelet.

FIGURE 8–2. Energy source wavelet (modified from Hardage, 1983)

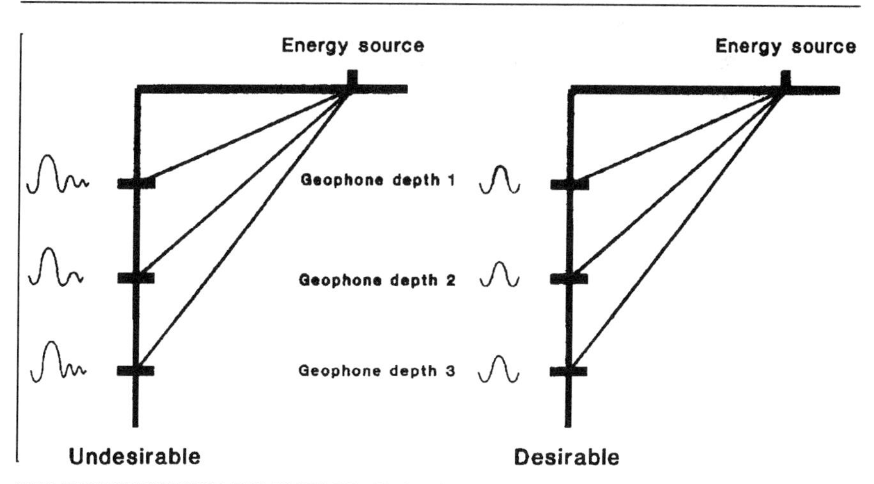

THE ENERGY SOURCE FOR VERTICAL SEISMIC PROFILING SHOULD GENERATE A CONSISTENT AND REPEATABLE WAVELET. ONE SURVEY MAY REQUIRE SEVERAL HUNDREDS OF SHOTS. THE ILLUSTRATION AT THE RIGHT SHOWS THE DESIRED CONDITION, AS THE WAVELET SHAPE REMAINS CONSISTENT FOR ALL OTHER GEOPHONE DEPTHS.

2. The output level of VSP energy should be carefully selected for optimum response without overkill. Many geophysicists have learned that it is *not* true that "the bigger the energy source, the better the response."
3. In all VSP surveys, the downgoing events are much stronger than the upgoing. As the output strength of the VSP source increases, more downgoing events will be created due to numerous reverberating layers in the near-surface part of the stratigraphic section. The increase in the number and amplitude of the downgoing events must be greater than the gain of the amplitude of the upgoing events. It is not unusual to record a decent VSP survey using an energy source of modest strength.
4. Surface energy sources used in VSP surveys are:
   a. Dynamite

      Buried dynamite charges are particularly effective producers of seismic body wave energy, and they are widely used as the surface energy source for VSP. Heven & Lynn (1958) found that using dynamite charges suspended in air, downhole signal amplitude was reduced by a factor of 30 compared to the signals generated by buried shots, even though the air charge size was reduced from the buried charge size by only a factor of 2.

      One objection concerning dynamite sources frequently expressed by VSP users is that it is too difficult to shoot several tens of shots and maintain consistent wavelet shape. However, if great care is exercised in the field, reasonably invariant shot wavelets can be obtained. To accomplish this the diameter and depth of the shot hole must be maintained throughout the course of the survey. The charge must be located below the weathered layer. A small charge (.5–1.5 kg) may be used, sometimes even smaller. Thirty-inch or thirty-six inch casing may be used to protect the formation. Figure 8–3 illustrates proposed shot hole configurations to produce invariant VSP shot wavelets.
   b. Mechanical Impulse Source

      Many varieties of sources exist that can apply a vertical impulsive force to generate seismic energy. Such sources are acceptable as VSP energy sources, but they should be tested for an area before being used. Some surface sources are characterized by limited frequency bandwidth and a tendency to generate severe shallow reverberations. It is advisable to conduct actual field tests at each onshore position to determine if valid information can be recorded.
   c. Vibrators

      A wide variety of vibratory energy sources are available that are attractive for use in VSP work. They are quite mobile, which

FIGURE 8–3. Shot hole design for repeatable wavelet (modified from Hardage, 1983)

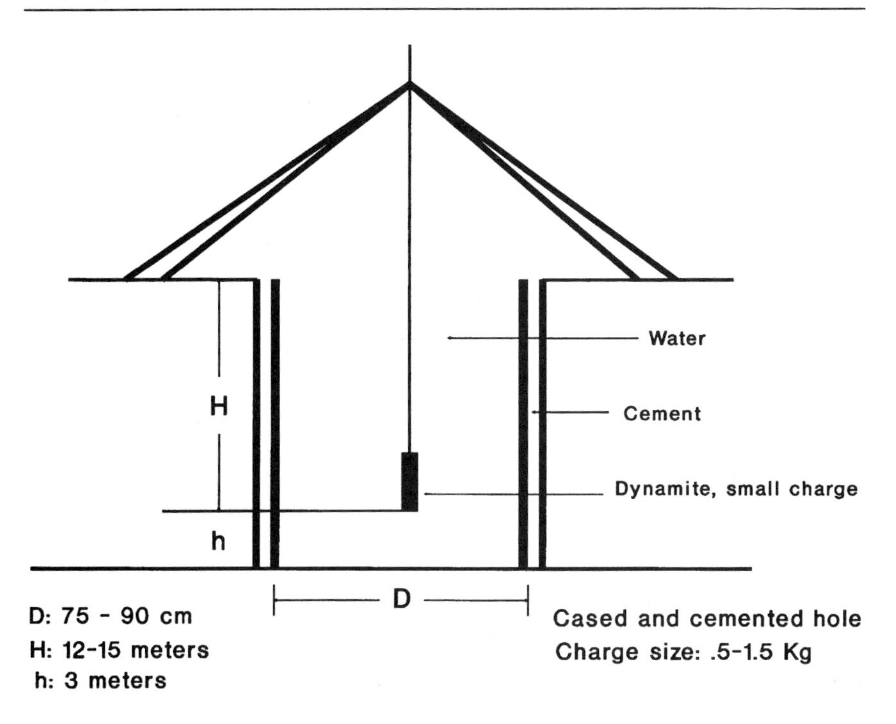

permits VSP experiments to use many different source locations. The Vibroseis® system is engineered to adjust the input pilot signal to the ground to meet resolution requirements needed in a particular VSP recording. The amount of the energy can be designed for optimum signal-to-noise ratio by varying the size or number of vibrators. If the site is populated by random noise, Vibroseis units are an excellent energy source because cross-correlation of Vibroseis data enhances signal-to-noise ratio by discriminating against noise outside of the sweep frequency range. On the other hand, coherent noise having frequencies within the sweep band width will be enhanced by the correlation operation, but these can be attenuated in the data processing stage.

Vibroseis is preferable for VSP surveys because its signals can be designed to suit the resolution of an area and its wavelet is repeatable and consistent.

®Registered trademark of Conoco Oil Company.

FIGURE 8–4. Air gun used as a stationary energy source in marine VSP (modified from Hardage, 1983)

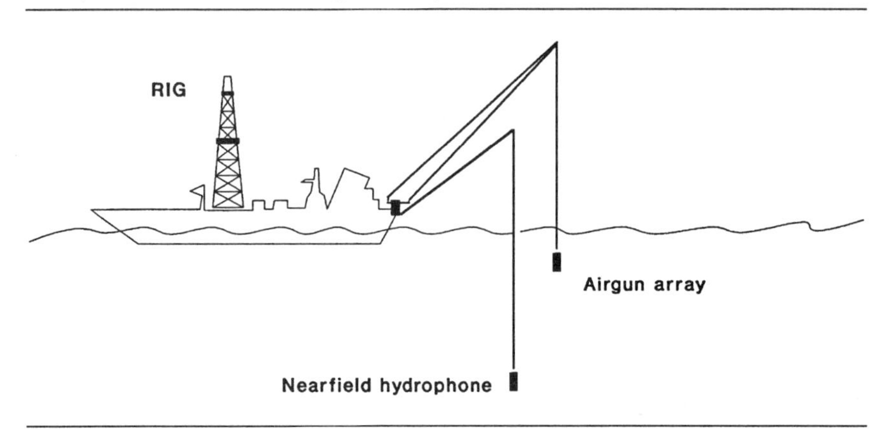

d. Air Guns

Air guns are by far the dominant source used in offshore vertical-seismic profiling. If an offshore well is vertical, then many VSP objectives can be obtained with the air gun at a fixed location near the well head. The shooting is very simple, because the air gun is suspended from a work crane. The air gun can be operated from the high-capacity air compressor that is standard equipment on the rig. An air gun in operation may cause the rig to vibrate, but it will not cause any structural damage. It is much safer than explosives for use on offshore rigs. Figure 8–4 illustrates the placement of the air gun in marine vertical seismic profiling.

Marine air guns have several features that make them attractive for onshore use. They are small and portable, they can be fired at intervals of a few seconds, and they create highly repeatable wavelets. They must be submerged in water in order to function properly. Figure 8–5 illustrates the marine air gun as an onshore VSP energy source.

THE DOWNHOLE GEOPHONE

There is a major difference between the shape and construction of a geophone used for surface recording and the borehole geophone used to record a VSP survey, as shown in Figure 8–6.

The downhole geophone is carried within a massive housing that is designed to withstand the high pressures and temperatures encountered in deep wells. The same housing contains the mechanical deployment system that anchors the geophone to the borehole wall and electronic amplifying and telemetering circuits.

FIGURE 8–5. Using a marine air gun as an onshore VSP energy source (modified from Hardage, 1983)

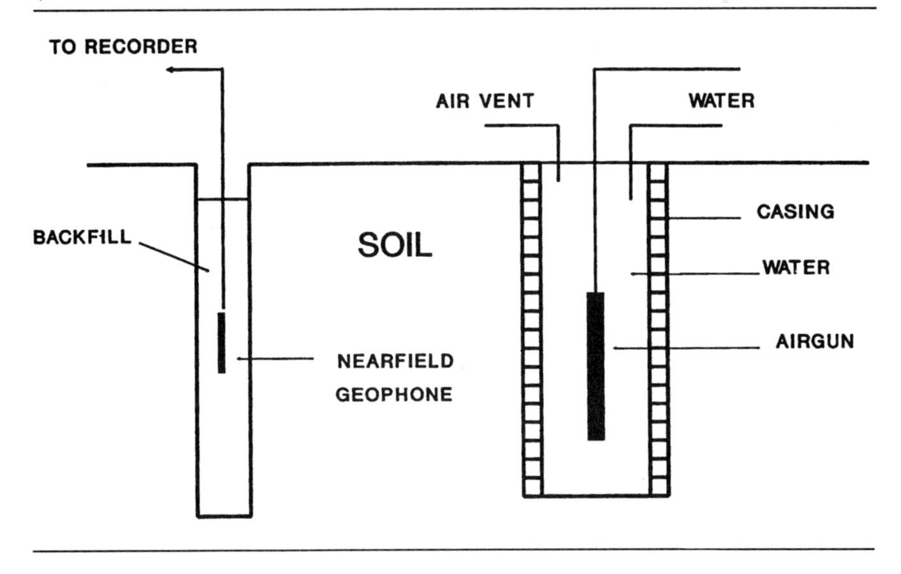

FIGURE 8–6. A typical land geophone and a VSP geophone (modified from Hardage, 1983)

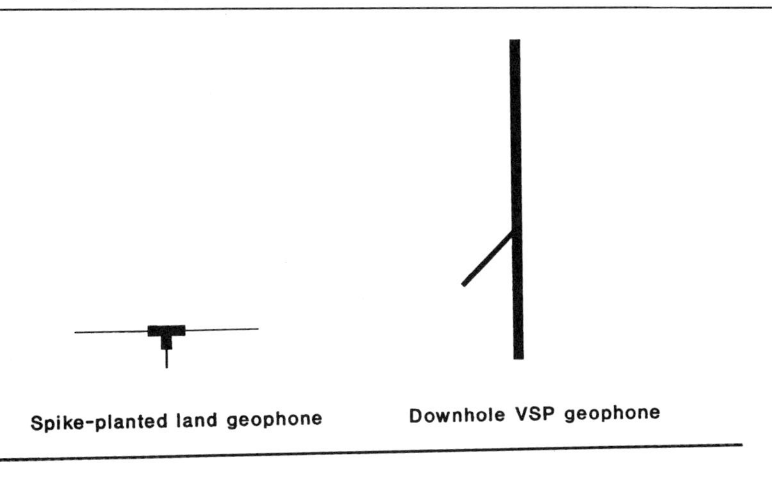

Parameters

Spike-planted land geophone	Downhole VSP geophone
Length: 10 cm (4 Inches)	Length: 3 m (9.8 Feet)
Diameter: 3 cm (one Inch)	Diameter: 10 cm (4 Inches)
Weight: 200 gm (.45 Pound)	Weight: 100 Kg (225 pounds)

RECORDING SYSTEM

VSP recordings should meet rigid standards regarding resolution, dynamic gain, and recording format. Both the downhole geophone data and near-field monitor geophone responses should be recorded with enough resolution (at least 12 bits, including sign) in order to capture high-resolution wavefronts. It is important to record the near-field wavelet in all marine VSP surveys, especially when performing source-signature deconvolution with an energy source, such as an untuned air gun, that creates a long wavelet.

NOISE ENCOUNTERED IN VSP

RANDOM NOISE

In some wells, formation irregularities, fluid movement behind casing or in the well may cause background noise. (These random noises will not be discussed in this text.)

GEOPHONE COUPLING

Some of the seismic noise recorded in surface geophones is caused by the poor planting of the geophone. A loosely planted geophone is noisier than one which has a good coupling with the earth. The same principle applies to the downhole geophone. Figure 8–7 shows the effect of coupling on signal.

CABLE WAVES

The propagation velocity of acoustic waves along cables such as those used in VSP surveys is a variable and depends on specific cable design. Propagation velocities are in the range of 2,500 m/sec to 3,500 m/sec.

In shallow holes and low-velocity stratigraphic sections, a cable wave can be the first arrival measured by the borehole geophone. If not correctly recognized, this wave can cause erroneous determinations of formation velocity. Cable-borne events are generally caused by wind or machinery vibrating the cable, but this type of noise can be reduced by slacking the cable after the tool is locked downhole. Figure 8–8 demonstrates how cable slack can dramatically reduce cable noise.

RESONANCE IN MULTIPLE CASING STRINGS

It is difficult to obtain usable VSP data inside multiple casing strings because one or more of the casings may not be bonded to another string or to the formation. Therefore, poor quality data are usually recorded near

FIGURE 8–7. VSP geophone coupling (courtesy Geophysical Press, from Hardage, B.A.: "Vertical Seismic Profiling, Part A: Principles," 1983)

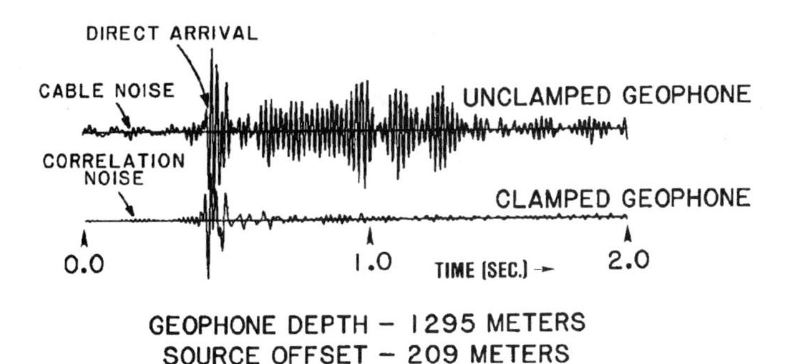

FIGURE 8–8. Effect of cable slack on geophone signal (courtesy Geophysical Press, from Hardage, B.A.: "Vertical Seismic Profiling, Part A: Principles," 1983)

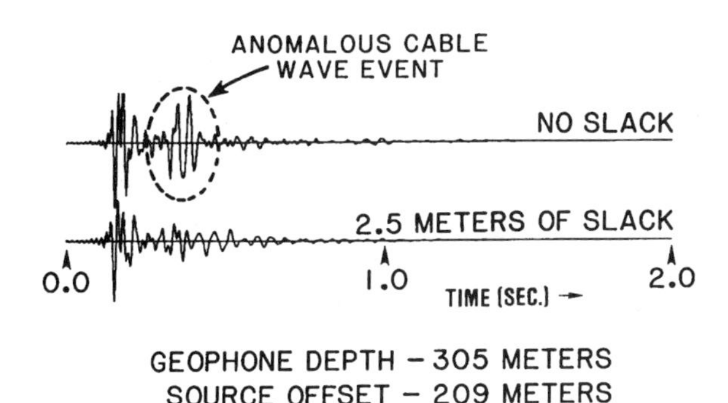

the surface where multiple casing are encountered. Processing can attenuate some of the noise patterns recorded, but to record a good VSP in a multiple-cased hole, all strings must be well cemented to each other. A good cement bond between the strings is the key to good VSP data inside multiple casing strings.

BONDED AND UNBONDED CASING

There is no difference in the shape of the wavelet recorded in cemented casings and in open hole, because cement is the medium best capable of transmitting seismic energy from the formation to the geophone. Data quality will deteriorate in unbonded single casing.

SURFACE WAVES

In land exploration, Rayleigh waves and Love waves propagate along the earth's surface in all directions away from the energy source, interfering with body waves from deep reflectors as they are recorded by geophones located at or near the surface. These waves are undesirable signals, and they prevent optimum imaging of stratigraphic and structural conditions. However, the VSP does not record Rayleigh or Love waves, because they do not reach the depth of the geophone.

TUBE WAVES

The fluid-filled borehole constitutes a cylindrical discontinuity that serves as a medium for propagating undesirable waves called *tube waves*. This is a disturbance that propagates along steel casings with a velocity of approximately 5.5 km/sec, Gal'Perin (1974). See Figure 8–9.

This is one of the most damaging of noise patterns, because it is coherent noise. As such, it cannot be reduced by repeating and summing shots. It is, in fact, usually amplified by summing, since its character is consistent for all records being summed. However, careful data processing procedures may attenuate the tube wave effectively.

SURFACE CULTURAL NOISE

This is the noise that exists because of the presence of people or running machinery near the VSP site. Examples of such noise sources are diesel engines, air compressors, electrical generators, and drillsite activity such as welding, pipe handling, and so forth. The amount of noise recorded depends upon whether VSP data are recorded while the drilling rig is still in place or after the rig has been removed. The only way to avoid severe cultural noise problems during a VSP survey is to create as quiet an environment as possible.

VSP FIELD PROCEDURES

It is essential to have efficient field operations because the longer a VSP survey is in progress, the more equipment failure or borehole deterioration.

Correct VSP field procedures should begin and end with a set of instrument tests. This should be done in order to confirm that the recording system itself does not attenuate some signal frequencies, the amplifier gains are correct, cross-feed has not been introduced into the data, and recording intervals are accurately determined.

FIGURE 8–9. Tube wave (modified from Hardage, 1983)

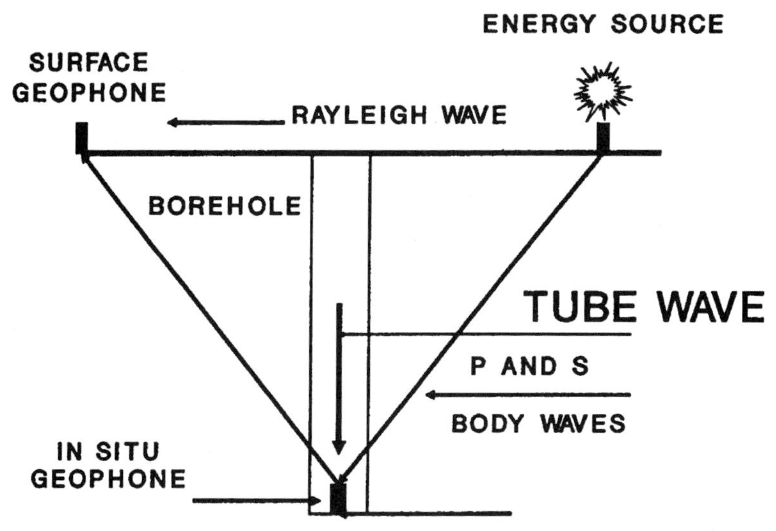

ELASTIC WAVE MODES INVOLVED IN SUBSURFACE SEISMIC RECORDING.

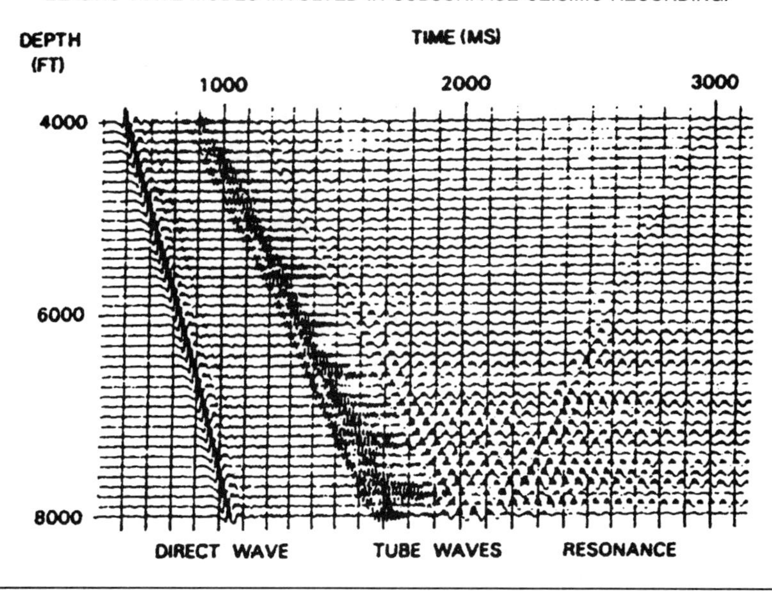

Geophone Tap Test

A geophone tap test determines the polarity of its output signal when its case moves in a specified direction in the borehole. This test is essential for correlation between VSP measurements and seismic data and to check that the geophone is functioning before it is lowered into the borehole.

ENERGY INPUT

As mentioned before, the fundamental difference between velocity and VSP surveys is that a velocity survey focuses only on the first breaks, whereas first breaks, downgoing, and upgoing events must be recorded by VSP.

The energy input required can be completely opposite for these two measurements. In the velocity survey, more energy can be input to downgoing wavelets as the geophone depth increases to maintain strong first break amplitude.

For a VSP survey, more input energy is sometimes needed when the geophone is at shallow depths so that weak late arrivals from deep reflectors can be recorded. In some cases, the energy input needed may be two or three times that required when the geophone is at the bottom of a deep hole. To get the required energy, multiple shots can be taken at the same level to maintain the same wavelet character. Increasing the strength of the input energy often creates a wavelet with a completely different character from the wavelet already recorded.

VSP DATA PROCESSING

It is extremely difficult for an interpreter to work with a set of unprocessed VSP field data. There is first the realization that the downgoing event is dominant. It is difficult to do any interpretation on the upgoing events. In addition, both random and coherent noises are present in the data. Extensive data processing is required to achieve the maximum utilization of the VSP recorded data.

FIELD DATA RECORDING

As the geophone is lowered into the well, exploratory data should be recorded at intervals of 300 to 500 meters (1,000–1,700 ft.). These measurements allow quality control personnel to select the appropriate recording parameters, such as source energy and the number of shots to be summed to enhance S/N ratio.

The VSP survey is recorded as the borehole geophone is raised in increments of a few meters from the bottom of the hole to the surface. The depth increment around the target horizon is normally small—100 feet (30 meters) or less—and becomes progressively longer as the geophone is pulled out of the hole. Multiple records are normally recorded at each geophone depth and vertically summed in the data processing stage.

Figure 8–10 shows the raw data of a VSP survey after the demultiplexing stage. One can see that the horizontal scale represents the geophone depth in the hole in unit distance, and the vertical scale represents the time in seconds.

FIGURE 8–10. Vertical seismic profile

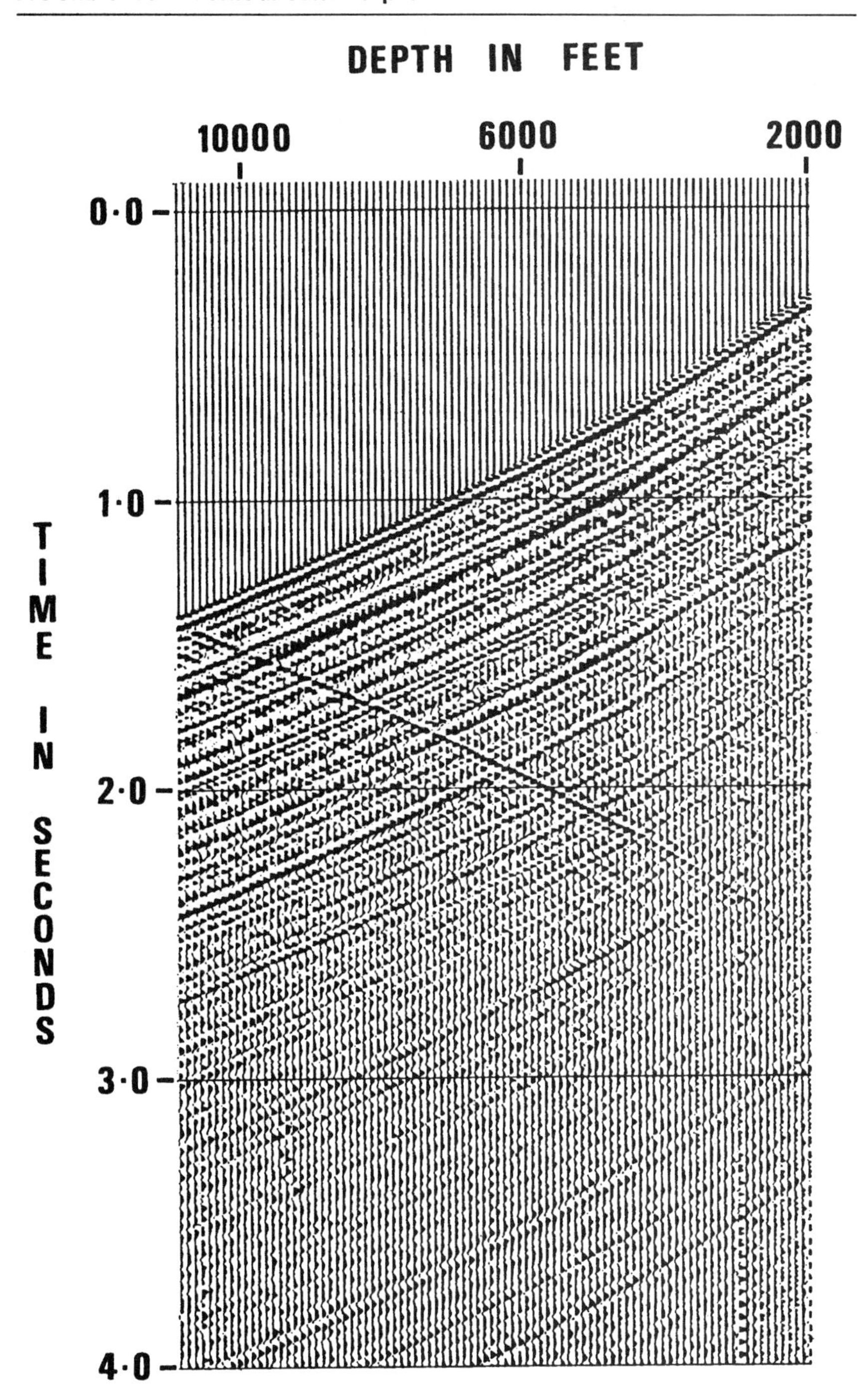

The first strong amplitude events are called the *direct arrivals* or *downgoing waves,* and their orientation trend from shallow time to the deeper times is from right to left. These are followed by downgoing multiples along the recorded VSP. Another event with an orientation trend that goes from deep time at the right to shallower time at the left is called an *upgoing wave.* It is a mirror image of the downgoing events. These waves must be separated before we can utilize the information from each type.

SEPARATING UPGOING AND DOWNGOING WAVES

Figure 8–11 represents the raypaths by which seismic energy can propagate from an energy source at the surface to a geophone in a borehole. Both primary and multiple reflections are shown, all upgoing events, and primaries and multiples. The reflectors in this analysis are assumed to be flat and horizontal. The source is located near the well so that the rays travel vertically. In Figures 8–12 and 8–13, the horizontal distance is exaggerated so that the rays are separated for visual clarity. T_1, T_2, and T_G are one-way vertical travel times to reflectors 1 and 2, and geophone depth, respectively.

FIGURE 8–11. Upgoing primaries and multiples

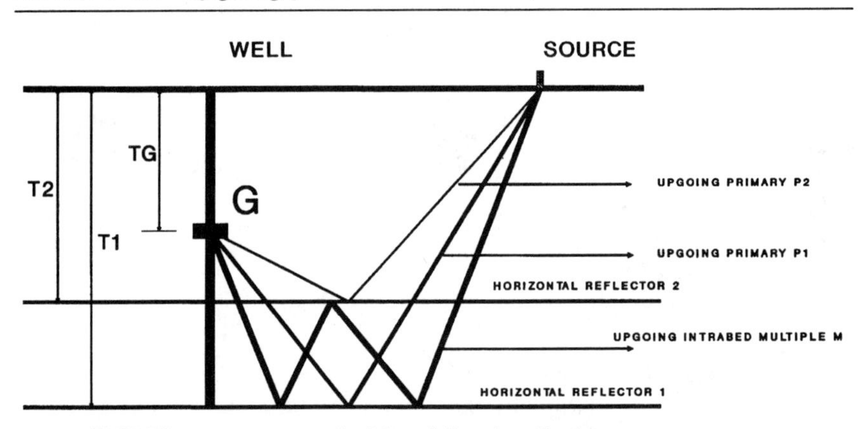

T1,T2,TG are one way vertical travel time to reflectors 1,2 and geophone G. Large offset is assumed for better visual clarity. Reflectors 1,2 are assumed to be horizontal.
t1:Upgoing travel time from reflector 1 to geophone.
t1=T1+(T1-TG)=2T1-TG
t2:Upgoing travel time from reflector 2 to geophone.
t2=T2+(T2-TG)=2T2-TG.
tM:Upgoing multiple travel time from reflector 1 to geophone
tM=T1+3(T1-T2)+T2-TG =2T1+2(T1-T2)-TG.

FIGURE 8–12. Downgoing surface and intrabed multiples

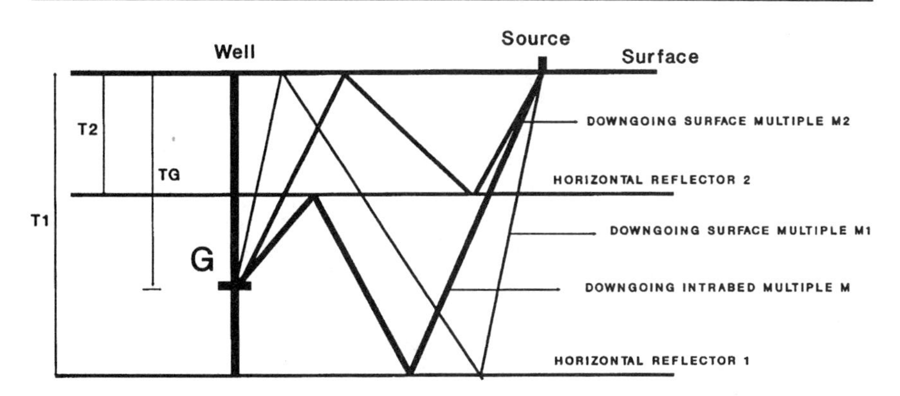

RAYPATHS DESCRIBING THE DOWNGOING SURFACE MULTIPLES AND INTRABED MULTIPLES THAT ARRIVE AT THE VSP GEOPHONE. ASSUME THAT REFLECTORS 1, 2 ARE HORIZONTAL

$t1 = 2T1 + TG$

$t2 = 2T2 + TG$

$$tM = T1 + (T1 - T2) + (TG - T2)$$
$$= T1 + T1 - T2 + TG - T2$$

$$tM = 2(T1 - T2) + TG$$

From the traveltime equations for upgoing events, the time of an upgoing event at the geophone depth is equal to the two-way time to the surface minus the one-way time to geophone depth. Similarly, the raypath describing the propagation of downgoing events and intrabed multiples is illustrated in Figure 8–12. From the traveltime equations for downgoing events, the time of a downgoing event at geophone depth is equal to the two-way time when the event is reflected downward plus one-way time to geophone depth.

From the traveltime equations, a static shift of either $+T_G$ or $-T_G$ will position VSP events at the same times at which they would be recorded by a surface geophone at position G.

By definition, T_G is the first-break time for the VSP trace recorded at geophone position G, so if we add $+T_G$ on both sides of the upgoing traveltime equation and $-T_G$ on both sides of the downgoing equation, we can separate the upgoing and downgoing events. See Figure 8–13.

DATA ENHANCEMENT

VSP data contain both downgoing and upgoing wave modes which overlie each other in varying degrees of complexity. The analysis of upgoing

FIGURE 8–13. Separating downgoing and upgoing events (courtesy of Seismograph Service Corporation)

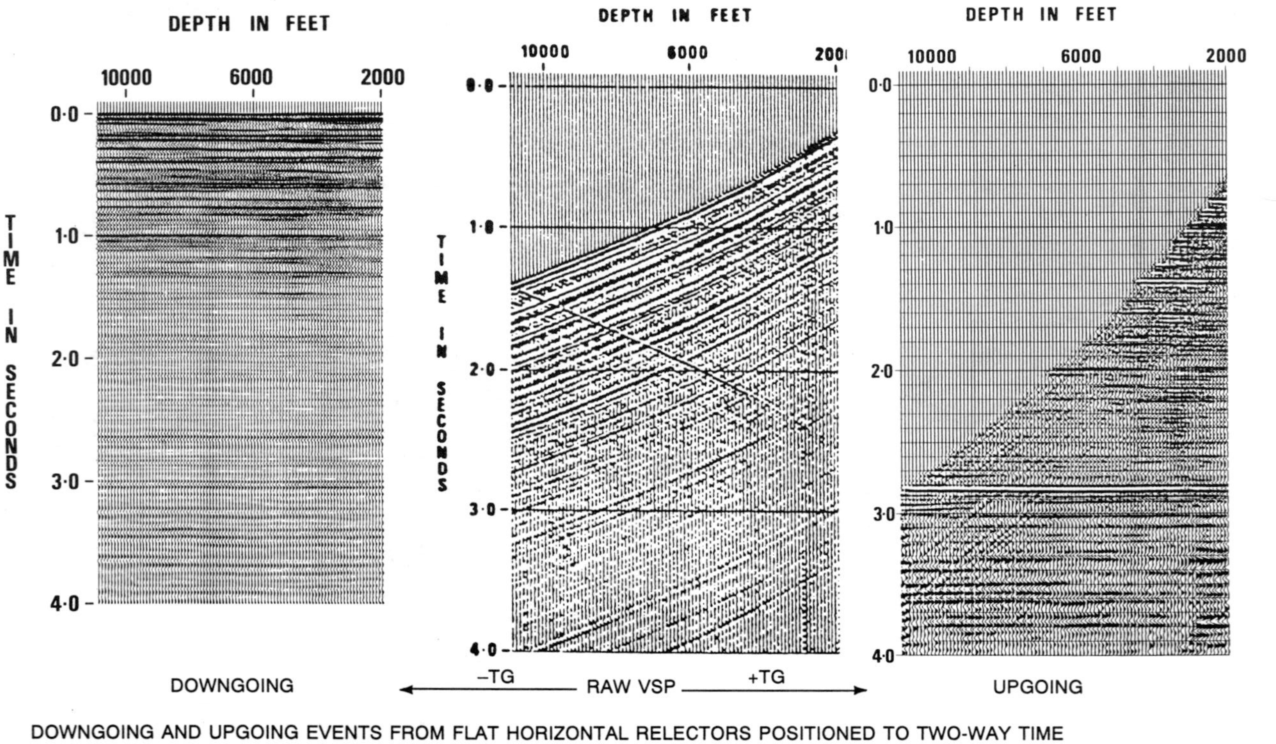

DOWNGOING AND UPGOING EVENTS FROM FLAT HORIZONTAL RELECTORS POSITIONED TO TWO-WAY TIME

modes is particularly important because these events are the ones recorded by surface seismic measurement.

Thus, numerical procedures that can attenuate the downgoing modes without seriously affecting upgoing events are very important in VSP data processing. The most common technique applied to remove a selected VSP wave mode is the F-K velocity filter.

F-K VELOCITY FILTER

This approach involves the design of velocity filters in frequency-wavenumber (F-K) space. The construction of these filters requires that VSP data be recorded at uniform increments in both time and space.

Both horizontal and vertical sampling must satisfy the Nyquist sampling theorem, which requires that at least two sample points be recorded within the shortest wavelength contained in the data.

Figure 8–14 illustrates the F-K velocity filtering applied in order to enhance the desired wave modes.

VSP APPLICATIONS

Some applications of vertical seismic profiling are listed below in two main categories. Some of these applications will be discussed in detail.

FIGURE 8–14. Velocity filtering (courtesy Geophysical Press, from Hardage, B.A.: "Vertical Seismic Profiling, Part A: Principles," 1983)

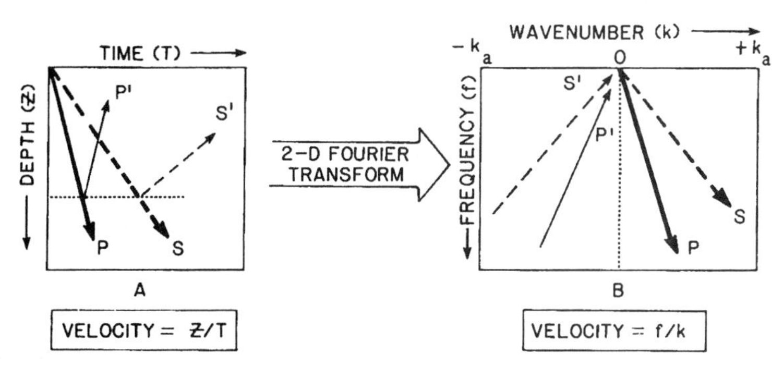

THE SEPARATION OF VSP WAVE MODES BY VELOCITY FILTERING CAN BE ACCOMPLISHED BY TRANSFORMING THE DATA FROM A FUNCTION OF TIME AND SPACE TO A FUNCTION OF FREQUENCY AND WAVENUMBER. THESE HYPOTHETICAL VSP DATA SHOW HOW DOWNGOING COMPRESSIONAL (P) AND SLOWER SHEAR (S) BODY WAVES AND UPGOING COMPRESSIONAL (P') AND SHEAR (S') REFLECTIONS TRANSFORM INTO F-K SPACE. DOWNGOING MODES APPEAR IN THE POSITIVE WAVENUMBER HALF PLANE AND UPGOING MODES IN THE NEGATIVE WAVENUMBER HALF PLANE. THE WIDTH OF EACH LINE IS PROPORTIONAL TO THE AMOUNT OF ENERGY CONTAINED IN THAT WAVE MODE.

EXPLORATION APPLICATIONS

- Determining reflection coefficients.
- Identification of seismic reflectors.
- Comparison of VSP with synthetic seismogram.
- Fresnel zone and VSP horizontal resolution.
- Seismic amplitude studies.
- Determining physical properties of the rocks.
- Seismic wave attenuation.
- Thin bed stratigraphy.

RESERVOIR ENGINEERING AND DRILLING APPLICATIONS

- Predicting depths of seismic reflectors.
- Predicting rock conditions ahead of the bit.
- Defining reservoir boundaries.
- Locating faults.
- Monitoring secondary recovery processes.
- Seismic tomography and reservoir description.
- Predicting high-pressure zones ahead of the bit.
- Detection of man-made fractures.

EXPLORATION APPLICATIONS OF VSP

REFLECTION COEFFICIENTS

Seismic waves are reflected at boundaries that separate rock layers of different acoustic impedances. The polarity, amplitude, and phase characteristics of these reflected wavelets are determined by the reflection coefficients at these boundaries.

Upgoing reflected waves contain much of the subsurface information that is recorded in using VSP. A correct understanding of the mathematics of a seismic wave-reflection coefficient is important when interpreting VSP, because the data contains both upgoing and downgoing waves, which are reflected from both the top and bottom of the boundary.

As has been shown, the reflection coefficient is:

$$R = \left(\rho_2 V_2 - \rho_1 V_1\right) / \left(\rho_2 V_2 + \rho_1 V_1\right)$$

This expression is valuable when making stratigraphic and lithologic interpretations of VSP.

IDENTIFICATION OF A SEISMIC REFLECTOR

A good interpreter will attempt to relate the surface-acquired reflection seismic data to subsurface stratigraphy and depositional facies. When doing this, one should keep in mind that good quality VSP data are capable of defining the depth at which each upgoing primary reflection is created in a stratigraphic section near the borehole. Therefore, by the use of VSP the correct stratigraphic interpretation of a surface-acquired reflection seismic section can be reached.

Using VSP data with a high signal-to-noise ratio, an interpreter can answer some questions such as:

1. Is a reflection generated at a stratigraphic boundary or at a time-stratigraphic boundary such as unconformities?
2. Which rock boundaries can be seen with seismic data and which cannot?
3. How reliably do synthetic seismograms made from well log data identify primary and multiple reflections?

Figure 8–15 shows VSP data recorded in a well where the stratigraphic and lithological conditions that create seismic reflections can be identified.

COMPARISON OF VSP WITH SYNTHETIC SEISMOGRAMS

The conventional tool to correlate between the subsurface stratigraphy and surface-measured seismic data is the synthetic seismograph. From the previous discussion, VSP can be used to identify lithology and subsurface stratigraphy with a high degree of accuracy. In contrast, a synthetic seismogram is only a *synthetic* representation of seismic measurements. In vertical seismic profiling, one can use the same type of source, a similar geophone, and the same instrumentation used to record the surface seismic data. In synthetic seismograph calculations, one can only approximate these aspects of the total seismic-recording process.

Figure 8–16 illustrates a comparison of surface seismic data crossing VSP well "P" and well "Z" with a synthetic seismogram.

FRESNEL ZONE AND VSP HORIZONTAL RESOLUTION

Figure 8–17 shows that a first-order Fresnel zone defines the smallest lateral dimension of a subsurface anomaly that can be resolved by surface-recorded seismic data. VSP data provide a better lateral resolution of the subsurface than do surface reflection measurements because the Fresnel zones involved in VSP measurements are smaller.

PREDICTING INTERVAL VELOCITY

One output of VSP data processing is a plot of acoustic impedance, ρV, versus depth. The change in sedimentary rocks densities is very small compared to their interval velocities; therefore, the density term can be dropped. Density is often calculated from velocity-density relationship using Gardner's equation. A plot of interval velocity versus depth can be

FIGURE 8–15. Identification of a seismic reflector (courtesy Geophysical Press, from Hardage, B.A.: "Vertical Seismic Profiling, Part A: Principles," 1983, adapted from Balch et al., 1981)

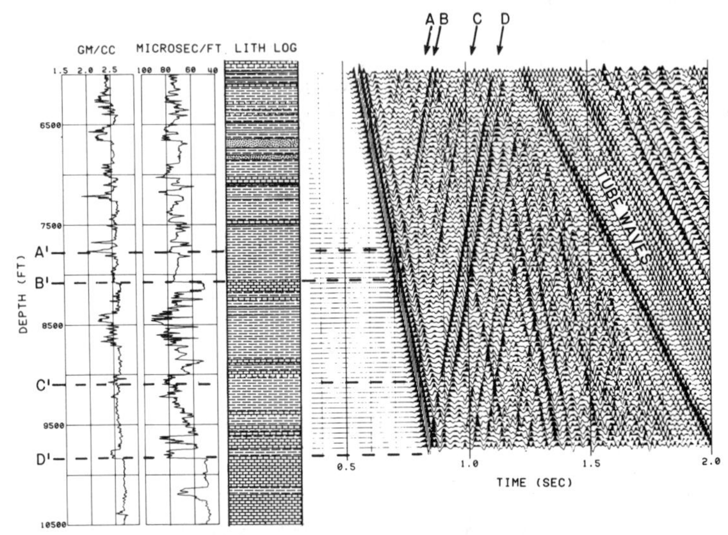

AN EXAMPLE OF THE RELIABILITY WITH WHICH VSP DATA CAN OFTEN IDENTIFY PRIMARY SEISMIC REFLECTORS. FOUR UPGOING PRIMARY REFLECTIONS ARE SHOWN BY THE LINEUP OF BLACK PEAKS LABELED A, B, C, D. THE SUBSURFACE DEPTH OF THE INTERFACE(S) THAT GENERATED EACH REFLECTION CAN BE DEFINED BY EXTRAPOLATING THE APICES OF THE BLACK PEAKS DOWNWARD UNTIL THEY INTERSECT THE FIRST BREAK LOCI OF THE DOWNGOING COMPRESSIONAL EVENT. THESE DEPTHS ARE LABELED A', B', C', D'. THESE ARE RAW FIELD DATA. NO PROCESSING HAS BEEN DONE OTHER THAN A NUMERICAL AGC FUNCTION HAS BEEN APPLIED TO EQUALIZE ALL AMPLITUDES.

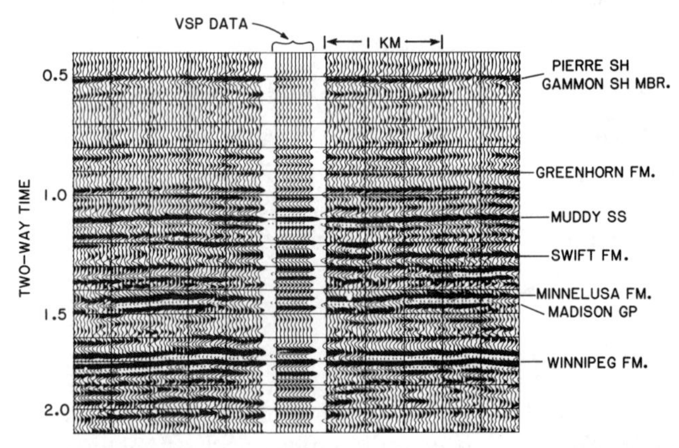

COMPARISON BETWEEN SURFACE-RECORDED REFLECTION DATA AND PROCESSED VSP DATA AT THE USGS MADISON LIMESTONE TEST WELL NO. 2. (ALTERED FROM BALCH ET AL., 1981B).

FIGURE 8–16. Comparison of VSP with synthetic seismogram (courtesy Geophysical Press, from Hardage, B.A.: "Vertical Seismic Profiling, Part A: Principles," 1983)

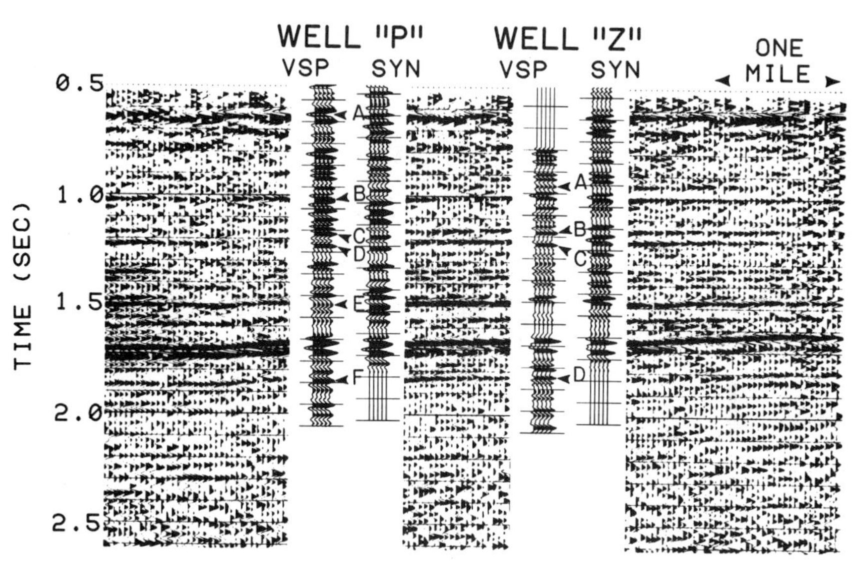

COMPARISON OF SURFACE SEISMIC DATA CROSSING VSP STUDY WELLS "P" AND "Z" WITH SYNTHETIC SEISMOGRAMS AND VSP DATA RECORDED IN THE WELLS. THE LETTERED ARROWHEADS SHOW WHERE THE VSP DATA ARE A BETTER MATCH TO THE SURFACE DATA THAN ARE THE SYNTHETIC SEISMOGRAM DATA.

obtained in addition to the velocity value in the interpretation. Interval velocities can be used in engineering applications, as will be demonstrated later in this section.

Figure 8–18 illustrates an acoustic impedance versus depth display, derived from a VSP survey.

ENGINEERING APPLICATIONS OF VSP

1. Predicting Depth of a Seismic Reflector

    Predicting drilling depth to key seismic markers is a common practice in oil and gas exploration. Many geophysicists are able to make accurate depth estimates from surface determinations of seismic velocities. These estimates become more difficult and less accurate when a well is in a wildcat area where there is little drilling history or in a poor quality seismic recording area. Using VSP to predict reflector depth will yield the greatest benefits in those areas where seismic reflection is poor. An important factor that works in favor of VSP is that the geophone is located deep in the borehole away from the surface noise.

FIGURE 8–17. Fresnel zone and VSP resolution (courtesy Geophysical Press, from Hardage, B.A.: "Vertical Seismic Profiling, Part A: Principles," 1983)

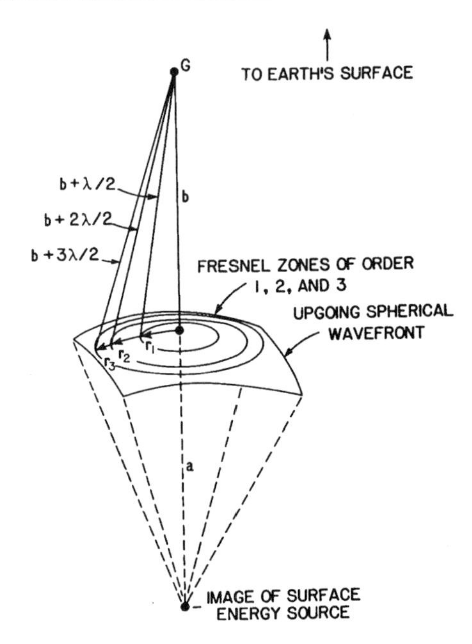

THE ABILITY OF SEISMIC MEASUREMENTS TO RESOLVE SUBSURFACE GEOLOGICAL FEATURES IN A HORIZONTAL SENSE IS CONTROLLED BY THE FIRST ORDER FRESNEL ZONE. THIS ILLUSTRATION SHOWS A VSP GEOPHONE, G, APPROACHING A SUBSURFACE REFLECTOR, WHICH IS BEING ILLUMINATED BY A SPHERICAL WAVEFRONT CREATED BY A SURFACE SOURCE ABOVE G (I.E., $a > b$). THE RAYPATH PICTURE IS SIMPLIFIED BY USING VIRTUAL RAYPATHS FROM THE SOURCE TO THE REFLECTOR.

Figure 8–19 shows this application. The intersection between the downgoing wave and the upgoing wave indicate the top of the reflector (A) at 9,850 ft depth.

2. Looking Ahead of the Bit

A way that VSP data has been used to predict the distance from the drill bit to a deeper formation is illustrated in Figure 8–20.

It is assumed that the well has been drilled to 8,000 feet and that VSP data are recorded from that depth upward far enough so that deep reflection events can be seen and interpreted. This means that data should be recorded from the bottom of the hole to about 2,000 feet above bottom with constant depth increment.

The question to be answered is: "How far below the current drilling depth is Reflector A?" As a first approximation, it can be assumed that the downgoing first arrival wavelet continues below 8,000 feet with the same depth curvature as the recorded data intervals from 6,000 to

FIGURE 8–18. Predicting velocity ahead of the bit (courtesy of Seismograph Service Corporation)

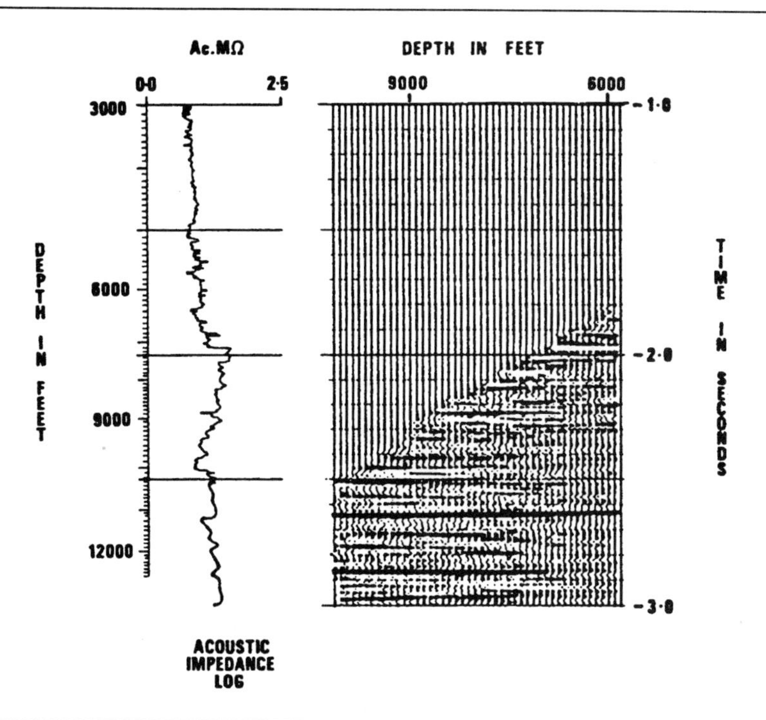

FIGURE 8–19. Predicting depth of a seismic reflector (courtesy Geophysical Press, from Hardage, B.A.: "Vertical Seismic Profiling, Part A: Principles," 1983)

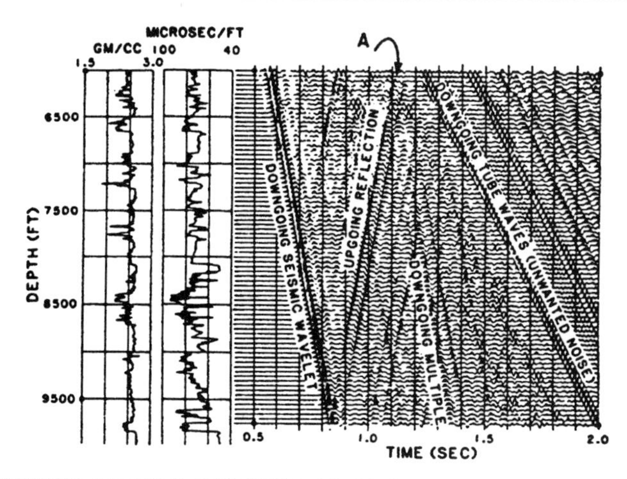

VSP DATA SHOWING A STRONG REFLECTOR, "A", ORIGINATING AT A DEPTH OF 9,850 FEET.

FIGURE 8–20. Looking ahead of the bit (courtesy Geophysical Press, from Hardage, B.A.: "Vertical Seismic Profiling, Part A: Principles," 1983)

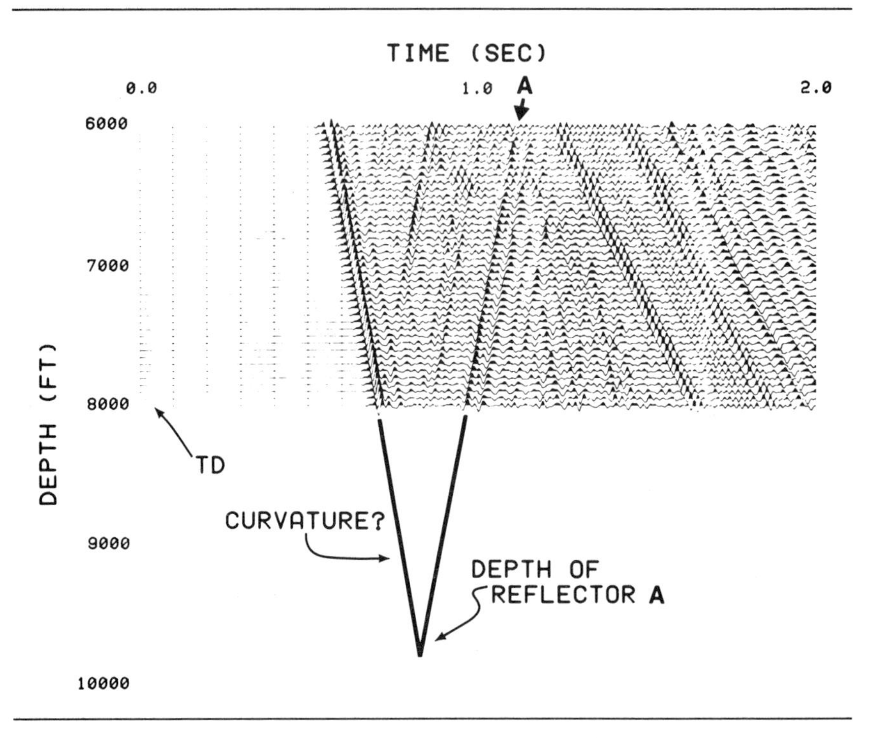

8,000 feet. The intersection of the downgoing extension line and upgoing event A will intersect at the depth of Reflector A.

3. Predicting Pore Pressure and Porosity Ahead of the Bit

Vertical seismic profiling is a technique that provides a common ground for engineers, geologists, and geophysicists.

One of the most interesting applications of the VSP is to predict conditions beneath the drilling bit. Stone (1982) and others have used the technical advantages of the VSP to predict interval velocity and depth beneath the bit. Depth and interval velocity are the basis of the computation of pore pressure and porosity in well log analysis. There is a potential for calculation of these important rock properties using VSP data.

VSP can be used to provide information for:

- Amplitude correction.
- Extraction of waveforms and multiples.
- Design of deconvolution parameters.
- Predicting velocities ahead of the bit.

- Predicting transit time versus depth ahead of the bit, which can be used to calculate other petrophysical properties of the rock.
- Predicting abnormal pressure zones ahead of the bit, allowing the drilling engineer to plan for pressure control.
- Predicting porosity by utilizing the predicted transit time and knowing the transit times of other needed rock and fluid in the borehole.
- Identifying stratigraphy and lithology on the surface-acquired seismic data.

4. Defining Reservoir Boundaries—Offset Source VSP

Karus, et al. (1975) showed a plot for first arrival amplitudes from VSPs shot from two sides of a well, which is just off the edge of a producing field. The sources were arranged so that one goes progressively down the well (see Figure 8–21). Direct arrival paths from SP-2 eventually must pass through the indicated producing horizons. An amplitude loss is observed when source-receiver paths encounter "hydrocar-

FIGURE 8–21. Offset-source VSP defining reservoir boundaries (modified from Karus et al., 1975)

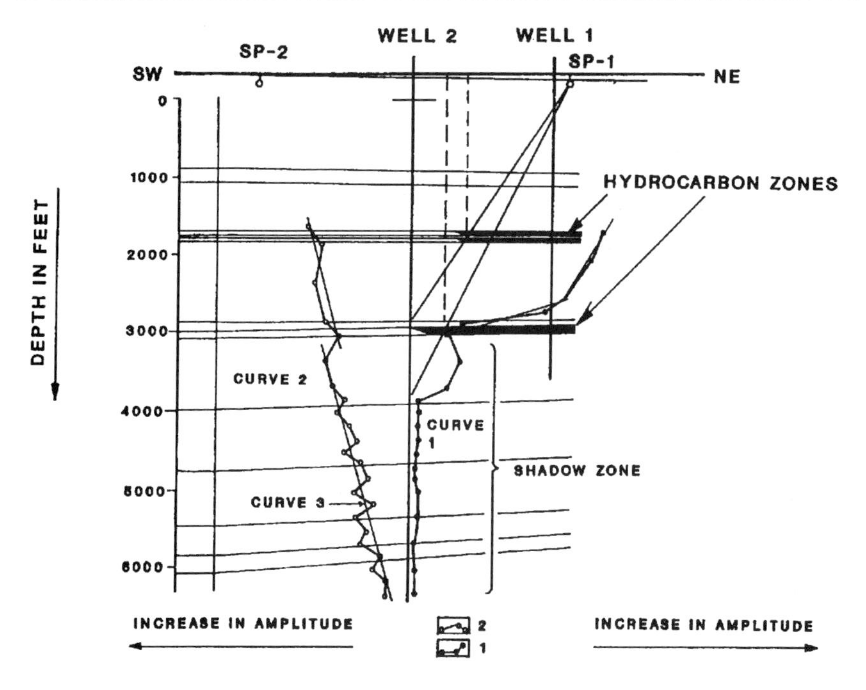

CURVE 1: AMPLITUDE MEASURED FROM SP-1. NOTE LOSS OF AMPLITUDE OF FIRST ARRIVAL UNDER THE SHADOW ZONE.
CURVE 2: AMPLITUDES MEASURED FROM SP-2.
CURVE 3: AVERAGED AMPLITUDE CURVES USED TO CALCULATE ATTENUATION.

FIGURE 8–22. VSP transmission amplitude anomaly and defining reservoir boundaries (modified after Mustafayev, 1967)

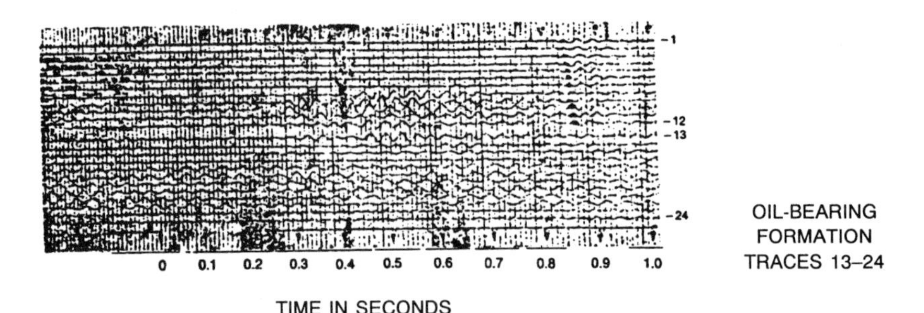

VSP RECORD SHOWS TRANSMISSION AMPLITUDE ANOMALY ASSOCIATED WITH OIL-BEARING ZONE IN KARLA FIELD, AZERBAYDZHAN, USSR. NOTE THAT TRACES 1–12 DO NOT INTERSECT THE FIELD AND SHOW PRONOUNCED AMPLITUDE AT .850 SECOND. TRACES 13–24 INTERSECT THE OIL-BEARING ZONE: NO AMPLITUDE IS APPARENT.

bon accumulation." The amplitude loss observed when raypaths encounter the hydrocarbon pay zone is very pronounced.

K. A. Mustafayev (1967, 1969) described similar investigations in the Chakhnaglar and Kala fields in Azerbaydzan. Figure 8–22 illustrates the field configuration and presentation. It was obtained by using a single borehole source at a depth greater than the oil-bearing formation. A line of detectors was laid out at the surface in such a manner that 12 receivers intersected the pay zone while the other 12 receivers did not. The VSP shows a transmission amplitude anomaly associated with the oil-bearing formation in the field. Trace 1 through 12, which do not intersect the zone, show a distinct first arrival at 0.85 second. On Traces 13 through 24, which intersect the pay zone, no first arrival is apparent. The suspected tendency for oil-and gas-bearing horizons to absorb seismic energy has been under investigation for decades, and these VSP investigations tend to support the concept.

STRATIGRAPHIC APPLICATIONS OF VSP

Cramer (1988) showed the use of multi-offset VSPs to develop a "D" Sand field in the Denver-Julesburg basin of Colorado. "D" Sand production in Wattenburg Field was established by drilling Well 34–3 (see Figure 8–23). Three successive wells were drilled and failed to determine the extent of the "D" Sand in the field. A stratigraphic model suggested that the offset VSP technique could be used to delineate the extent of the "D" Sand reservoir.

On the basis of this study, a five-offset VSP was conducted in the discovery well, and a zero-offset VSP was run in this well to complete the delineation of the reservoir and to confirm the second location.

FIGURE 8–23. "D" Sand field and geologic cross-section (copyright © 1988, Society of Petroleum Engineers, from Cramer, P.N.: "Reservoir Development Using Offset VSP Techniques in the Denver-Julesburg Basin," *Journal of Petroleum Technology* (February 1988))

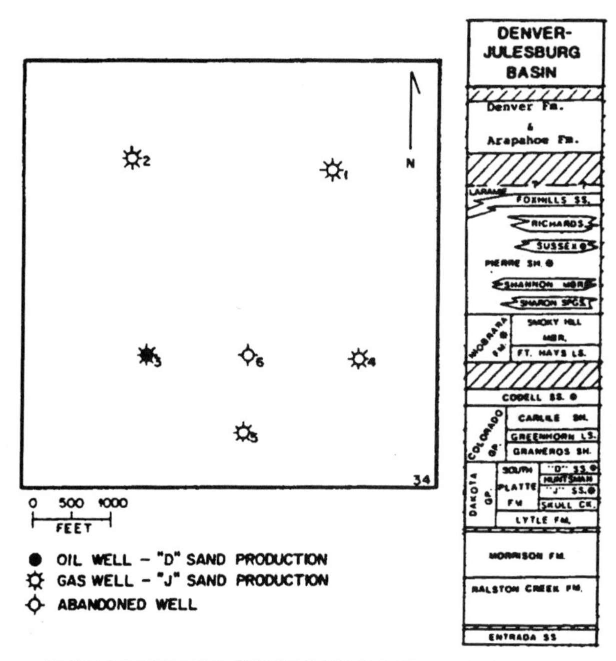

PLOT OF ACREAGE AND WELLS DRILLED IN SECTION 34

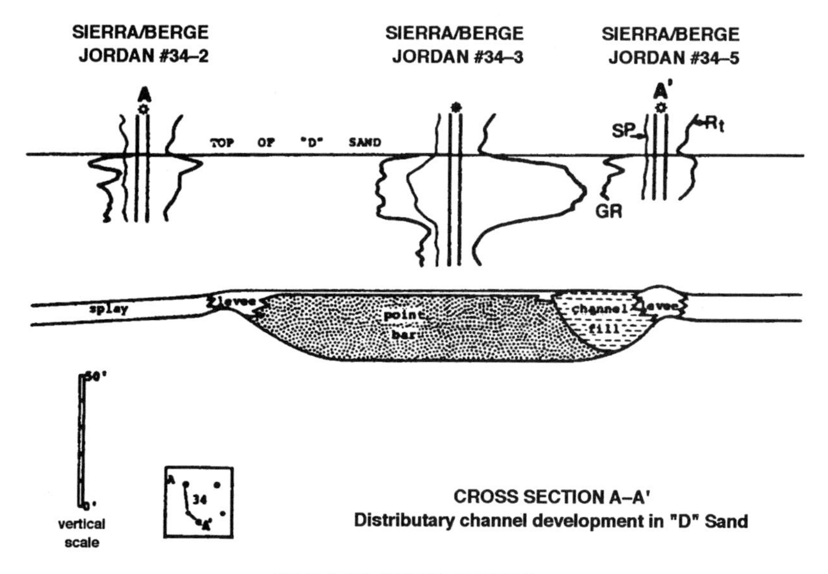

GEOLOGIC CROSS-SECTION

SURVEY MODELING

Figure 8–24 illustrates the geological model used to study VSP resolution and survey design, and the results of modeling synthetic offset VSP data correlated to Sand "D" thickness. Sand thickness was allowed to change from 7 to 30 feet (2–9 m). One can see a distinct character change that can be correlated with the known changes in the "D" Sand thickness. Modeling also demonstrated that reasonable survey parameters (i.e., source distances, geophone level spacing, number of levels, etc.) could be chosen to give a lateral profile length of 2,000 feet (600 m). From this model the decision was made to proceed with the VSP offset survey.

SURVEYING PLAN

The first decision was to determine which well would be used to do the VSP. Well 34–3 was selected for the following reasons: (a) its location would allow VSP data to be collected over most of the acreage held by the operator, and (b) Sand "D" was known to exist in Well 34–3, which would allow reliable correlation and calibration of the offset VSP to well logs. Some concerns were expressed regarding the use of this well, because it would have to be shut in and cleaned out, resulting in loss of production and revenue for at least two days. A second concern was the lack of cement in the upper portion of the well, above 6,300 feet (1920 m). Without cement to couple the casing to the formation, a path of seismic energy to the well geophone may not exist. After some brainstorming, Well 34–3 was selected as the site for the VSP survey.

DATA ACQUISITION

The original and modified plans for the VSP survey are illustrated in Figure 8–25. A two-ms sampling interval was chosen for recording, and the vibrator sweep frequency was 10 to 80 Hz. A record length of 17 seconds and listen time of 5 seconds were selected. Six vibrators were used for the survey. Two vibrators were used at each location source point, so two runs of the well geophone into the hole were needed, shooting from three offset locations on each run.

The survey was conducted as planned. According to the plan, the northwest, north, and northeast offsets were the first to be acquired. Some problems arose when the well geophone was pulled shallower than 6,300 feet, because no signal could be observed. This depth coincided with the top of the cement behind casing. The only possible solution was to adjust the source offset so that the desired lateral coverage of 2,000 feet (600 m) could be achieved without raising the well geophone above the top of the cement.

To change the configuration at night was difficult, but the decision was made to continue, hoping that data processing would be able to solve

FIGURE 8–24. Survey modeling (copyright © 1988, Society of Petroleum Engineers, from Cramer, P.N.: "Reservoir Development Using Offset VSP Techniques in the Denver-Julesburg Basin," *Journal of Petroleum Technology* (February 1988))

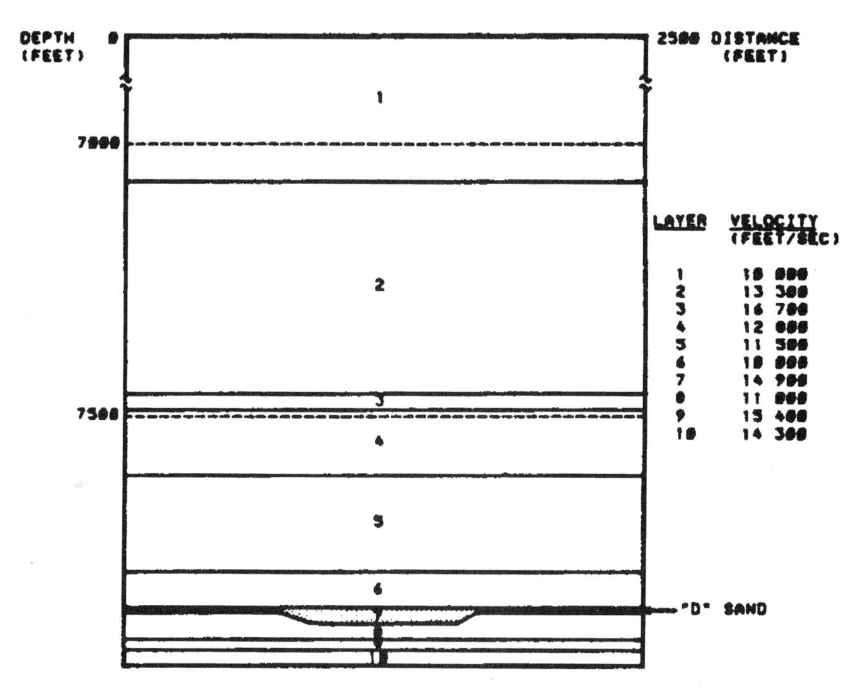

GEOLOGIC MODEL USED TO STUDY VSP RESOLUTION AND SURVEY DESIGN

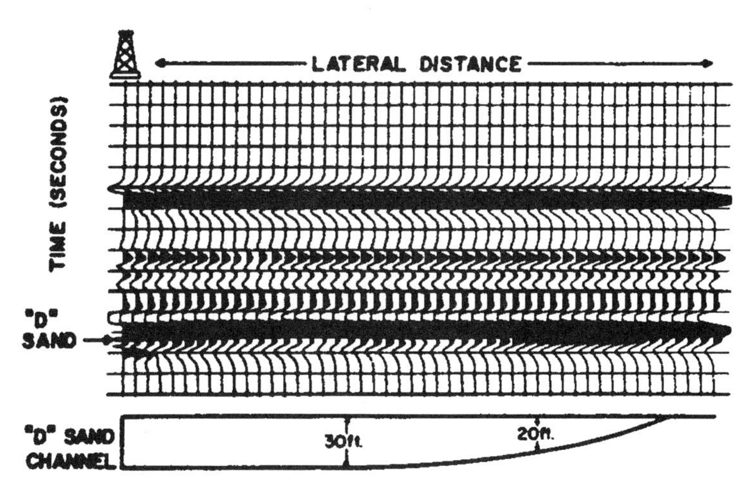

RESULT OF MODELING, SYNTHETIC OFFSET VSP DATA CORRELATED TO SAND "D" THICKNESS

FIGURE 8–25. Multi-offset VSP field plan (copyright © 1988, Society of Petroleum Engineers, from Cramer, P.N.: "Reservoir Development Using Offset VSP Techniques in the Denver-Julesburg Basin," *Journal of Petroleum Technology* (February 1988))

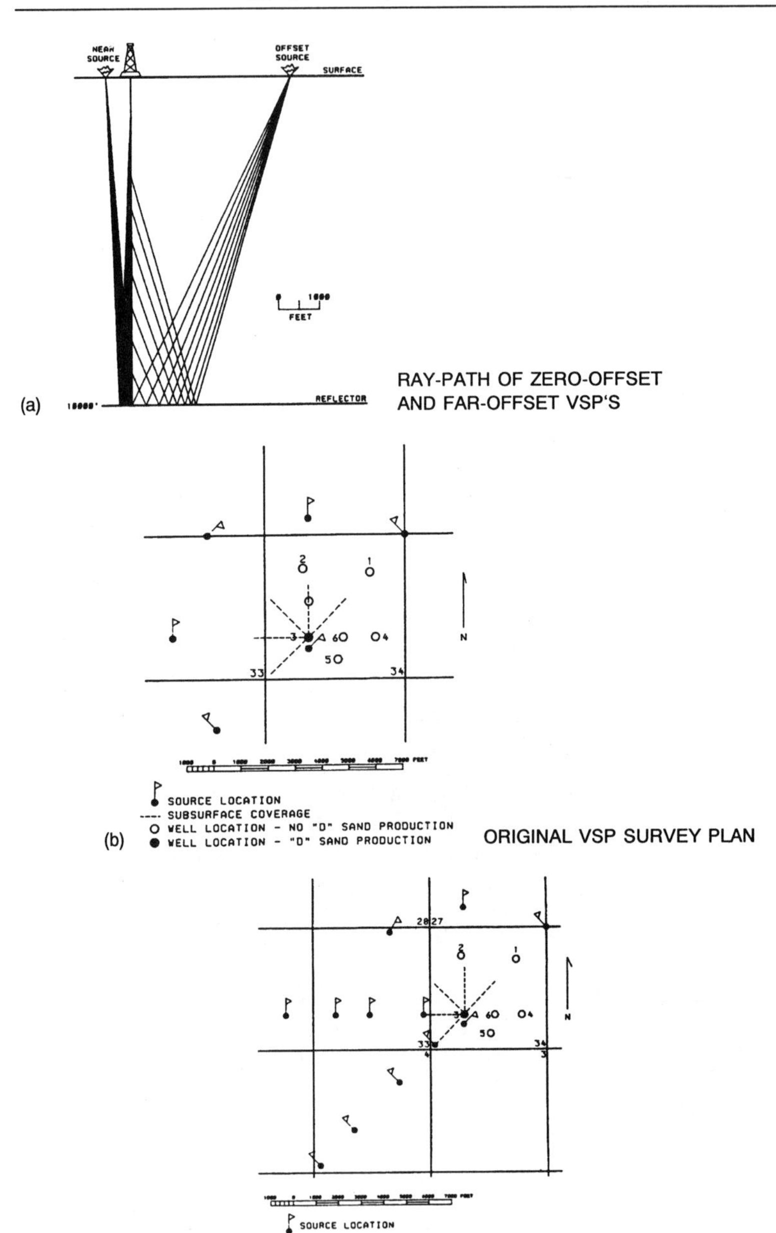

FIGURE 8–26. Correlation between model data and zero-offset VSP (copyright © 1988, Society of Petroleum Engineers, from Cramer, P.N.: "Reservoir Development Using Offset VSP Techniques in the Denver-Julesburg Basin," *Journal of Petroleum Technology* (February 1988))

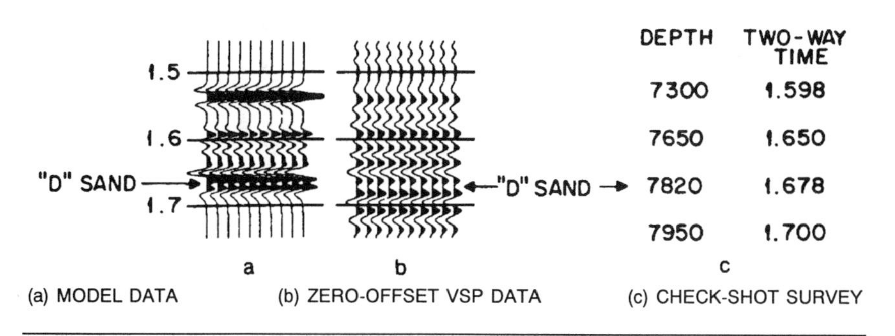

some data problems and enhance the quality. In the morning, the remaining two far offsets and near offsets were acquired. The survey plan for the remaining far offsets was redesigned with four source points in each direction. In addition, the spacing between geophone levels was tightened to 50 feet (15 m) to provide closer spacing of reflection points in order to preserve the lateral resolution.

Figure 8–26 shows a correlation of model data and zero offset VSP. Figure 8–27 and 8–28 show the final offset VSP.

DATA INTERPRETATION

The final display shows VSP data from each profile transformed to offset two-way time domain, or CDP seismic data. With a check shot, the survey was computed from zero-offset survey to include a time/depth chart. From the zero-offset VSP, the "D" sand was located after comparing it to the model, then identifying the "D" sand response of each far offset and mapping it away from the well (see Figure 8–29).

Two problems complicated the interpretation: (a) the poor data recorded in the northwest, north, and northeast in uncemented casing. No reliable interpretation could be made through the zone, although Sand "D" on these offsets shows that it is disappearing midway in the record. It cannot be stated reliably that the edge of the buildup is where Sand "D" disappeared, but only where the sand extends to at least at this point. It may extend further, but this cannot be determined from the data.

SUBSEQUENT LOCATION

Well 34–7, located 1,650 feet (500 m) northeast of Well 34–3, was drilled and completed in Sand "D" with 19 feet (6 m) of oil pay, compared to 26 feet (8 m) of pay in Well 34–3. To continue delineation of the reservoir

FIGURE 8–27. Final offset VSP data displays (copyright © 1988, Society of Petroleum Engineers, from Cramer, P.N.: "Reservoir Development Using Offset VSP Techniques in the Denver-Julesburg Basin," *Journal of Petroleum Technology* (February 1988))

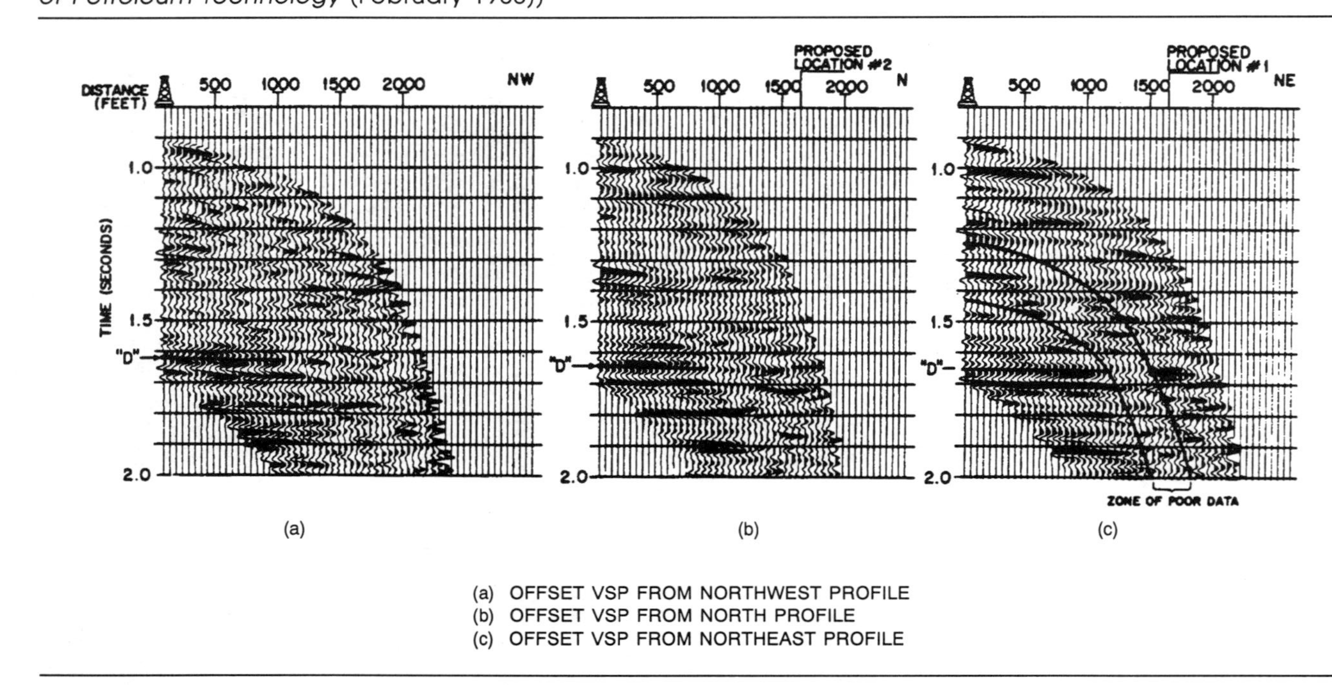

FIGURE 8–28. Final offset VSP data displays (copyright © 1988, Society of Petroleum Engineers, from Cramer, P.N.: "Reservoir Development Using Offset VSP Techniques in the Denver-Julesburg Basin," *Journal of Petroleum Technology* (February 1988))

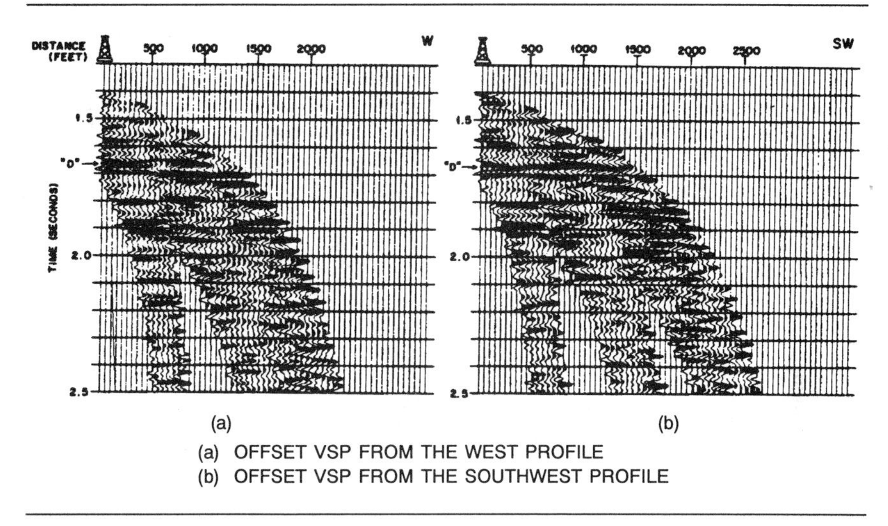

(a) OFFSET VSP FROM THE WEST PROFILE
(b) OFFSET VSP FROM THE SOUTHWEST PROFILE

extent and to locate a new development location, another survey was planned using Well 34–7 for the survey. This survey was conducted in the open hole immediately after the well was logged and before casing was set. The survey was conducted according to the original plan, and it was completed in 48 hours.

RECOMMENDATIONS FOR USING MULTI-OFFSET VSP

1. Seismic modeling of the proposed survey should be conducted before the survey to verify that the resolution required to solve the problem can be obtained, and to assist in designing survey parameters such as source offset and geophone level increment, and to assist the interpreter in understanding the record area.
2. Surveys should be conducted in either a completely cased hole or in the open hole before casing is set, because poorly cemented casing causes serious degradation of the VSP data quality.
3. Multi-offset VSP surveys should be designed to utilize all the possible existing well control to confirm modeled results.
4. A near offset VSP should be run with the far offset to establish velocity control to aid in correlating data with the well logs.
5. VSP is one of the geophysical methods that can be applied in developing a field economically.

FIGURE 8–29. Results of interpretation of the VSP data (copyright © 1988, Society of Petroleum Engineers, from Cramer, P.N.: "Reservoir Development Using Offset VSP Techniques in the Denver-Julesburg Basin," *Journal of Petroleum Technology* (February 1988))

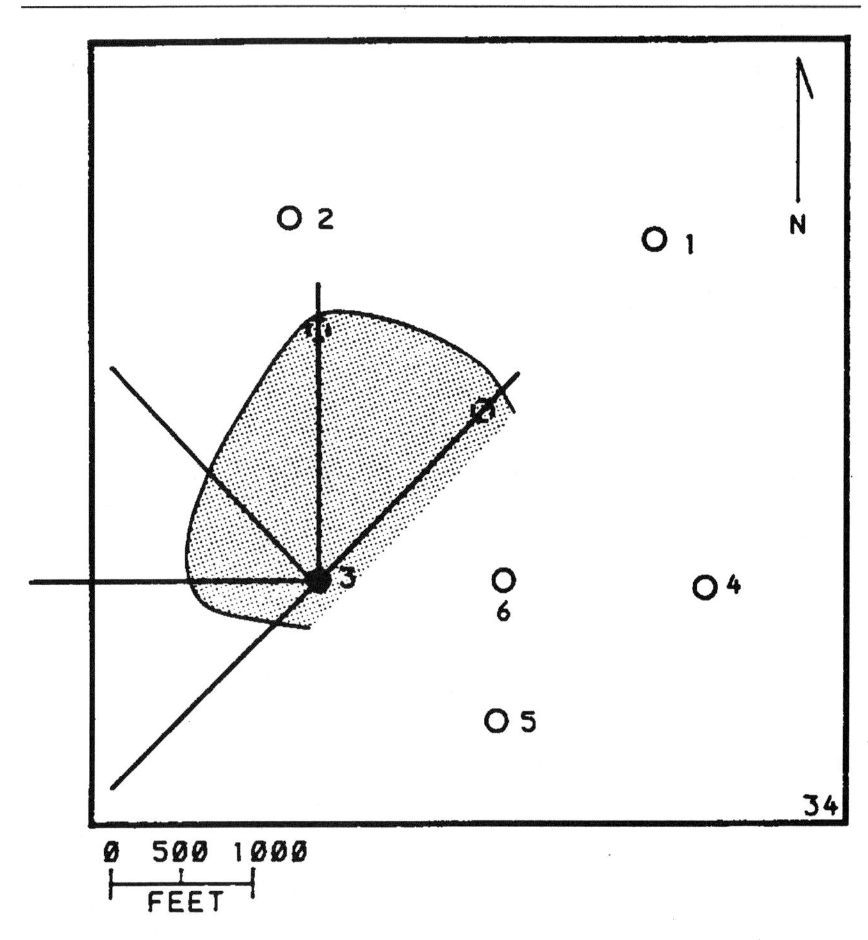

SUMMARY AND DISCUSSION

Vertical seismic profiling (VSP) is a technique that has proved its value and applications in petroleum exploration and development.

VSP has its limitations like any other tool, but its applications are numerous, and it is open for your innovation to think of an application that will help you solve a particular problem. You should check your idea by modeling the problem and find out if you can get the desired seismic resolution by using the VSP technique.

In the past, engineers and managers were concerned when they found out that a VSP survey would be run at a certain time during the drilling of the well. Their concerns were about possible mechanical problems and the rig idle time (72 hours).

A few years ago, the cost to run a VSP survey in a 12,000-foot well with 50 to 75 levels was well over $35,000 and not less than $10,000 to process the survey. The turnaround time was three to four weeks and the information obtained was velocity-depth information.

Today, VSP surveys cost substantially less. You will find below estimated costs per area in the United States and Canada for approximately 70 levels. The cost is in thousands of dollars and includes the data processing. The first number is for the vertical survey (ZVSP), and the second number is for the offset VSP in addition to the (ZVSP).

Area	**(ZVSP)**	**(ZVSP)+Offset VSP**
Midcontinent	25	40
Offshore Gulf of Mexico	50	75
West Coast	30	50
Rocky Mountains	30	46
Alaska	60	90
Gulf Coast	40	57
Canada	30	46

The turnaround time is a few days and sometimes overnight in case of emergency.

The survey is done routinely as any logging tool. In a vertical survey eight to ten levels per hour can be surveyed. In land surveys, perhaps six to eight levels per hour can be taken because it takes more time to inject the energy source. In offshore surveys, it may take more time, four to five levels per hour.

The VSP survey is going to provide the geophysicist with the seismic velocity, seismic time to geological depth conversion, and the next seismic marker. It provides the geologist with well prognosis. It will tell the engineer the location of the drilling bit or at what depth he can expect a high-pressure zone. If he can predict the high-pressure zone ahead of time, he can take action to head off problems.

A VSP survey will cost from $25,000 and will take two to three days turnaround time to get information that will help in evaluating a well and avoid high-pressure zones. With minimal rig idle time, the survey is definitely more economical than a blowout.

The VSP will play an important role in borehole geophysics, reservoir characterization, and transmission tomography as we will discuss in Chapter 11.

KEY WORDS

Casing
Cementing
Caliper log
Downgoing wave (Direct arrivals)
First break time
Intrabed multiples
Multioffset VSP
Pilot signal (waveform)
Upgoing wave
Velocity survey

BIBILOGRAPHY

Angeleri, G. P. and E. Loinger. "Amplitude and Phase Distortions Due to Absorption in Seismograms and VSP." Paper presented at 44th annual meeting of EAEG, 1982.

Balch, A. H., M. W. Lee, J. J. Miller and R. T. Ryder. "The Use of Vertical Seismic Profiles in Seismic Investigations of the Earth." *Geophysics* 47 (1982): 906–918.

Balch, A. H. and M. W. Lee. "Some Considerations On the Use of Downhole Sources in Vertical Seismic Profiles." Paper presented at 35th Annual SEG Midwestern Exploration Meeting, 1982.

Barton, D. C. "The Seismic Method of Mapping Geologic Structure." *Geophy. Prosp. Amer. Inst. Min. and Mat. Eng* 1 (1929): 572–624.

Bilgeri, D. and E. B. Ademeno. "Predicting Abnormally Pressured Sedimentary Rocks." *Geophysics* 30 (1982): 608–621.

Biot, M. A. "Propagation of Elastic Waves in a Cylindrical Bore Containing a Fluid." *J. Appl. Physics* 23 (1952): 997–1005.

Blair, D. P. "Dynamic Modeling of In-Hole Mounts for Seismic Detectors." *Geophys. Jour. Roy. Astron. Soc.* 69 (1982): 803–817.

Bois, P., M. Laporte, M. Laverne and G. Thomas. "Well-to-Well Seismic Measurements." *Geophysics* 37 (1972): 471–480.

Brewer, H. L. and J. Holtzscherer. "Results of Subsurface Investigations Using Seismic Detectors and Deep Bore Holes." *Geophys. Prosp.* 6 (1958): 81–100.

Butler, D. K. and J. R. Curro Jr. "Crosshole Seismic Testing—Procedures and Pitfalls." *Geophysics* 46 (1981): 23–29.

Cheng, C. H. and M. N. Toksoz. "Elastic Wave Propagation in a Fluid-Filled Borehole and Synthetic Acoustic Logs." *Geophysics* 46 (1981): 1042–1053.

Cheng, C. H. and M. N. Toksoz. "Tube Wave Propagation and Attenuation in a Borehole." Paper presented at Massachusetts Institute of Technology Industrial Liaison Program Symposium, Houston, 1981.

Cheng, C. H. and M. N. Toksoz. "Generation, Propagation and Analysis of Tube Waves in a Borehole." Paper P, Trans. SPWLA 23rd Annual Logging Symposium, vol. I., 1982c.

Chun, J., D. G. Stone and C. A. Jacewitz. "Extrapolation and Interpolation of VSP Data." Tulsa, OK: *Seismograph Service Companies Report*, 1982.

Crawford, J. M., W. E. N. Doty and M. R. Lee. "Continuous Signal Seismograph." *Geophysics* 25 (1960): 95–105.

DiSiena, J. P. and J. E. Gaiser. "Marine Vertical Seismic Profiling: Paper OTC 4541, Offshore Technology Conference, Houston, TX, 1983, p. 245–252.

Douze, E. J. "Signal and Noise in Deep Wells." *Geophysics* 29 (1964): 721–732.

Gaizer, J. E. and J. P. DiSiena. "VSP Fundamentals That Improve CDP Data Interpretation." Paper S12.2, 52nd Annual International Meeting of SEG, Technical Program Abstracts, 1982a, p. 154–156.

Gal'perin, E. I. "Vertical Seismic Profiling." *Soc. Expl. Geophys. Special Publ.* 12 (1974): 270.

Hardage, B. A. "An Examination of Tube Wave Noise in Vertical Seismic Profiling Data." *Geophysics* 46 (1981b): 892–903.

Hardage, B. A. "A New Direction in Exploration Seismology is Down." *The Leading Edge* 2, no. 6 (1983): 49–52.

Jolly, R. N. "Deep-Hole Geophone Study in Garvin County, OK." *Geophysics* 18 (1953): 662–670.

Karus, E. V., L. A. Raybinkin, E. I. Gal'perin, V. A. Teplitskiy, Yu B. Demidenko, K. A. Mustafayev and M. B. Raport. Detailed Investigations of Geological Structures by Seismic Well Survey. 9th World Petroleum Congress *PD* 9 (4), V. 26 (1975): 247.

Kennett, P., R. L. Ireson and P. J. Conn. "Vertical Seismic Profiles—Their Applications in Exploration Geophysics." *Geophys. Prosp.* 28 (1980): 676–699.

Lang, D. G. "Downhole Seismic Technique Expands Borehole Data." *Oil and Gas Jour.* 77, No. 28 (1979a): 139–142.

Lang, D. G. "Downhole Seismic Combination of Techniques Sees Nearby Features." *Oil and Gas Jour.* 77, No. 29 (1979b): 63–66.

Lash, C. C. "Investigation of Multiple Reflections and Wave Conversions By Means of Vertical Wave Test (Vertical Seismic Profiling) in Southern Mississippi." *Geophysics* 47 (1982): 977–1000.

Lee, M. W. and A. H. Balch. "Theoretical Seismic Wave Radiation From a Fluid-Filled Borehole." *Geophysics* 47 (1982): 1308–1314.

Lee, M. W. and A. H. Balch. "Computer Processing of Vertical Seismic Profile Data." *Geophysics* 48 (1983): 272–287.

Levin, F. K. and R. D. Lynn. "Deep Hole Geophone Studies." *Geophysics* 23 (1958): 639–664.

McCollum, B. and W. W. Larue. "Utilization of Existing Wells in Seismograph Work." *Early Geophysical Papers* 12 (1931): 119–127.

Mustafayev, K. A. "Increased Absorption of Seismic Waves in Oil and Gas Saturated Deposits." *Prikladnaya Geofizika* 47 (1967): 42.

Pucket, M. "Offset VSP: A Tool for Development Drilling." *TLE* 10 (1991): 18–24.

Quarles, M. "Vertical Seismic Profiling—A New Seismic Exploration Technique." Paper presented at the 48th Ann. Internat. Mtg of SEG, 1978.

Rice, R. B., et al. "Developments in Exploration Geophysics, 1975–1980." *Geophysics* 46 (1981): 1088–1099.
Riggs, E. D. "Seismic Wave Types in a Borehole." *Geophysics* 20 (1955): 53–60.
Stewart, R. R., R. M. Turpening and M. N. Toksoz. "Study of a Subsurface Fracture Zone by Vertical Seismic Profiling." *Geophy. Res. Lett.* 9 (1981): 1132–1135.
Stone, D. G. "VSP—The Missing Link." Paper presented at the VSP Short Course Sponsored by the Southeastern Geophysical Society in New Orleans, 1981.
Stone, D. G. "Prediction of Depth and Velocity on VSP." Paper presented at the 52nd Ann. Mtg. of SEG, Dallas, Texas, 1982.
Stone, D. G. "Predicting Pore Pressure and Porosity From VSP Data." Paper presented at the 53rd Ann. Mtg. of SEG, Las Vegas, Nevada, 1983.
van Sandt, D. R. and F. K. Levin. "A Study of Cased and Open Holes for Deep Seismic Detection." *Geophysics* 28 (1963): 8–13.
Wyatt, K. D. "Synthetic Vertical Seismic Profile." *Geophysics* 46 (1981a): 880–991.
Wyatt, S. B. "The Propagation of Elastic Waves Along a Fluid-Filled Annular Region." Master of Science Thesis. University of Tulsa, Tulsa, OK, 1979.
Zimmermann, L. J. and S. T. Chen. "Comparison of Vertical Seismic Profiling Techniques." *Geophysics* 58 (1993): 134–140.

CHAPTER 9

AMPLITUDE VERSUS OFFSET ANALYSIS

INTRODUCTION

The amplitude of a reflected seismic signal normally decreases with the increase of the distance between source and receiver. This decrease is related to the dependence of reflectivity on the angle at which the seismic wave strikes the interface, spreading, absorption, near surface effects, multiples, geophone planting, geophone arrays and instrumentation.

In certain depositional environments, the amplitude variation can also be an important clue to the lithology or to the presence of hydrocarbons. An increase in amplitude with increased offset, resulting in a "bright spot" on the section, may indicate a gas sand reservoir. A decrease in amplitude with offset may indicate a carbonate reservoir. However, these amplitude anomalies are masked in the common midpoint stack (CMP), as every trace of the stack section represents an over-all average of offsets in the common midpoint gather.

AMPLITUDE VERSUS OFFSET METHODOLOGY

Amplitude versus offset analysis is designed to retrieve the variation in amplitude with angle of incidence by conducting the analysis on the normal moveout corrected gathers *before* stack.

REFLECTION COEFFICIENT

The amplitude of a seismic reflection is related to three rock properties.

1. V_p = compressional wave velocity
2. V_s = shear wave velocity
3. ρ = density

The interpretation of a stacked seismic section is restricted to the zero-offset model. Accordingly, an incident plane wavefront of amplitude A_0 on a horizontal interface will produce a reflected plane wavefront of an amplitude of A_1. The ratio of A_1 to A_0 is defined as the ***reflection coefficient*** (R) of this interface, and it is expressed by the following relation:

$$R = \left(\rho_2 V_2 - \rho_1 V_1\right) / \left(\rho_2 V_2 + \rho_1 V_1\right)$$

POISSON'S RATIO

Poisson's Ratio is defined as the ratio of transverse strain to longitudinal strain of a material under stress. For example, if a piece of rubber is squeezed it is shortened, but it also becomes wider as the volume remains approximately constant. The ratio of the change in width to the change in length is the Poisson's ratio of that material. In seismic applications, it is the ratio between the velocities of P and S waves:

$$\sigma = \left[.5 - \left(V_s / V_p\right)^2\right] / \left[1 - \left(V_s / V_p\right)^2\right] \quad \text{(Sheriff, 1973)}$$

For liquids, V_s vanishes, and σ is 0.5.

REVIEW OF AVO DEVELOPMENT

THE ZOEPPRITZ EQUATION

Zoeppritz derived a relationship governing the reflection and transmission coefficients for plane waves as a function of angle of incidence and six parameters, three on each side of the reflecting interface. These are V_p, V_s, and density. The equation is complex, and its solution is laborious.

SHUEY'S SIMPLIFICATION

Shuey in 1985 simplified the Zoeppritz equation to the following:

$$R(\theta) = R_0\left[A_0 R_0 + \Delta\sigma / \left(1 - \sigma\right)^2\right] \sin^2\theta + \Delta V_p / V_p \left(\tan^2\theta - \sin^2\theta\right) / 2$$

where $R(\theta)$ = the compressional wave coefficient
A_0 = the normal, gradual decrease in amplitude with offset
R_0 = is the amplitude at normal incidence ($\theta = 0$)
(at normal incidence, amplitude and reflection coefficient are the same)

The first term on the right side of the equation, R_0, gives the reflectivity at normal incidence ($\theta = 0$).

The second term characterizes $R(\theta)$ at an intermediate angle. The coefficient of the second term is a combination of elastic properties that can be determined by analyzing the offset dependence of event amplitude.

If the amplitude of the event is normalized to its value for normal incidence, then:

$$A = A_0 + \left[1/(1-\sigma)^2\right]\left[\Delta\sigma/R_0\right]$$

A_0 specifies the normal, gradual *decrease* of amplitude with offset. Its value is small enough that the main information conveyed is in the second term, in which $\Delta\sigma$ is the contrast in Poisson's ratio at the reflecting interface.

HILTERMAN'S MODIFICATION OF SHUEY'S EQUATION

Hilterman modified Shuey's equation and established a linear relationship between incident angle and reflection coefficient.

$$R(\theta) = R_0 \cos^2\theta + 2.25\Delta\sigma \sin^2\theta$$

Where R_0 = reflection coefficient at normal incidence

$= (\rho_2 V_{p_2} - \rho_1 V_{p_1})/(\rho_2 V_{p_2} + \rho_1 V_{p_1})$

$\Delta\sigma = \sigma_2 - \sigma_1$

θ = angle of incidence

This approximation for amplitude behavior with angle of incidence is valid if $\theta < 30°$, $R_C(\theta) < 0.15$, and $V_p \approx 2V_s$.

For plane waves only, by using linear regression, it is possible to get estimates of R_0 from normal moveout corrected CDP gathers. Essentially, the following are known or can be estimated:

$R(x,t)$ = $R(\theta)$, the seismic trace amplitude

$\theta(x,t)$ = angle of incidence as a function of offset and time

For each CDP, R_0 and $\Delta\sigma$ can be computed for each time sample. The result is two seismic sections, one called the *normal incidence* section and the other the *delta sigma*, or Poisson's ratio, section.

If $R(\theta)$ is normalized by dividing by $\cos^2\theta$, the following equation is obtained:

$$R(\theta)/\cos^2\theta = R_0 + 2.25\Delta\sigma \tan^2\theta$$

Which is a linear equation of the form

$$y_i = b + mx_i$$

CONCEPTS AND INTERPRETATION OF AVO

Refer to Figure 9–1. For a given interface, acoustic and elastic properties are given. Both the Zoeppritz and Shuey equations are applied to obtain a relationship between reflection coefficient and incident angle (in degrees).

The two equations give the same results up to 10° angle of incidence, and they do not differ significantly up to 45°. One can see the increase of reflection coefficient (amplitude) with the increase of angle of incidence or offset.

FIGURE 9–1. Typical concept

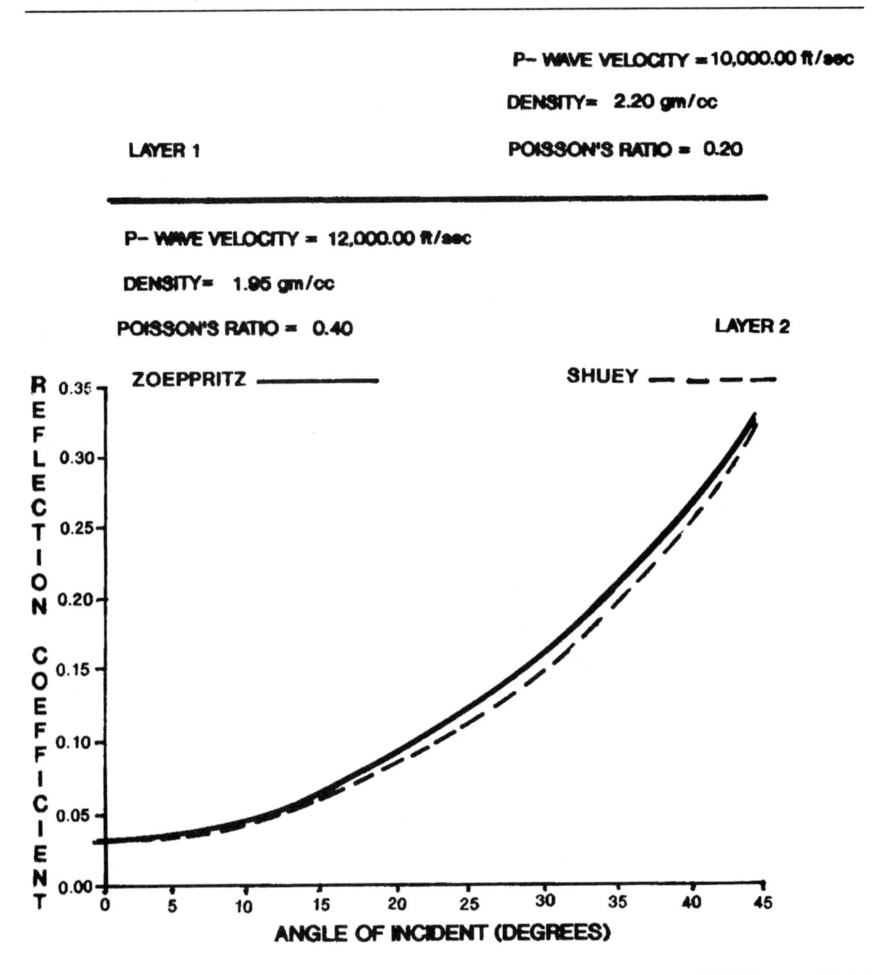

Figure 9–2 is the reflection coefficient versus angle of incidence for a typical Gulf Coast gas sand. The reflection coefficients are the troughs and have negative signs, since the plane wavefront is passing from high velocity and density to lower velocity and density. Notice the decrease of Poisson's ratio and the increase in absolute amplitude with angle of incidence. Figure 9–3 shows little change of reflection coefficient with the angle of incidence. Shuey's curve shows a slightly smaller value of reflection coefficient than Zoeppritz for the same angle of incidence between

FIGURE 9–2. Gulf Coast gas sand

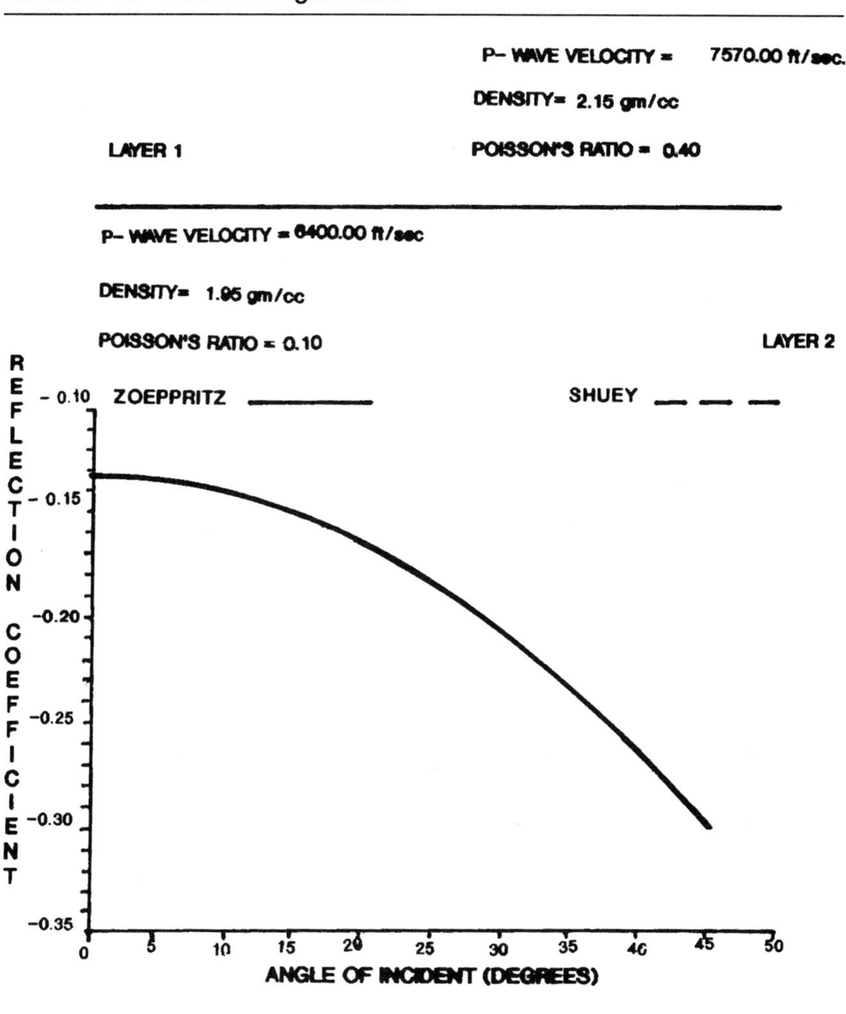

FIGURE 9–3. Little change—possible brine sand

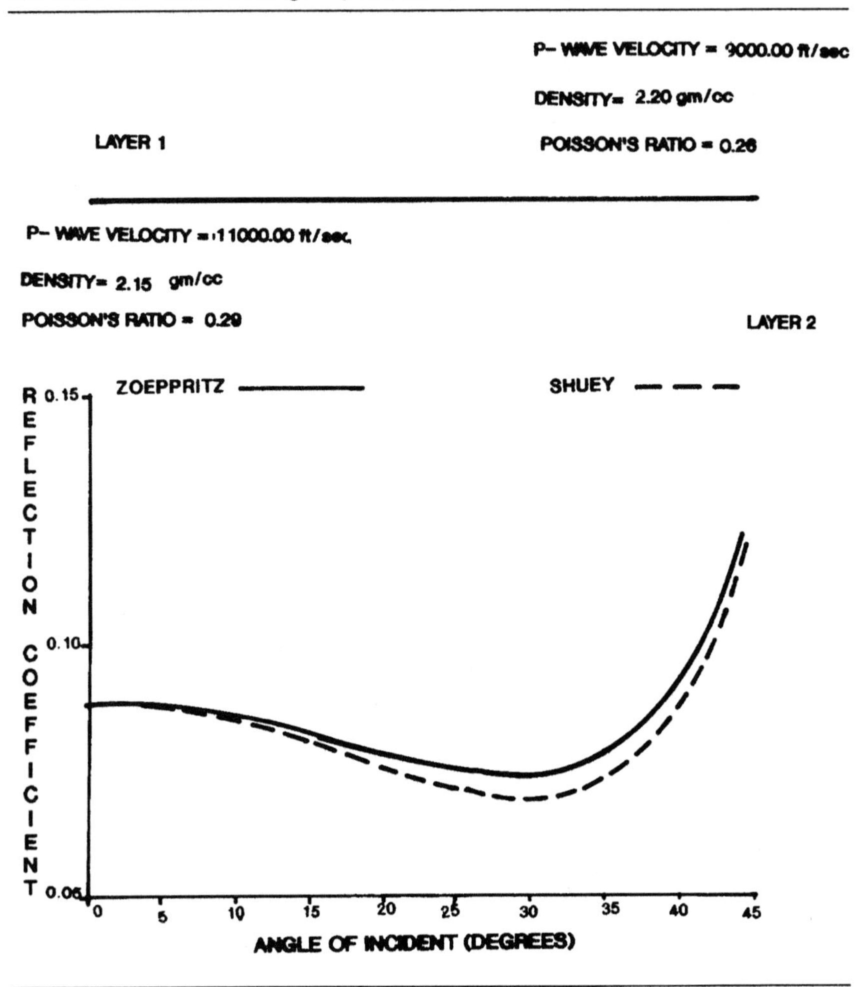

(10–40°). Otherwise, the two curves are essentially identical and may indicate brine (salt water) sands.

Figure 9–4 illustrates a decrease of amplitude with increase of angle of incidence or offset. This is a typical *dim* spot as observed in carbonate rocks. The curve was derived from the modeling of the Austin Chalk formation in the Texas Gulf Coast.

Notice the departure of the curves from the two methods of computing the angle of incidence. Yet, both show the same trend of the relationship between reflection coefficient and the angle of incidence.

FIGURE 9–4. Dim spot in carbonate rock—Austin chalk

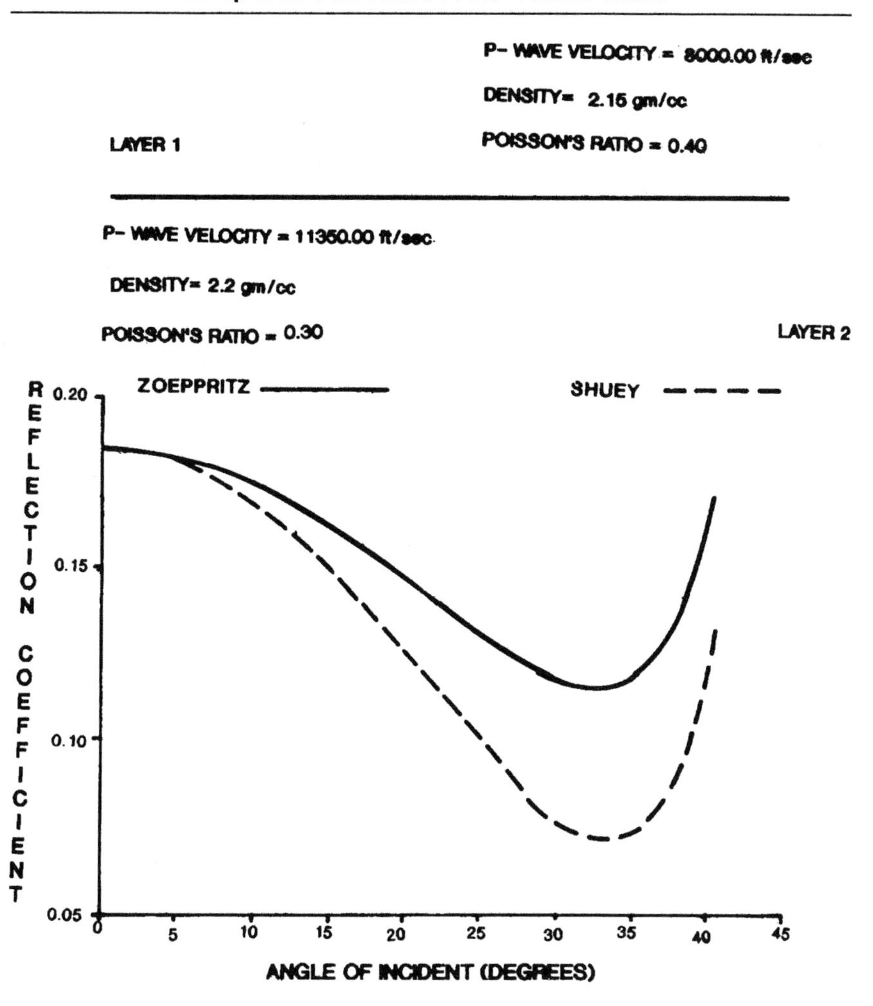

Figure 9–5 shows a slight decrease of amplitude with the angle of incidence, and the trends of the two curves match up to 30° of angle of incidence. The Shuey curve suggests it is consistent for all angles, while the Zoeppritz curve suggests an increase in amplitude. Above 30°, the calculations will be sensitive due to the NMO stretch on far offsets within the CMP gather.

Figure 9–6 shows a decrease of amplitude with increase of angle of incidence, which suggests a dim spot anomaly. The two curves have the

FIGURE 9–5. Little change—shale to sand

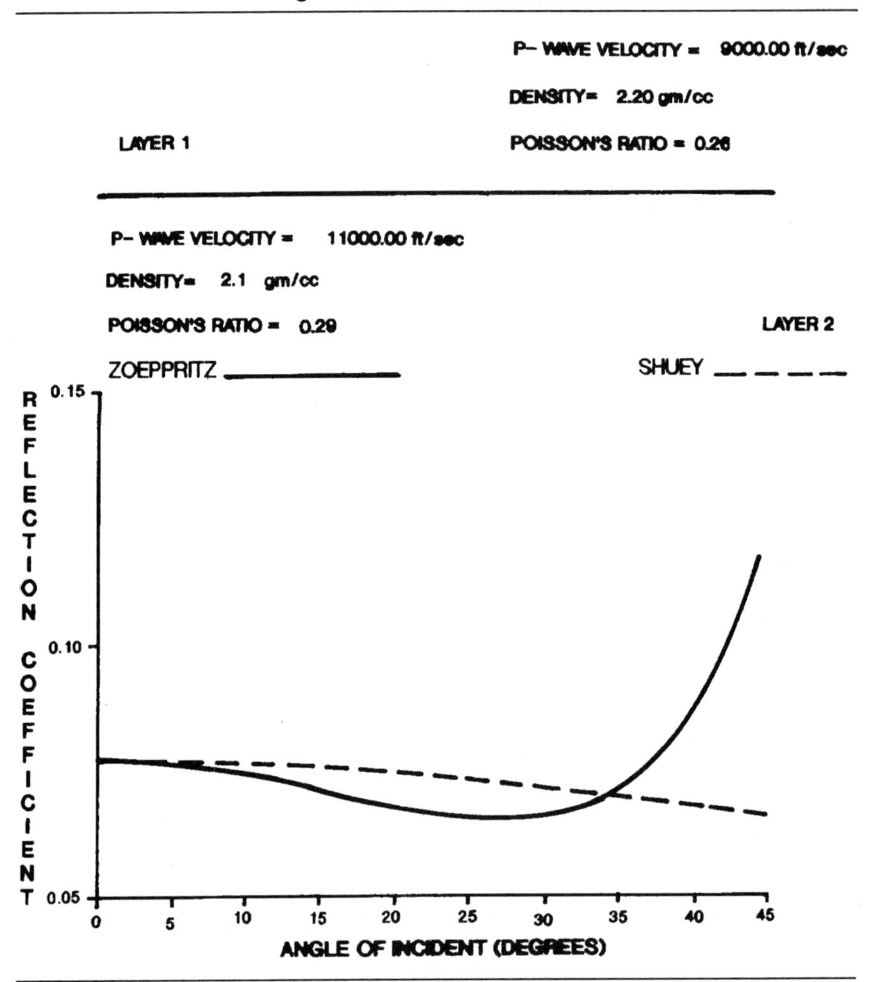

same trend, even though the Zoeppritz curve may show higher reflection coefficient values above 35° angle of incidence.

Zoeppritz equations are the complete solution, which relates the change in amplitude with the angle of incidence. The other approximations such as Shuey's are acceptable to a certain extent for most lithologies. Other approximations are suitable for some localized and specific areas.

GEOPHONE ARRAY CORRECTION

The data will be handled, for the most part, in the data reduction and setting up the lines geometry, as we discussed briefly in Chapter 5. A

FIGURE 9–6. Decreased amplitude with increased angle—dim spot. (carbonate)

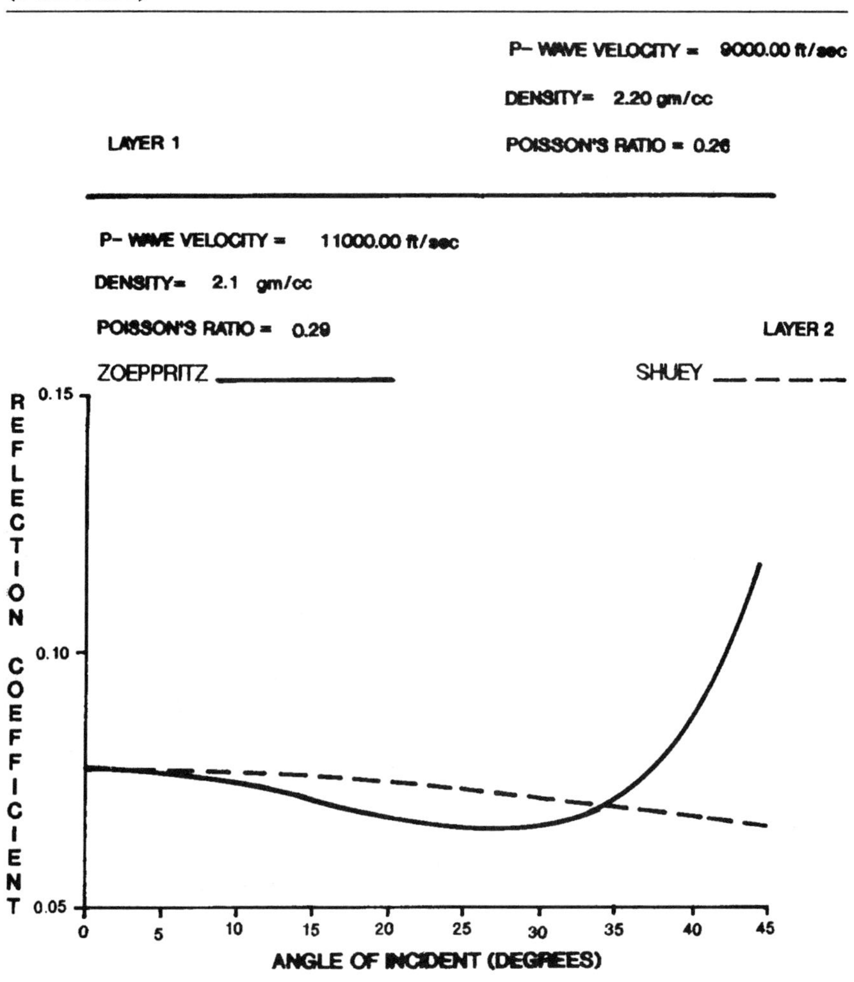

geophone array correction must be applied to compensate for the time differential within the array from the first to the last geophone in the pattern.

Figure 9–7 and Figure 9–8 show the theory and a synthetic example to illustrate the need to correct for the geophone array spread. Figure 9–8 shows 12 geophones planted in line on the ground over 220 feet. Observing from left to right on the first reflector, the time differential between geophone 1 and geophone 12 is about 10 ms. The individual geophone signals are summed together in one trace on the extreme right.

FIGURE 9–7. Geophone array correction

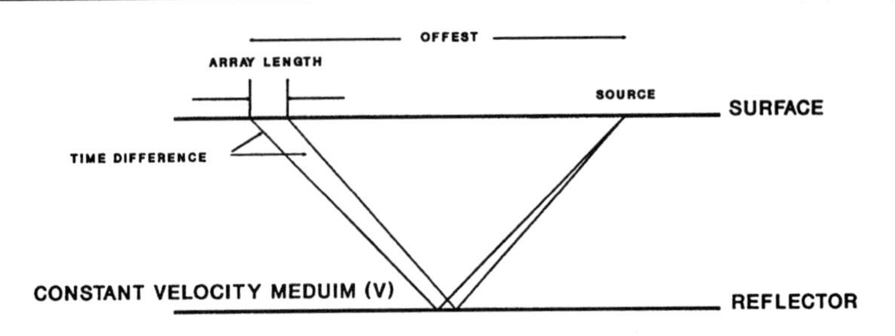

TIME DIFF.= (OFFSET X ARRAY LENGTH)/ TRAVEL TIME X VEL. SEQ.

FIGURE 9–8. Array of 12 geophones (After Fouquet, courtesy of Seismograph Service Corporation)

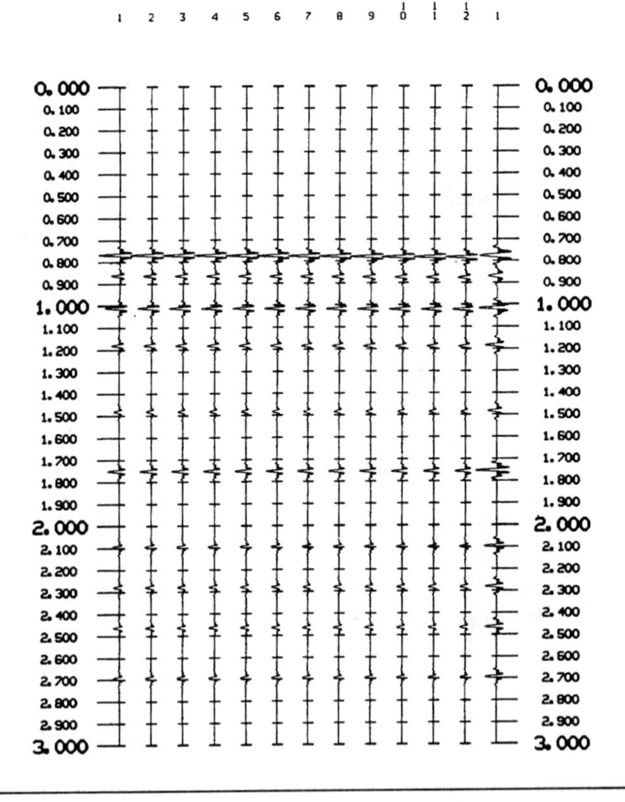

Each trace has the same high-frequency component but the summed trace, recorded as the array response, lacks some of the high frequencies because of the time shift from geophone to geophone within the array.

The differential time within the array from the first geophone to the last geophone decreases with depth, as the angle of incidence decreases with depth. It is critical to maintain the high-frequency component up shallow, especially if the data is recorded for shallow targets.

DATA PROCESSING FLOW CHART

Figure 9–9 illustrates the flow of data processing designed to preserve and enhance the true amplitude of each trace within the CMP.

Scaling is a critical step, and it should be done in a surface-consistent manner. Figure 9–10 shows the scale factor display, which is used

FIGURE 9–9. Recommended AVO processing flowchart

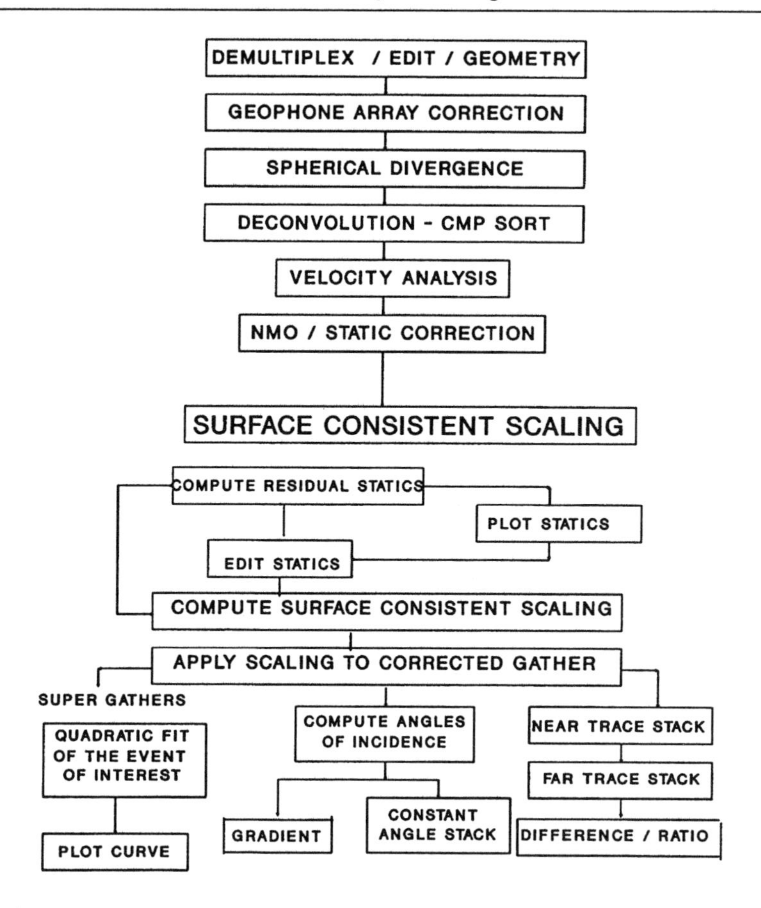

FIGURE 9–10. Surface consistent scaling display (courtesy of Seismograph Service Corporation)

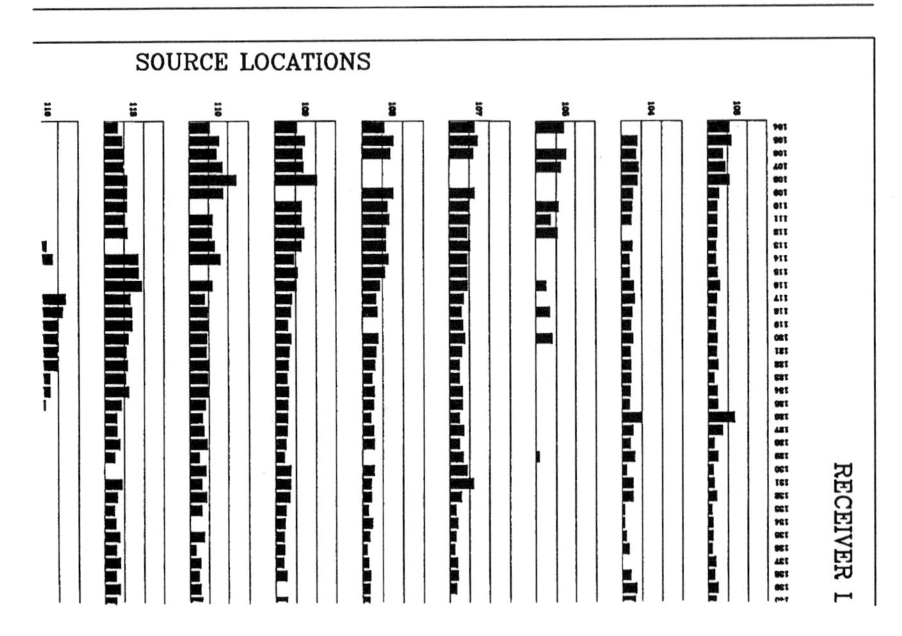

for editing and modifying to derive the scale factor for each source and receiver on the line. It is used as a quality control display; the final surface-consistent scaling is achieved when all the bars on the display are approximately equal in hight.

Figure 9–11 illustrates an AVO stack with all the offsets in the CMP gather. The amplitude anomaly at the event between 1.600–1.700 seconds at CMP 305–337 represents a *bright* spot. It stands out in the section and is normally a direct indicator of gas sand reservoirs.

Figure 9–12 is the same line. A range of the near traces in every CMP gather corrected for NMO are stacked to form this line. The bright spot anomaly has disappeared.

Figure 9–13 is the stack generated from a set of far offset traces in every CMP gather stacked. The bright spot stands out.

Figure 9–14 is the differential amplitude between far trace and near trace amplitudes. One can see that the bright spot is still anomalous on the section.

From this discussion, an increase in the amplitude of a seismic event with the increase of distance from the source to receiver (related to increased angle of incidence) represents a geological marker. In this case, the bright spot is associated with a gas sand reservoir.

FIGURE 9–11. AVO stack (courtesy of Seismograph Service Corporation)

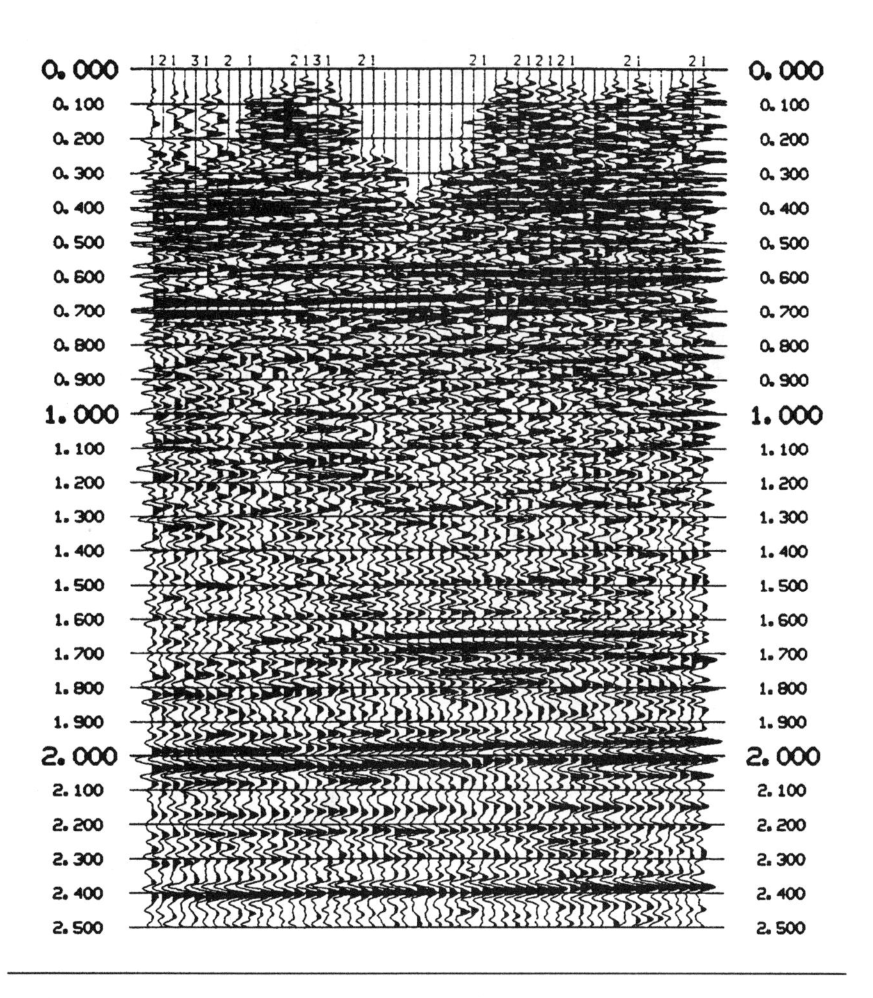

Figure 9–15 is a display of the amplitude versus offset, or angle of incidence, run of CMP gathers 310, 314, 318 from Figure 9–11. Notice the time slices between 1.600–1.700 seconds. One can see the increase of the amplitude as the offset increases on all three, but it is most pronounced on CMP 318. A curve of RMS or maximum amplitude can be plotted to define this anomaly.

FIGURE 9–12. Near traces stack (courtesy of Seismograph Service Corporation)

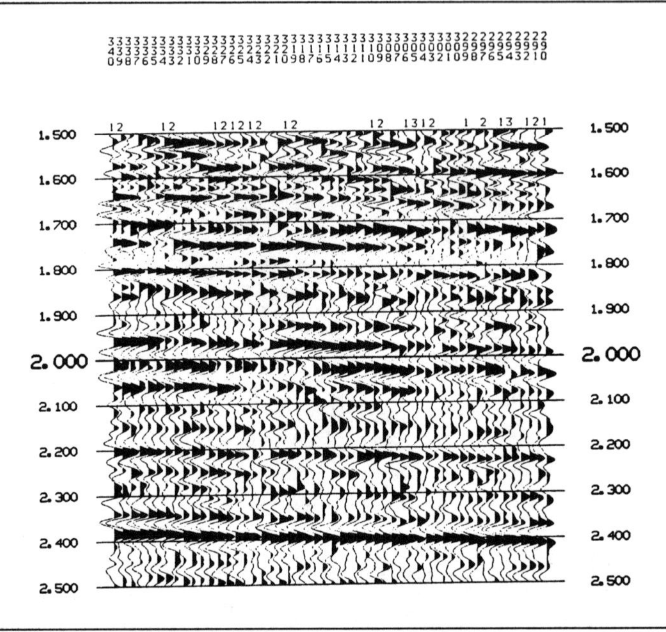

FIGURE 9–13. Far traces stack (courtesy of Seismograph Service Corporation)

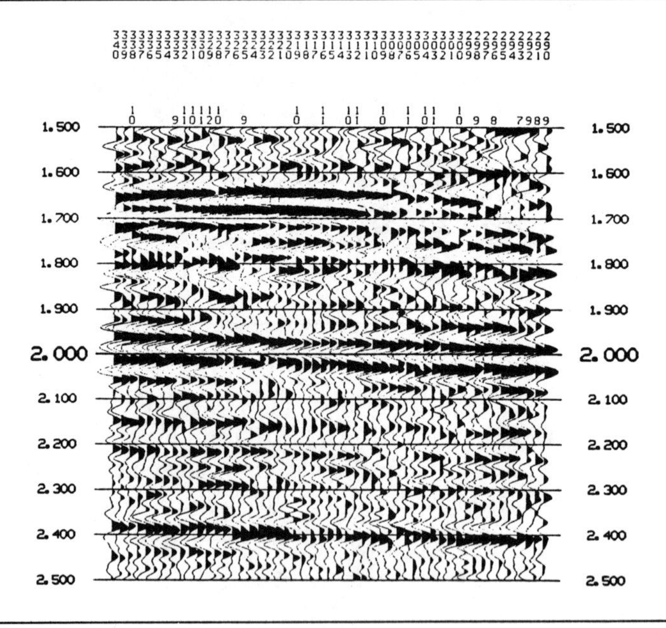

FIGURE 9–14. Difference stack (courtesy of Seismograph Service Corporation)

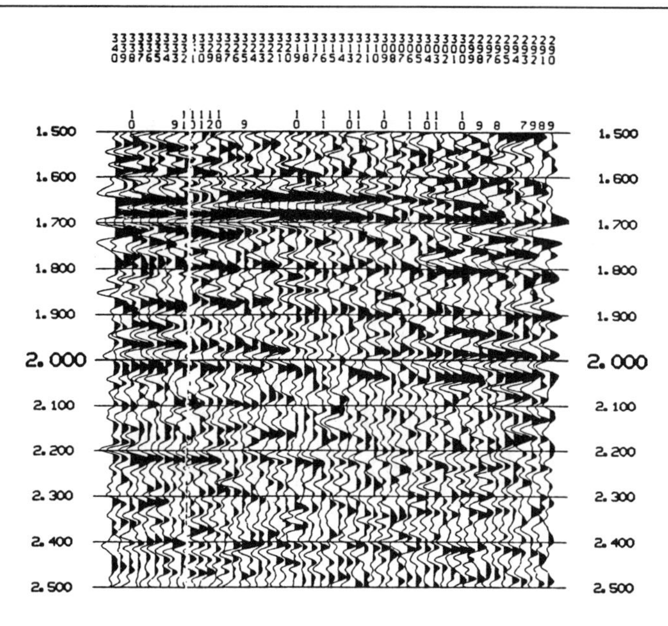

CONSTANT-ANGLE STACK

As we reviewed in the AVO analysis, AVO provides the interpreter with tools to observe and measure variation in the amplitude with either offset or angle of incidence. We discussed how observation of the change of amplitude with offset can be done on a corrected NMO gather or CMP.

To observe how amplitude varies with reflection angle, however, it is convenient to transform traces recorded at fixed offset to traces characterized by a fixed (or a limited range) of angle of incidence. The distinction between fixed offset traces or fixed angle traces is illustrated in Figure 9–16.

In the constant angle stack gather, each angle trace is generated by a partial stacking of traces in an NMO corrected CMP gather. The extent of partial stacking is by an angle range width or window beam. The annotated angle represents the central angle of the range.

Figure 9–17 is a computer printout of a common midpoint, normal moveout corrected, gather. The near trace is trace 1 at the left, and the far trace is trace 22 at the right. The data has a 4 ms sampling rate, and a total of 450 samples scanned for this illustration.

Every trace has a sample range from 54–450, and for each sample the angle of incidence was computed; a window beam of 5 degrees range

FIGURE 9–15. Amplitude vs angle of incidence (courtesy of Seismograph Service Corporation)

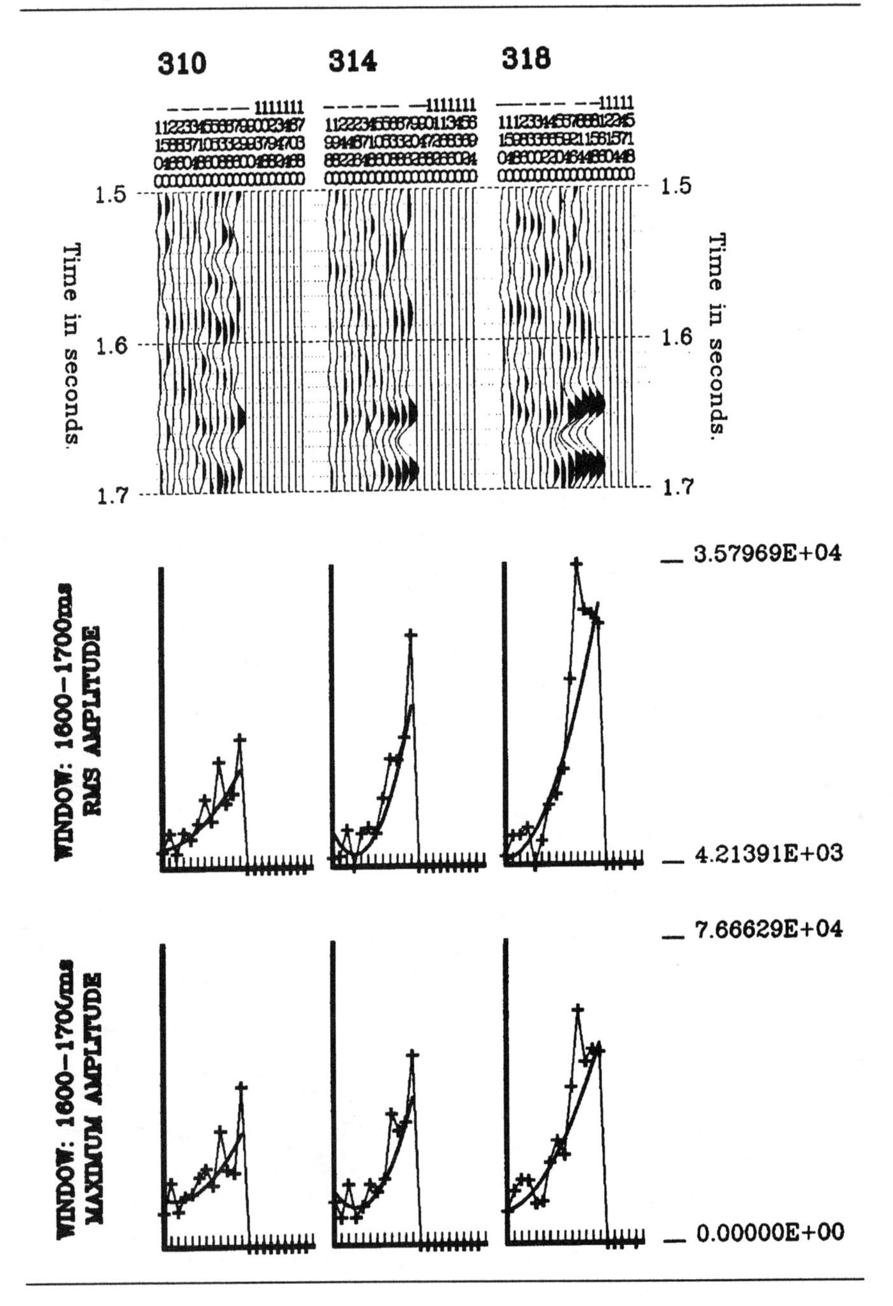

FIGURE 9–16. Constant offset and constant angle (courtesy of Seismograph Service Corporation)

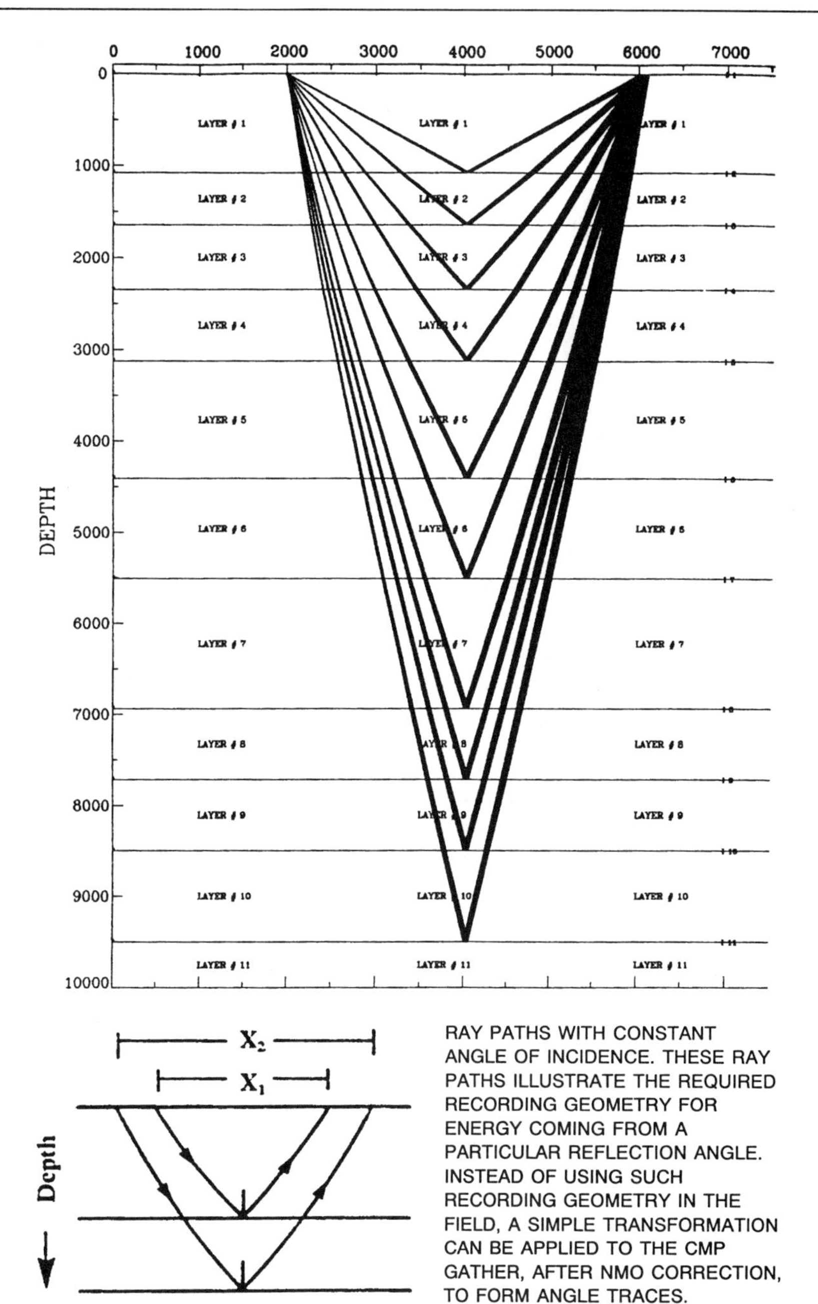

FIGURE 9–17. Computer printout, angle of incidence (courtesy of Seismograph Service Corporation)

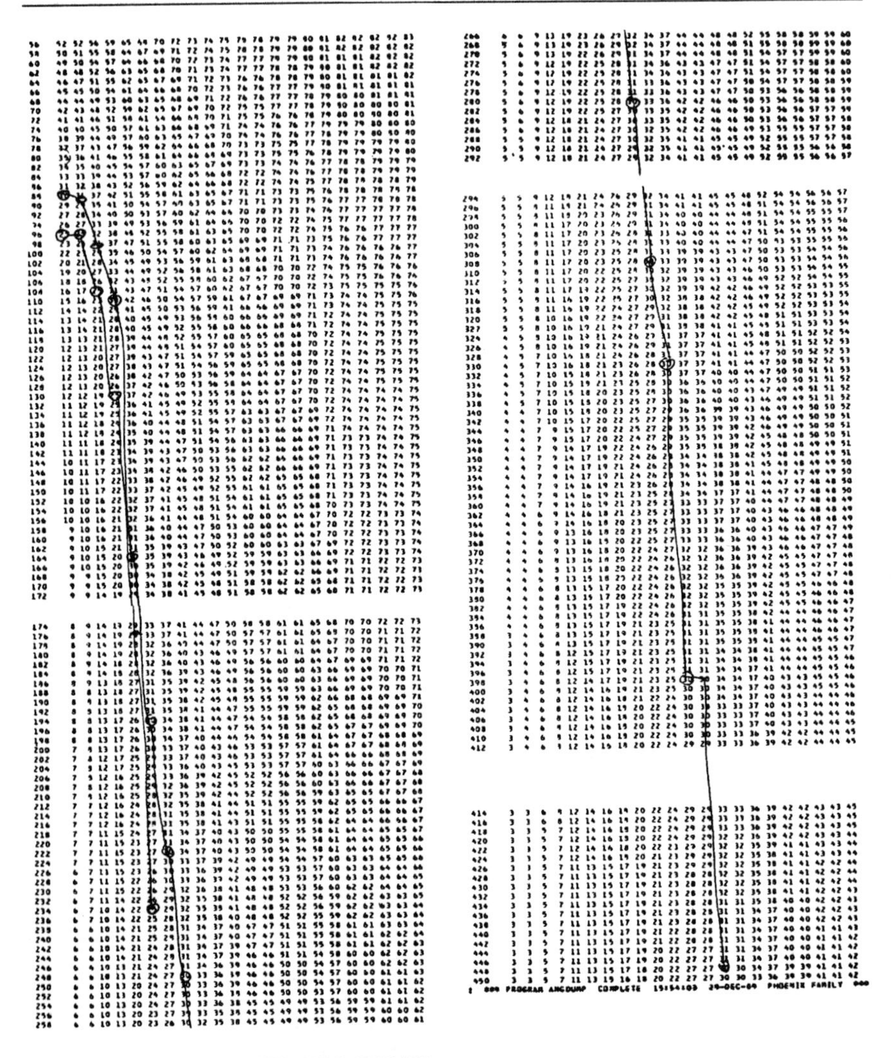

- CMP GATHER, NMO APPLIED
- 22 TRACES (FOLD)
- 4 MS SAMPLE RATE
- TRACE 1 TO LEFT, TRACE 22 TO RIGHT
- ANGLES LISTED FOR EVERY SAMPLE AND FOR EVERY TRACE

FIGURE 9–18. Constant angle stacks (courtesy of Seismograph Service Corporation)

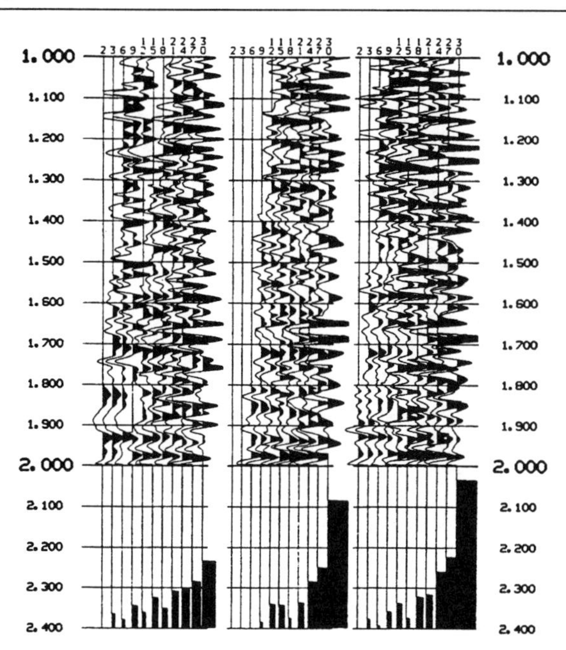

was chosen. All the partial traces with an angle from 1 to 5 degrees were used and stacked. The annotation was at the midpoint of this beam, or 3 degrees; the second beam will be 4 to 8 degrees, then all partial traces in the gather having an angle of incidence within this range will be stacked and annotated 6 degrees, and so on.

Figure 9–18 illustrates this approach; it was done on three common midpoint, normal moveout corrected, gathers. The constant angle range is from 2° to 30°. A bar graph representing the amplitude variation with the angle of incidence is plotted below each set of constant angle stack gathers. As we can see, the amplitude at the same window of investigation (1.6 to 1.7 seconds) increases with increased angle of incidence.

AVO ATTRIBUTES AND DISPLAYS

A number of other parameters can be displayed on sections in a manner similar to the conventional stack, such as a near-trace stack, which consists of short trace distances selected from each CMP gather, corrected for NMO, and stacked together. Likewise, a far-trace stack with selected far offset distance range can be stacked together to form a far-trace stack.

The observed variation (increase) of amplitude can be easily seen as the far trace offset stack section shows a pronounced increase in the amplitude. Also, the amplitude ratio between the near-and far-trace stacks can be displayed, as well as the gradient normal incidence amplitude.

These attributes are shown in Figures 9–19, 9–20, 9–21 and 9–22. Color displays may be used for easier interpretation.

Other useful displays that can be created include:

- Amplitude versus $\sin^2\theta$.
- P-wave reflection.
- Gradient of amplitude versus $\sin^2\theta$.
- Mimic of shear wave stack can be generated by assuming that the S-wave velocity is half of the P-wave velocity. The travel time of the section is governed by the P-wave velocity.
- Poisson's Ratio stack, using the same assumption that the shear wave velocity is half the P-wave velocity.

All these are shown in Figures 9–23 and 9–24.

PROCESSING DON'TS

Certain data processing operations, although possible, must be avoided in order to preserve the amplitude versus offset relation:

Multichannel operations such as:

- Mixing traces will remove the significance of true amplitude.
- Trace-to-trace scaling with small windows.
- F-K operations.
- Deconvolution derived from trace summation.

ADVANTAGES OF AVO

AVO is a proven tool in the verification of direct indicators, such as bright spots in gas sands and dim spots in carbonate reservoirs, and any related amplitude anomalies. An AVO analysis using NMO-corrected CMP is a two-dimensional analysis, whereas a stacked trace is a one-dimensional analysis. The amplitude variation with the angle of incidence is another tool used for confirmation of observed anomalies. For a given rock, we can estimate Poisson's ratio, which can be used to determine its elastic properties.

FIGURE 9–19. Far trace stack (courtesy of Seismograph Service Corporation)

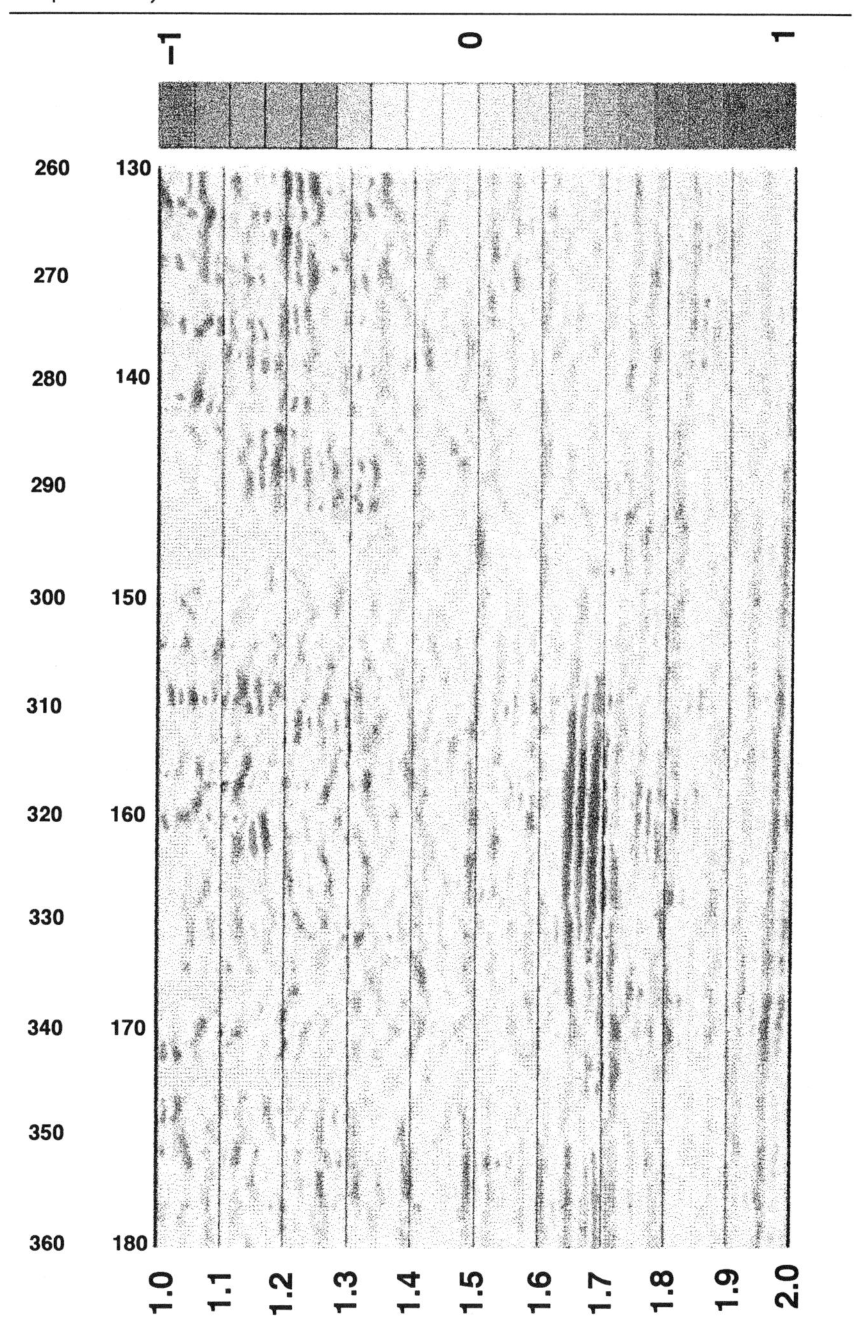

FIGURE 9–20. Near trace stack (courtesy of Seismograph Service Corporation)

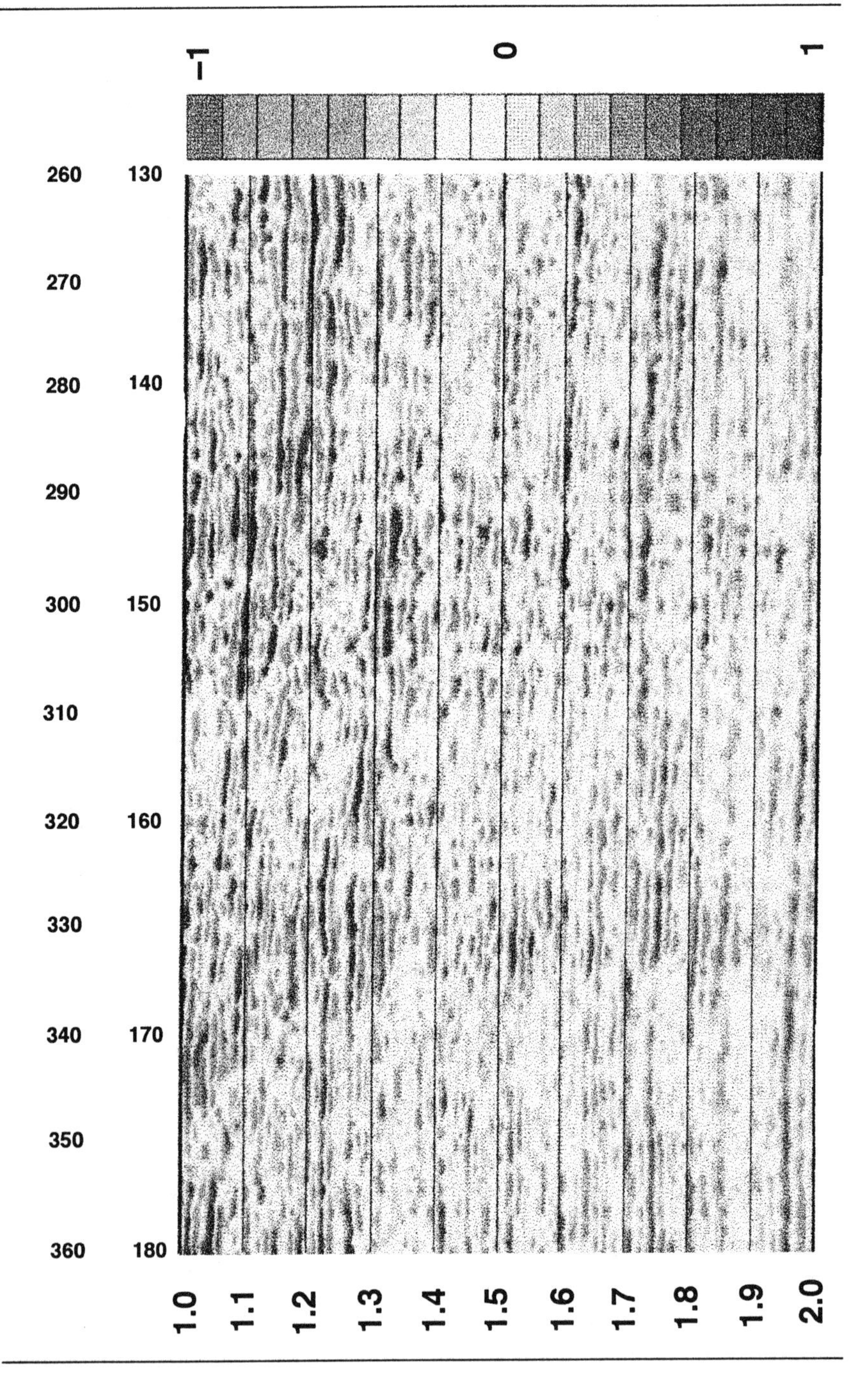

FIGURE 9–21. Gradient * normal incidence amplitude (courtesy of Seismograph Service Corporation)

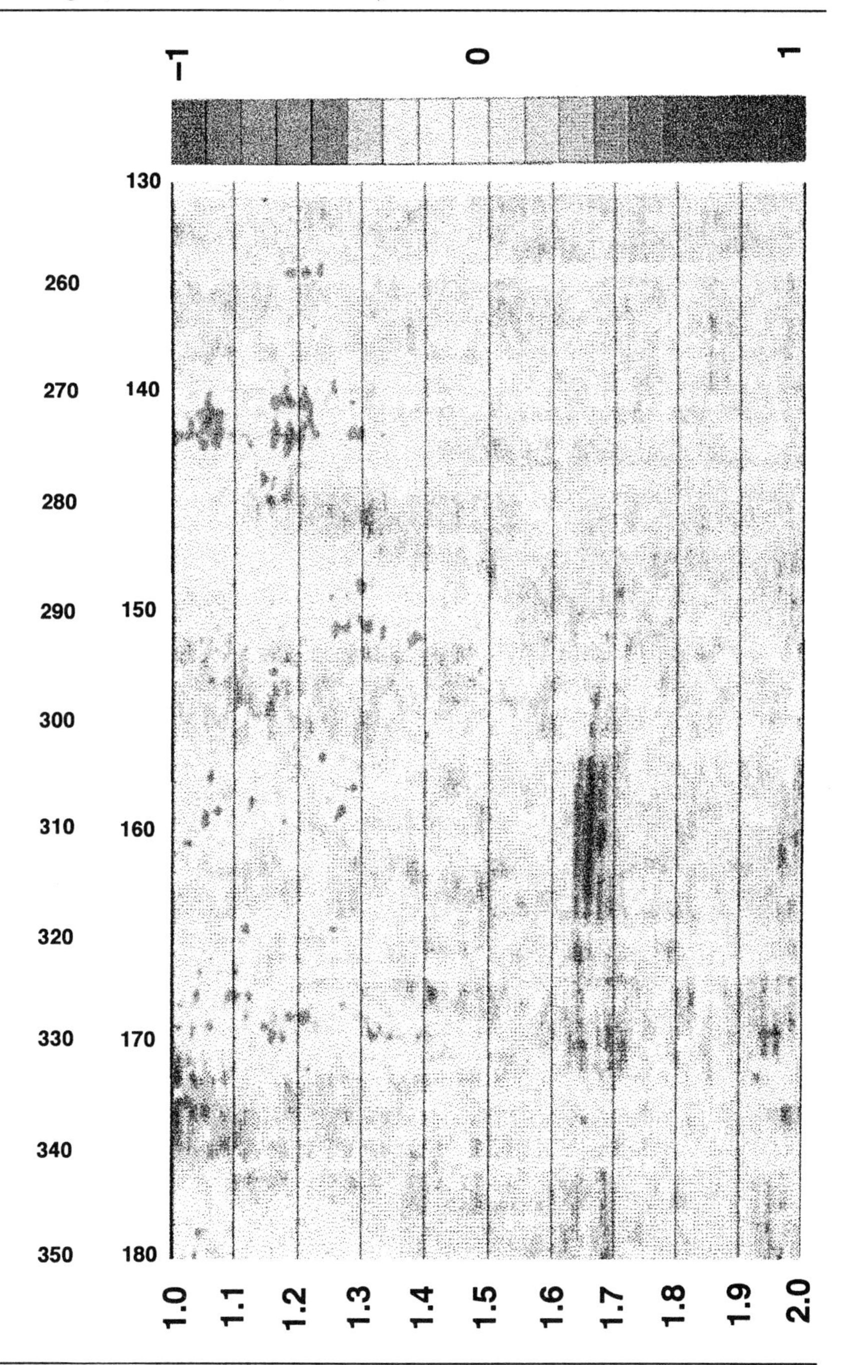

FIGURE 9–22. Far-near trace ratio (courtesy of Seismograph Service Corporation)

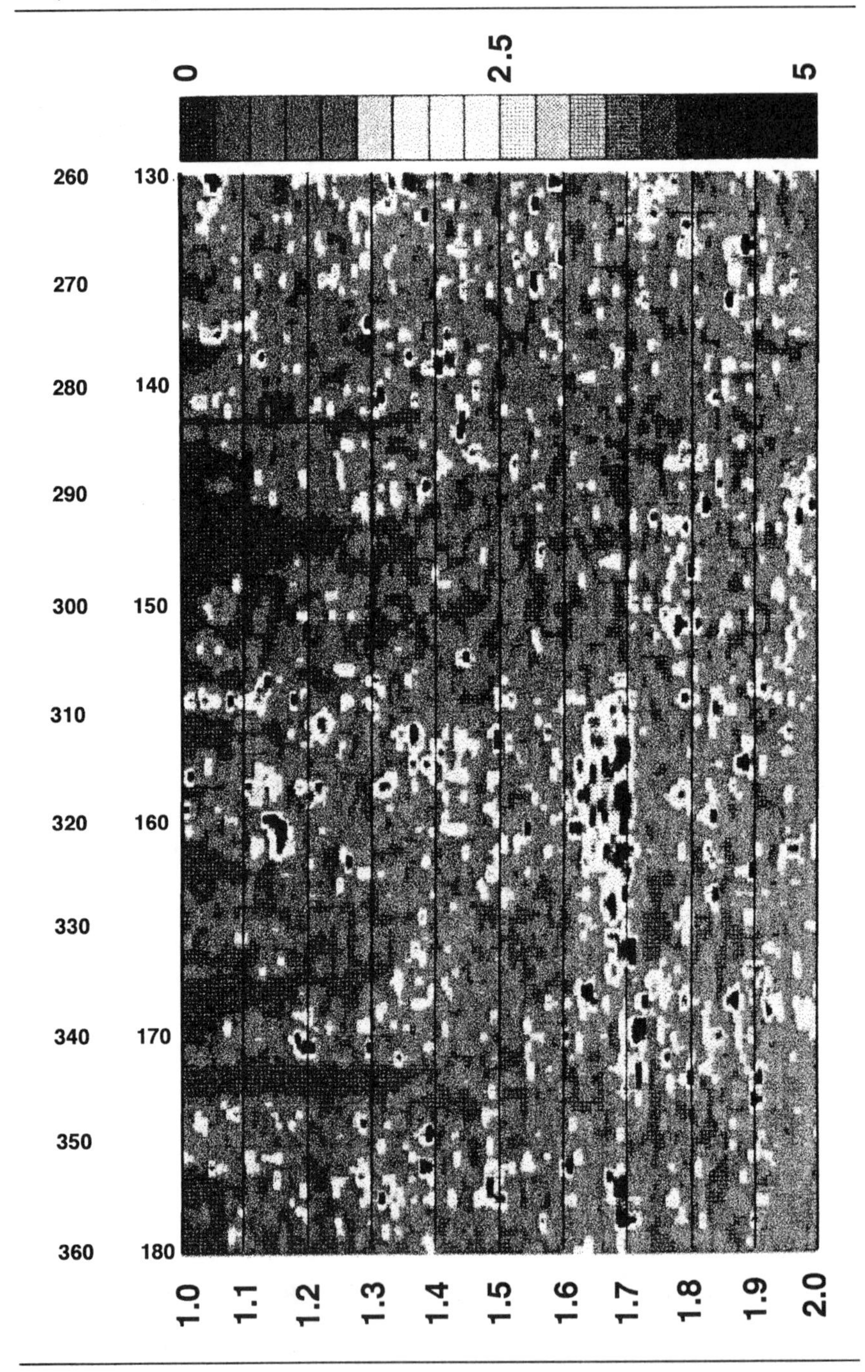

FIGURE 9–23. Bright-spot and P-wave stacks (courtesy of Western Geophysical)

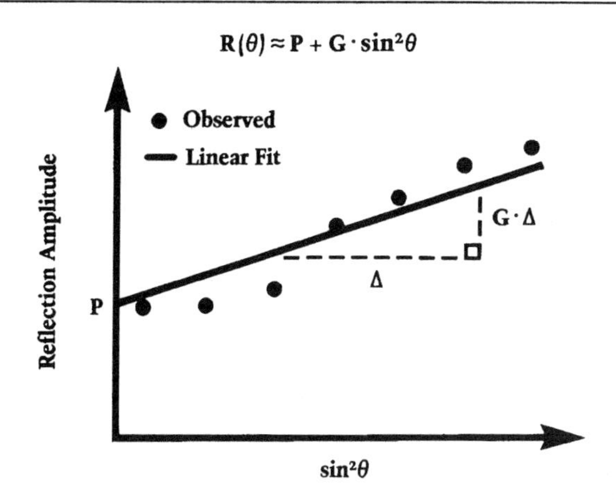

(a) LINEAR FIT OF AVO CURVE OF REFLECTION AMPLITUDE VERSUS $\sin^2\theta$. FOR ANGLE OF INCIDENCE θ LESS THAN 25 DEGREES, THE AMPLITUDE OF A P-WAVE REFLECTED FROM A PLANAR INTERFACE BETWEEN TWO CLASTIC MEDIA VARIES LINEARLY WITH $\sin^2\theta$. IN THE LINEAR FIT, P IS THE INTERCEPT, AND G IS THE GRADIENT (OR SLOPE).

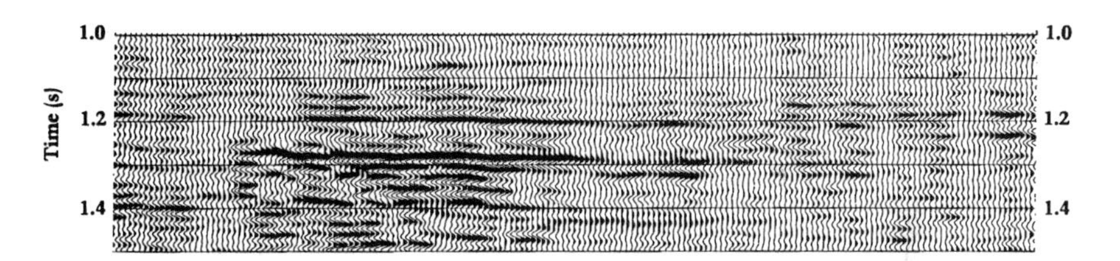

(b) CLOSE-UP OF CMP STACK SECTION. THE BRIGHT SPOT IS THE STRONG REFLECTION JUST ABOVE 1.25 S. A PEAK INDICATES A DECREASE IN IMPEDANCE.

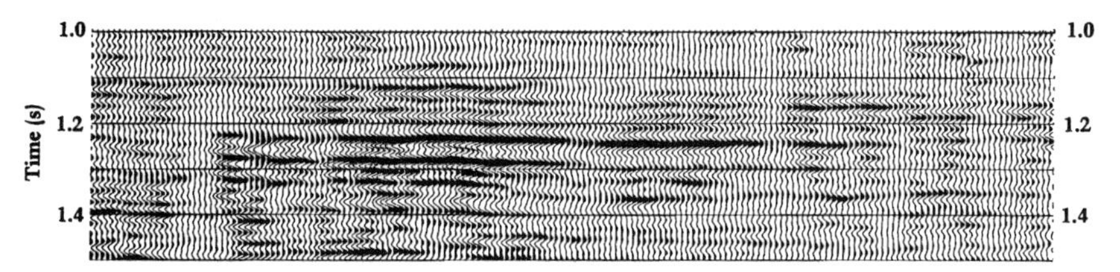

(c) P-WAVE STACK. EACH TRACE IS THE SEQUENCE OF INTERCEPTS, P, DERIVED FROM LINEAR FITS OF RELECTION AMPLITUDE VERSUS $\sin^2\theta$ FOR AN NMO-CORRECTED CMP GATHER.

FIGURE 9–24. S-wave and Poisson's ratio stacks (courtesy of Western Geophysical)

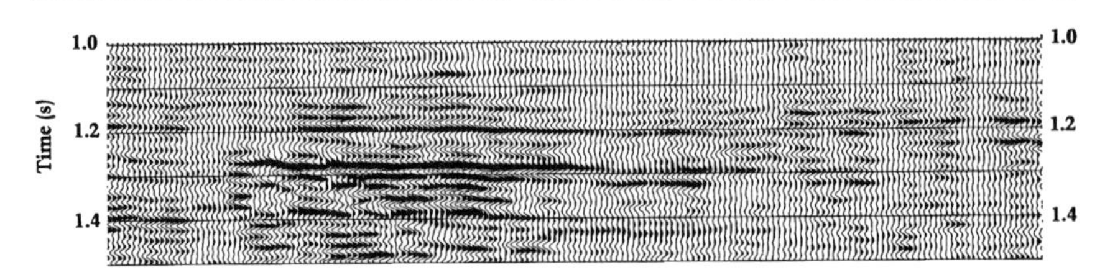

S-WAVE STACK. THESE DATA MIMIC WHAT WOULD HAVE BEEN RECORDED FROM A ZERO-OFFSET S-WAVE SURVEY, BUT WITH THE TRAVELTIME GOVERNED BY THE P-WAVE VELOCITY. EACH TRACE (OBTAINED BY COMBINING INTERCEPT P AND GRADIENT G [FIGURE 9–23(a)] AND ASSUMING S-WAVE VELOCITY IS HALF THE P-WAVE VELOCITY) REPRESENTS THE RESPONSE TO CONTRASTS IN S-WAVE IMPEDANCE.

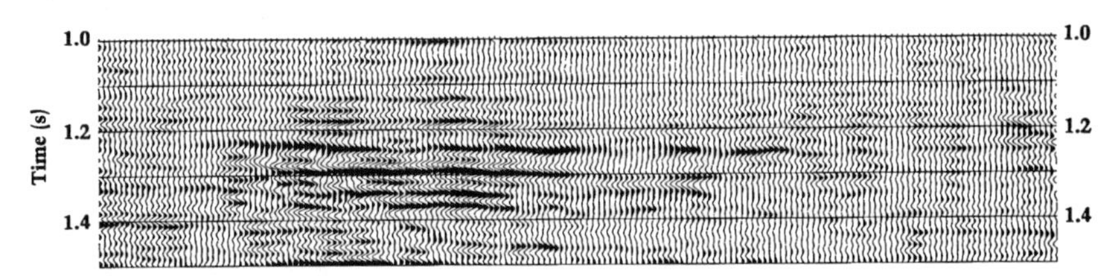

POISSON'S-RATIO STACK. EACH TRACE (OBTAINED USING THE SAME ASSUMPTIONS AS FOR THE S-WAVE STACK, BUT WITH A DIFFERENT COMBINATION OF THE INTERCEPT AND GRADIENT) SHOWS THE RESPONSES TO CHANGES IN POISSON'S RATIO. AS WITH THE S-WAVE STACK, THE TRAVELTIME IS GOVERNED BY THE P-WAVE VELOCITY.

APPLICATIONS OF AVO

RESERVOIR BOUNDARY DEFINITION

One of the most difficult tasks for a reservoir engineer or geologist is to estimate the hydrocarbons in place, since there are so many variables involved, especially in the exploration stage. One of these variables is the areal extent of the reservoir. Also, in the development stage of a field, the boundary of the reservoir must be defined in order to design the optimum spacing between development wells.

With adequate coverage of detailed seismic data, which has been processed for AVO analysis, we can distinguish the boundary of the reservoir with a great deal of accuracy. The preferred seismic survey is a 3-D or an extensive, detailed 2-D seismic survey.

FUTURE EXPLORATION OR EXPLOITATION

AVO analysis, especially on a proven anomaly that has been documented by drilling and successful completion, can be used to locate look-alike anomalies in other geological provinces that have the same or close geological and lithological settings. The AVO analysis should always be integrated with all other geological and engineering information.

PREDICTING HIGH-PRESSURE GAS ZONES

Bright spot anomalies are normally associated with gas sand formations or lenses. Some of these lenses in the shallow part of the seismic section may have abnormally high pressure. Where it is documented that such geological conditions are present in an area, an AVO analysis of seismic data will permit the drilling engineer to take precautions to account for this high-pressure zone and modify his drilling program before the bit hits the zone.

SUMMARY AND DISCUSSION

The amplitude versus offset technique, in which the changes of amplitude with the angle of incidence are analyzed, is widely used as a hydrocarbon indicator.

The analysis is done on the common midpoint (CMP) corrected gathers before stacking. It is a two-dimensional analysis compared to the stacked traces, which is one dimensional.

In the near future, it may be possible in a certain area to identify lithology of the rock from the shape of the reflection coefficient versus angle of incidence (Koefoed, 1955).

AVO techniques can be used in exploration to delineate look-alike features, to define depositional environments, delineate reefs, and identify gas sands to name few.

In the field data acquisition, a long spread is desirable in order to investigate the change of the amplitude with offset on the far traces (long distances from the source).

How long will the spread be? The spread length is a function of the target depth, velocity in the area, and structure and maximum frequency to be recorded. The spread length can be determined from field tests or from the contractor's experience in the area.

Two-dimensional seismic modeling could help in this regard. A special data processing sequence is needed to investigate the variation of amplitude with offset. This technique preserves the relative true amplitude of the seismic traces within the CMP. It is used to identify the rock lithology and its fluids and/or gas content. One should know the rock properties in a nearby field, velocity information, and stratigraphy in order to carry a sensible AVO interpretation. Data processing should be closely monitored, and always study the CMP gathers for amplitude changes. Other displays such as partial stacks and ratio sections are recommended for thorough investigations. It is essential to integrate the amplitude changes with other geological, geophysical, and petrophysical information via modeling in order to complement the interpretation.

Many software processing packages are available. Data can be manipulated to analyze a variation of one variable such as Poisson's ratio, normal incidence, and others.

The average cost per mile for 2-D standard seismic-data processing is about $400. A typical line configuration is 120 channels, 55 foot group interval, 110 foot shot interval, and 4 ms sampling interval.

For AVO analysis add an additional 40%. Color attributes are extra. Prices may change with time; it is always recommended to check on current prices. AVO processing is not costly compared to the benefits of upgrading the interpretation of a play or avoiding a blowout from a high-pressure zone.

KEY WORDS

Amplitude versus offset (AVO)
Constant angle stack
Quadratic fit
Dim spot
NMO stretch
Poisson's ratio

BIBLIOGRAPHY

Backus, M. M. "The Reflection Seismogram in a Layered Earth." *Bulletin of the American Association of Petroleum Geologists* 67 (1983): 416–417.

Castagna, J. P. "Petrophysical Imaging Using AVO." *TLE* 12 (1993): 172–178.

Domenico, S. N. "Effect of Brine-Gas Mixture on Velocity in an Unconsolidated Sand Reservoir." *Geophysics* 41 (1976): 882–894.

DeVoogd, N. and H. Den Rooijen. "Thin layer response and spectral bandwidth." *Geophysics* 48 (1983): 12–18.

Gassaway, G. S. and H. J. Richgels. "SAMPLE, Seismic Amplitude Measurement for Primary Lithology Estimation." *53rd SEG Mtg.*, Las Vegas, Expanded Abstracts (1983): 610–613.

Gelfand, V., P. Ng, and K. Larner. "Seismic Lithologic Modeling of Amplitude-Versus-Offset Data." *56th SEG Mtg.*, Houston, Expanded Abstracts (1986): 332–334.

Hindlet, F. "Thin Layer Analysis Using Offset/Amplitude Data." *56th SEG Mtg.*, Houston, Expanded Abstracts (1986): 332–334.

Hill, N. R., and I. Lerche. "Acoustic Reflections From Undulating Surfaces." *Geophysics* 51 (1986): 2160–2161.

Hilterman, F. "Is AVO the Seismic Signature of Lithology? A Case History of Ship Shoal-South Addition." *TLE* 9 (1990): 15–22.

Hilterman, F. "AVO: Seismic Lithology." *SEG, Course Notes,* 1992.

Koefoed, O. "On the Effect of Poisson's Ratios of Rock Strata on the Reflection Coefficients of Plane Waves." *Geophy. Prosp.* 3 (1955): 381–387.

Koefoed, O. and N. DeVoogd. "The Linear Properties of Thin Layers, with an Application to Synthetic Seismograms over Coal Seams." *Geophysics* 45 (1980): 1254–1268.

Meissner, R. and M. A. Hegazy. "The Ratio of the PP- to SS-Reflection Coefficient as a Possible Future Method to Estimate Oil and Gas Reservoirs." *Geophys. Prosp.* 29 (1981): 533–540.

Muskat, M., and M. W. Meres. "Reflection and Transmission Coefficients for Plane Waves in Elastic Media." *Geophysics* 5 (1940): 149–155.

Narvey, P. J. "Porosity Identification Using AVO in Jurassic Carbonate, Offshore Nova Scotia." *TLE* 12 (1993): 180–184.

Ostrander, W. J. "Plane Wave Reflection Coefficients for Gas Sands at Non-Normal Angles of Incidence." *Geophysics* 49 (1984): 1637–1648.

Schoenberger, M. and F. K. Levin. "Reflected and Transmitted Filter Functions for Simple Subsurface Geometries." *Geophysics* 41 (1976): 1305–1317.

Shuey, R. T. "A Simplification of Zoeppritz Equations." *Geophysics* 50 (1985): 609–614.

Wright, J. "Reflection Coefficients at Pore-Fluid Contacts as a Function of Offset." *Geophysics* 51 (1986): 1858–1860.

Young, G. B. and L. W. Braile. "A Computer Program for the Application of Zoeppritz's Amplitude Equations and Knott's Energy Equations." *Bulletin of the Seismological Society of America* 66 (6) (1976): 1881–1885.

Yu, G. "Offset-Amplitude Variation and Controlled-Amplitude Processing." *Geophysics* 50 (1985): 2697–2708.

CHAPTER 10

3-D SEISMIC SURVEYS

INTRODUCTION

The acquisition and processing of data for the purpose of detailed subsurface imaging present a set of new challenges to the mineral resources industry. 3-D seismic data acquisition and processing represent a step forward in the continuing development of imaging technology.

In the past, few 3-D surveys were run because of the high cost of acquisition and processing. Advances in acquisition hardware, such as telemetry systems with multistation and multiline capabilities, improvements in marine navigation systems which can accurately identify the position of every shot and receiver, and improvements in data processing hardware and software, have made it possible to process large volumes of 3-D seismic data within reasonable time and cost limits.

WHY, WHERE, AND WHEN TO USE 3-D

In areas where structural traps are being explored, the subsurface beds generally dip in many directions. A 3-D survey can lead to a high degree of reliability in the interpretation because the volume of data provides more accurate and detailed sections.

Drilling programs based on 3-D usually have a high success ratio. This tool has proved its effectiveness in development drilling in complex structure areas, defining small, subtle features such as narrow channel sands, surfaces of nonconformity, and stratigraphic traps, which the conventional 2-D survey fail to reliably locate and define.

The 3-D survey has proved its success in monitoring the progress of enhanced oil-recovery projects, where a change in the seismic wavelets is an indication of the advancement of the front. 3-D has many engineering applications, some of which are still under field observation, and for which final results have not been published.

As with any tool, 3-D has limitations that should be recognized. Every project must stand on its own merit, and the survey should be designed after studying all available information. This will be discussed later.

Figure 10–1 shows why 3-D surveys are preferred over 2-D surveys. Figure 10–1a is a subsurface model consisting of a single dipping plane interface between homogeneous media. Line 1 is in the direction of dip, line 2 is in the strike direction, and line 3 is in an arbitrary direction. After migration, data at the intersection mis-ties at point X.

Figure 10–1b shows: (1) migration along dip line 1 and (2) migration along strike line over depth model in Figure 10–1a. Point D, after migration, moves updip to D' along the dip line, but does not move along strike line 2. This causes a mis-tie, as indicated.

Figure 10–1c is a plan view of migration at intersection D along the three lines indicated. Point D moved up dip to D', its true subsurface location, along dip line 1. Point D did not move on strike line 2. The same point moves to D'' along line 3, which is of arbitrary direction.

Complete imaging is achieved by migrating the data again along the direction perpendicular to 3, to move the event from D'' to D'.

FIGURE 10–1. Why 3-D? (modified after Workman, 1984)

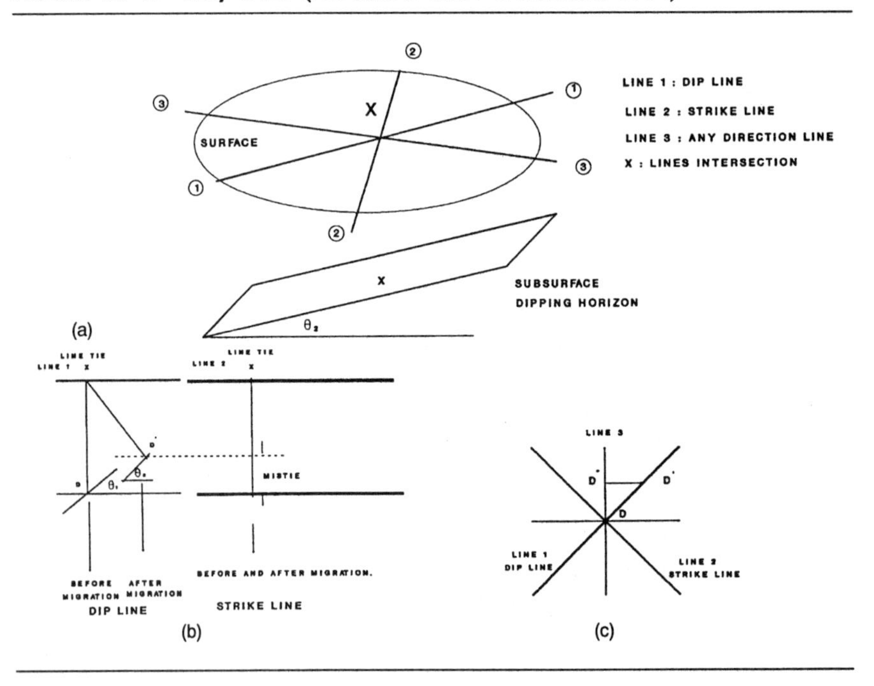

3-D DATA ACQUISITION

Field procedure and survey arrangements are always subject to optimization, because there is always a desire to get maximum quality for minimum cost.

Several distinct 3-D seismic methods have been developed in recent years. The various approaches are considered in terms of their attributes, such as resolution, target depth, and convenience of access. Marine surveys may be broadly categorized as exploration and development 3-D, recorded as a dense grid of 2-D surveys.

Marine 3-D methods may be conveniently grouped into three categories:

1. Exploration 3-D, in which a sparse grid of data is recorded, and the requirements of cross-line spatial sampling are met by interpolation.
2. Short-offset 3-D, in which a very short streamer close to the source is used to record a finely sampled set of traces of near-normal-incident data.
3. Conventional 3-D, the current technique, in which streamers of normal length record a finely sampled, multifold grid of data.

3-D data acquisition on land is commonly carried out by swath shooting in which receiver cables are laid out in parallel lines (in-line direction), and shots are positioned in a perpendicular direction (cross-line direction). Before designing the 3-D survey for an area, certain preparations should be made:

- Study all available data, maps, photos, and send permit man to the field to investigate the site of the survey.
- Study the geology, any available 2-D seismic, VSP, and logs, and identify the projected target.
- Get 2-D seismic data: geometry, offset, fold, dip, source type, and receiver.
- Establish basic parameter models, correlate logs for target.
- Allow for obstacles, environments and industries in your design.
- Test for azimuths, offsets, fold.
- Consider equipment, cost effectiveness, and time constraints.
- Establish 3-D template, receivers, and shots.

3-D SURVEY SYSTEM DESIGN OVERVIEW

After studying all available data, a target description and survey objectives become clear and can be defined. To obtain the required resolution from the 3-D, it is imperative to study the noise patterns in the area, either by

FIGURE 10–2. Recommended 3-D design considerations

1. AVAILABLE DATA—MAPS, PHOTOS, AND SCOUTING.
2. GEOLOGY, 2-D SEISMIC, VSP, LOGS, AND TARGET.
3. ALLOW FOR OBSTACLES, ENVIRONMENTS, AND INDUSTRIES IN DESIGN.
4. GET 2-D GEOMETRY—OFFSETS, FOLD, DIPS, SOURCE AND RECEIVER TYPES.
5. BASIC PARAMETERS—MODEL, CORRELATE LOGS FOR TARGET.
6. TEST FOR AZIMUTHS, OFFSETS, FOLD, ETC.
7. CONSIDER EQUIPMENT, COST EFFECTIVENESS, AND TIME CONSTRAINTS.
8. ESTABLISH 3-D TEMPLATE—SHOTS, RECEIVERS, ETC.

investigating any previously done test or by planning one before conducting the 3-D survey. All available 2-D seismic data, well logs, and VSP surveys should be studied for field geometry, source, fold, signal-to-noise ratio, subsurface dips, velocities, predominant frequency, maximum frequency, and vertical and horizontal resolution for the target. Information drawn from all these sources will help in the design of the survey.

Source and receiver arrays can be designed taking into consideration the cost effectiveness and logistics based on prospect description and operational constraints. Figure 10–2 illustrates the 3-D design system overview.

2-D SURVEY DESIGN

It is essential that you become familiar with designing the field parameters of 2-D surveys before learning how to design the 3-D survey parameters. In Chapter 4 we discussed some field geometry terminology such as source, receiver, group interval, shot interval, near offset, far offset, and fold coverage.

Now, it is time to talk about how to design some of these parameters and to discuss the factors that affect each parameter.

THE MODEL

Construct a model that will show the target horizon, its extent, dip, velocity, depth, and time, as well as all other seismic markers, velocities, times, dips, and approximate arrival time of each marker. Include the near-surface formation layer or layers and its velocity time.

Figure 10–3 shows the model and definition of the objectives. This model will furnish most of the parameters needed for design. It also is insurance on secondary survey objectives. The model can be extended to aid in determining more detail, such as the wavelengths needed for resolving the target.

FIGURE 10–3. Model

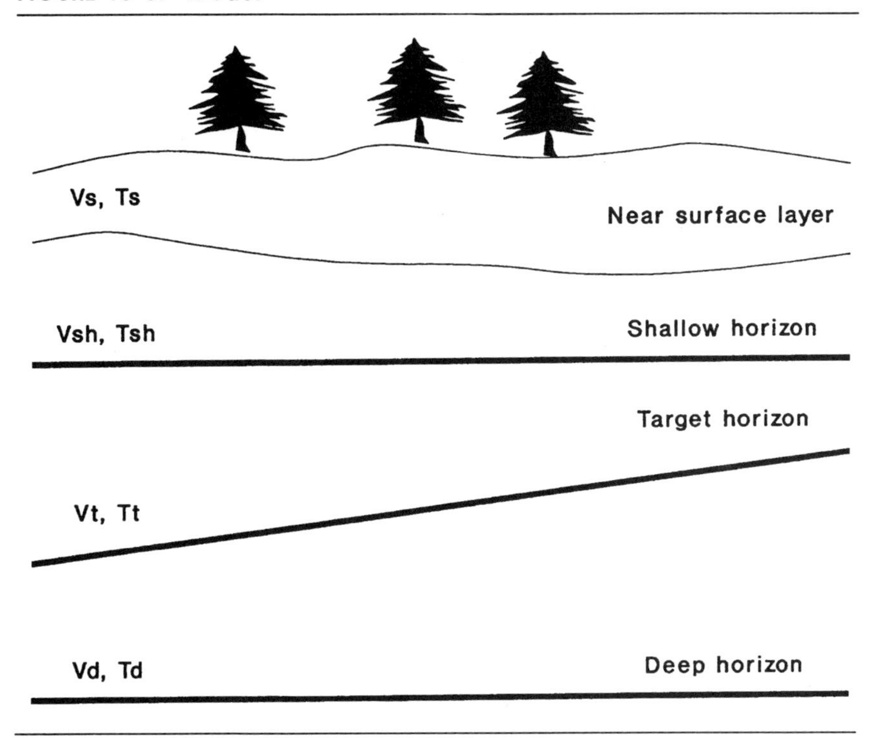

FIELD TERMINOLOGY

Figure 10–4 shows a 2-D field layout and the terminology that is used in this chapter to calculate each parameter.

2-D CALCULATIONS

The parameters we need to calculate are:

- Near and far offset (NO, FO).
- Vertical and horizontal resolution (dT & Gi).
- Shot and receivers interval (Si & Ri).
- Fold.

For the calculations, we need to know:

- Maximum frequency in the area.
- Frequency range of the targets.
- Noise patterns in the area.
- Model to image the target (velocities, depth, dips).

FIGURE 10–4. Field terminology

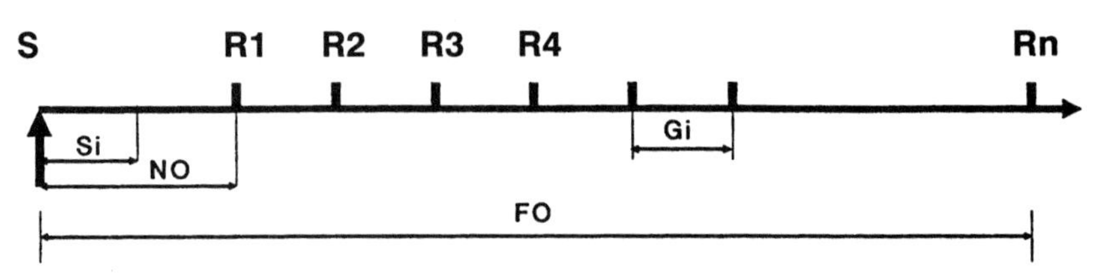

S: Source or Shot.
R1, R2, R3,....Rn: Receivers
Si: Source interval
Gi: Group interval
NO: Near offset
FO: Far offset
n: Number of channel

1. Design of Group Interval
   In designing the group interval, we must keep in mind the lateral extent of the target zone, its dip and velocity, and how it will be sampled to avoid migration and spatial aliasing.
2. Design of 2-D Near Offset
   Near offset design is a function of the shallowest horizon to be recorded, quality of data recorded on the near traces (normally noisy when they are very close to the source), velocity of the area, and previous experience in the area.
3. Design of 2-D Far Offset
   FO(max) is a function of the target depth in unit distance, target velocity in unit distance/sec, and near-surface velocity in unit distance per sec. In the case of a dipping event, a dip factor is added.
4. 2-D Geophone Pattern
   It is essential to design a geophone pattern or array that will attenuate undesirable signals and at the same time optimize signals from the target horizons. The design must take into account the noise frequency, depth of deepest and shallowest targets, and other factors.

3-D SURVEY DESIGN

Many basics of 3-D design are the same as in 2-D survey design—both have a target. To design:

- Vertical and horizontal resolution.
- Near and far offset.

- Surface energy power, filters, geophone arrays.
- Shot interval and fold coverage.

We need to know:

- Depth of target.
- Velocities, possible frequencies.
- Noise patterns in the area.
- Dips.

COMPUTATION OF VERTICAL RESOLUTION

Vertical resolution is defined as how closely two seismic events can be positioned vertically, yet be identified as two separate events.

Vertical resolution, or *tuning thickness,* is usually considered to be equal to one-fourth the dominant wavelength, which is a function of dominant frequency and velocity of the target horizon.

$$\text{Vertical resolution} = 0.25\left(V/F_0\right)$$

If the signal-to-noise ratio is greater than 1:1, vertical resolution is improved by improving the bandwidth. If the signal/noise ratio is less than 1:1, enhancing the ratio results in better vertical resolution.

COMPUTATION OF LATERAL RESOLUTION

Lateral (or horizontal) *resolution* refers to how closely two reflecting points can be situated laterally, yet be recognized as two separate points. It is a function of vertical wavelength λ_V and migration angle θ of the dipping horizon.

Figure 10–5 explains 2-D lateral resolution and the Fresnel zone. The reflecting portion AA′ is called the *half-wavelength Fresnel zone* or *first Fresnel zone.* A′O is the radius of the Fresnel zone, and it is a function of target depth, velocity, frequency, and wavelength.

MIGRATION APERTURE

As we have discussed, an unmigrated seismic section shows an anticlinal feature to be larger than it actually is. Moreover, the steeper the dip angle, the greater the exaggeration. It follows, then, that a 3-D survey must cover a larger area than the actual size of the feature. This area is called the *migration aperture.*

Figure 10–6a shows a depth profile that represents a surface model with dipping reflector CD buried in a homogeneous medium. Zero-offset modeling using normal incidence may yield Figure 10–6b. Although it is not shown in this figure, the time section should show diffraction off the edges of the dipping reflecting segment. Migration moves event C′D′ updip to its true position CD, which is overlaid in time section for comparison.

FIGURE 10–5. 2-D lateral resolution—fresnel zone

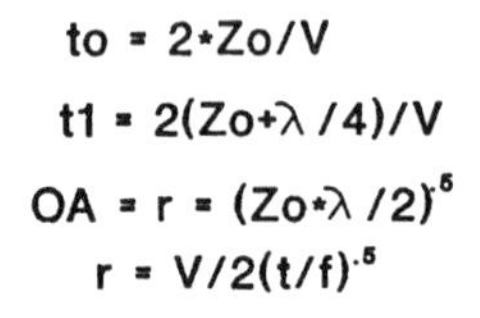

S X Zo + λ/4 Zo A O A' Z

WHERE:

Zo = DEPTH IN UNIT DISTANCE at time to

Z = (Zo+ λ/4) AT TIME t1, IN UNIT DISTANCE.

λ= IS WAVELENGTH IN UNIT DISTANCE.

V = VELOCITY OF THE REFLECTOR IN UNIT DISTANCE/SEC.

f = FREQUENCY OF REFLECTOR IN HZ.

LATERAL RESOLUTION REFERS TO HOW CLOSE TWO REFLECTING POINTS CAN BE SITUATED HORIZONTALLY, YET CAN BE RECOGNIZED AS TWO SEPARATE POINTS RATHER THAN ONE.

THE REFLECTING PORTION AA' IS CALLED THE HALF-WAVELENGTH FRESNEL ZONE (HILTERMAN,1982), OR FIRST FRESNEL ZONE (SHERIFF,1984)

FIGURE 10–6. Migration aperture (After Chun and Jacewitz, 1981. Courtesy Seismograph Service Corporation)

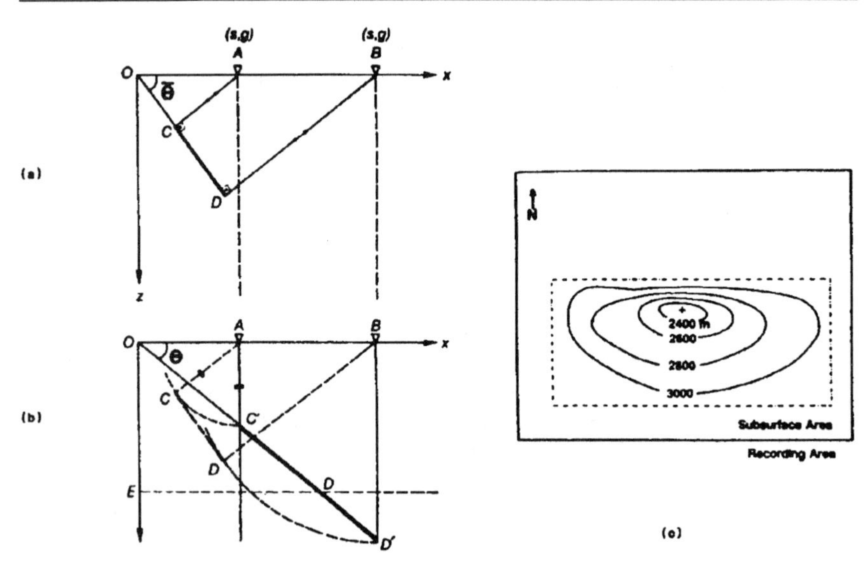

THE REFLECTION SEGMENT C'D' IN TIME SECTION (b), WHEN MIGRATED, IS MOVED UPDIP, STEEPENED, SHORTENED, AND MAPPED ONTO ITS TRUE SUBSURFACE LOCATION TO ACCOUNT FOR MIGRATION APERTURE. THE SIZE OF A 3-D SURVEY OVER A SUBSURFACE STRUCTURE NORMALLY IS GREATER THAN THAT OF THE STRUCTURE.

The horizontal extent of the zone of interest is the distance OA; if the line length had been limited to OA during recording, then the time section would be blank of the event. If, on the other hand, the recording had been limited to AB, then C′D′ would be absent from the migrated section. Although the target is confined to OA, the time section must be recorded over the longer segment OB.

The line must be long enough to include diffractions that may be present in the data. Additionally, the recording time must be long enough to include diffraction tails and all dipping events of interest.

The spatial (horizontal) and temporal (vertical) displacements of a point on a dipping event, resulting from migration, depend upon the velocities, depths, and dips of the events.

The same considerations apply to 3-D surveys. Figure 10–6c is a depth contour map of a structural high. The subsurface extent of the objective portion is indicated in the smaller, dashed rectangle. The actual survey size needed to define the target is outlined by the outer rectangle. The northern flank of the structure is the steepest part, so the survey area must be extended mostly in this direction. Extension in other directions can be determined accordingly. A typical 3-D survey of a subsurface anomaly of 3 × 3 kilometers may require a surface grid measuring 9 × 9 kilometers.

Migration aperture is a function of time (T) to the event, velocity (V), and dip angle (θ):

$$MA = TV^2\theta/4000 \quad \text{(for } \theta \text{ in ms/unit distance)}$$
$$= \tfrac{1}{2}TV\sin\theta \quad \text{(for } \theta \text{ in degrees)}$$

If a structure has dip in more than one direction, the aperture must be calculated for each dip. In the example of Figure 10–6c, four different calculations were necessary.

GROUP INTERVAL

The design of the group interval is based on the dip, depth, velocity, and frequency of the target horizon. Experience in the area and familiarity with the local noise patterns are helpful in designing the group interval.

The *ideal group interval* is a function of the average velocity, maximum expected frequency, and dip angle of the target horizon in the *direction of the survey.* Dip angle can be used in the computations in either ms/unit distance or in degrees.

$$GI = 1000/(3 \times f_{max} \times \theta) \quad \text{(dip angle } \theta \text{ in ms/unit distance)}$$
$$GI = V/(6 \times f_{max} \times \sin\theta) \quad \text{(dip angle } \theta \text{ in degrees)}$$

where GI is in unit distance
V is in unit distance/sec
f_{max} is in Hertz

For a line along the true strike direction, θ is equal to zero.

The *maximum group interval* may be more economical in the field layout than the ideal one. As long as the calculated maximum group interval will attenuate the undesirable signal and give good signal-to-noise ratio, it is the one to use in the 3-D survey.

The maximum group interval is a function of the same parameters as the ideal group interval, but it is longer (coarser) than the ideal one.

$$GI = 1000/(2 \times f_{max} \times \theta) \quad \text{(dip angle in ms/unit distance)}$$

$$GI = V/(4 \times f_{max} \times \sin\theta) \quad \text{(dip angle in degrees)}$$

3-D GEOPHONE ARRAYS

Just as in 2-D surveys, geophones can be arranged in arrays of different numbers and weighting, and in different configurations such as in-line, L-shape, cross-shape, bunched, star, square, and windmill.

Swath-type shooting results in directional effects on the array pattern. If there is no ground roll in the area, the bunched array is probably the best and most economical geophone array to use.

Figure 10–7 illustrates the different types of geophone arrays that may be used in a 3-D seismic survey.

FIGURE 10–7. Geophone arrays for 3-D

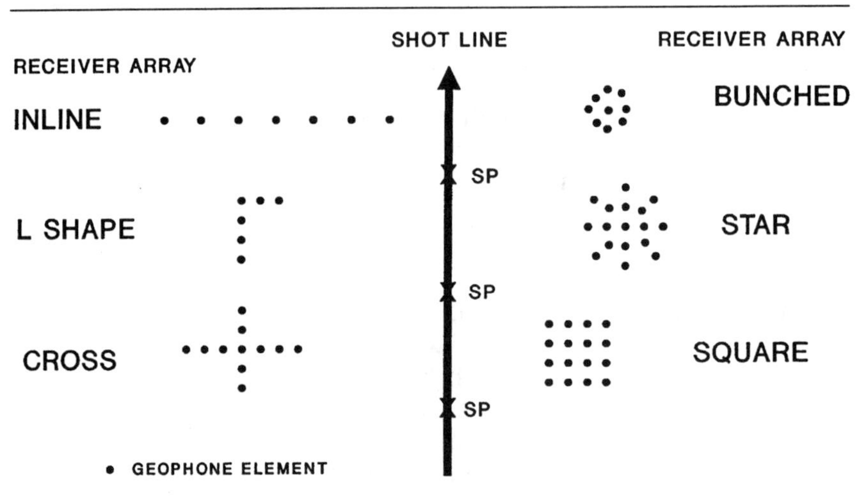

3-D HORIZONTAL RESOLUTION

3-D migration reduces the Fresnel zone in both the X and Y directions, increasing the horizontal resolution. Figure 10–8 illustrates the effects of 2-D and 3-D migration on a Fresnel zone.

If the group interval is taken to be one-half of the radius of the Fresnel zone, the calculation will give a larger value of group interval.

$$G_i = V / 4\sin\alpha (t/f)^{0.5}$$

Given the following parameters:

D = 2,000 m
t = 2 sec (one-way time)
f = 60 Hz
α = 45° (dip angle)
$V = D/t$ = 1,000 m/sec.
G_i = 1,000/(4 × .7071) × (2/60)$^{.5}$ = 64 meters

Trace interval is one-half the group interval, or 32 meters.

Trace spacing from the 3-D calculation is 24 meters, corresponding to a 48 m group interval. If a 64 m interval will give the same resolution as a 48 m interval, it is more economical in time and money to use the longer interval.

FIGURE 10–8. Effects of migration on fresnel zone

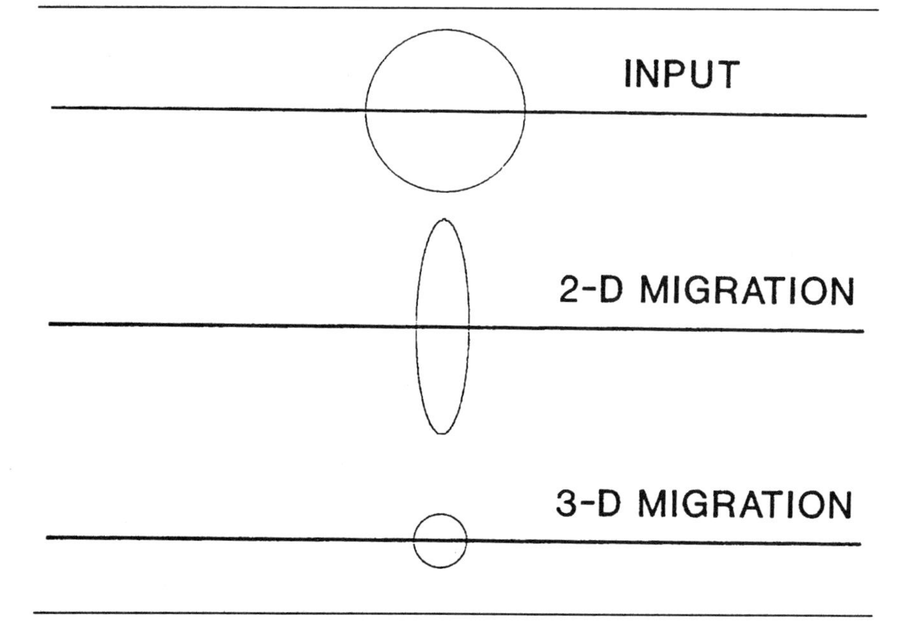

SWATH SHOOTING

3-D data acquisition on land is commonly carried out by swath shooting, in which receiver cables are laid out in parallel lines (in-line direction) and shots are positioned in the perpendicular direction (cross-line direction). As one shot is completed, the receiver cables are moved along the swath a number of stations (equivalent to the shot-line spacing), and the shooting is repeated. After one swath is completed, another swath parallel to it is recorded. This procedure is repeated over the entire area of the survey. Figure 10–9 shows a swath shooting layout with five receiver cables and a cross-line of shot points.

The swath-shooting method yields a wide range of source-receiver azimuths. (The source-receiver azimuth is the angle between a reference line such as dip line or receiver line, and the line that passes through the source and receiver stations).

FIGURE 10–9. Swath-shooting layout

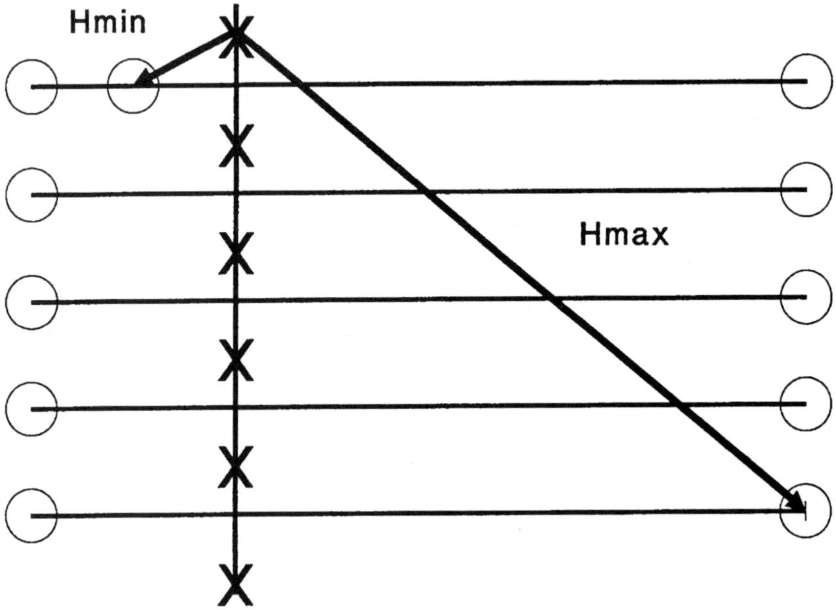

Hmin

Hmax

Hmin: DISTANCE FROM THE SOURCE TO THE NEAREST GEOPHONE ARRAY
Hmax: DISTANCE FROM THE SOURCE TO THE FARTHEST GEOPHONE ARRAY

ROLLALONG IN LONG DIRECTION OR COMPLETE
LINE ROLL IN OTHER DIRECTION

Swath shooting is economical. Better velocity analyses are obtained in data processing, since velocity analysis is reduced to one direction. Some advantages of swath shooting, if there are no environmental restrictions which will cause skips, are a uniform fold and good offsets. This type of shooting is best suited to simple structures and allows easy rectangular bins.

3-D Swath Design

3-D swath design is operationally simple over a large survey area, and it usually requires multiline equipment and a high number of channels. Each line of the swath is the same dimensions as in a 2-D spread. Thus, a five-line swath needs about five times the channel capacity of the conventional 2-D line.

Sources and receivers are one-directional. This is very important in calculating the angles and determining the offset to get better velocity analysis.

Bins

A *bin* consists of cells with dimensions one-half the receiver group spacing in the in-line direction and nominal line spacing in the cross-line direction (same as CDP in 2-D data processing). Traces that fall within a cell make up a common-cell gather. In marine 3-D surveys, not all of these traces are from the same shot line because of cable feathering.

Sorting of the data into cells is called *binning*. Bin size is a function of the velocity of the target, maximum frequency of the area, and dip angle of the target horizon. Figure 10–10 illustrates this relationship.

Figure 10–11 illustrates the attribute analysis for each bin, which are:

- Subsurface: target imaging.
- Fold: its effect on S/N ratio and attenuating multiples.
- Azimuth: which will effect the velocity analysis and give better imaging for the dipping structure.
- Offset range: will affect the velocity analysis to get the best stacking velocities and be able to get enough dynamic correction discrimination to attenuate multiples.

Full Range 3-D

A full azimuth range 3-D is well suited to small surveys, where source and receiver azimuths are multidirectional. Since the number of rollalong station moves is large, this type of shooting is especially wellsuited for complex structures. Because of the great volume of data collected, the processing cost is fairly high.

FIGURE 10–10. Bins

SAMPLING BASIC UNIT

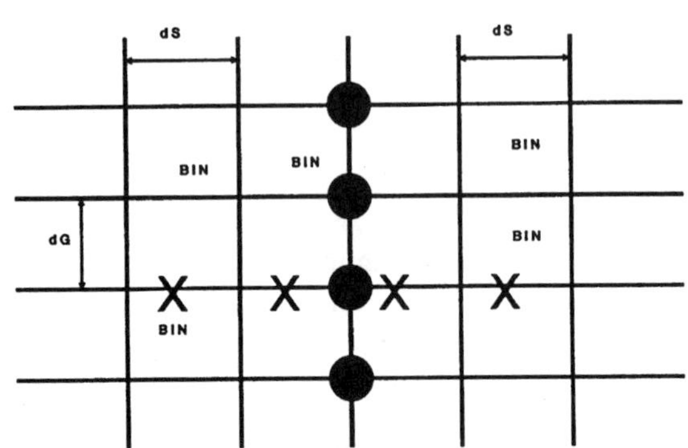

dG = GEOPHONE ARRAY INTERVAL

dS = SHOT INTERVAL

$$B = V/(2 * Fm * \sin \theta)$$ WHERE:

B = BIN SIZE
V = VELOCITY OF THE TARGET
Fm = MAXIMUM FREQUENCY
θ = DIP ANGLE OF THE TARGET

FIGURE 10–11. Attribute analysis for each bin

SUBSURFACE	FOLD
AZIMUTH	OFFSET RANGE

- SUBSURFACE - TARGET
- FOLD - S/N, MULTIPLES
- AZIMUTHS - STRUCTURE, DIP
- OFFSET-VELOCITIES, MULTIPLES

Design of Maximum Offset

The design of maximum offset, for both the nondipping and dipping case, is discussed in detail in Appendix B.

3-D DATA PROCESSING

Land Data Processing

Almost all concepts of 2-D seismic data processing apply to 3-D data processing, although some complications may arise in 3-D geometry, quality control, statics, velocity analysis, and migration. Editing of very noisy traces, spherical divergence correction for amplitude decay with depth and offset, deconvolution, trace balancing, and elevation statics are done in the preprocessing stage before collecting traces in common cell gathers. Sorting in common cell gathers introduces a problem if there are dipping events, because there are azimuthal variations in the NMO application within the cell.

Marine Data Processing

After preprocessing, the data is ready for common-cell sorting, and a grid is superimposed on the survey area. This grid consists of cells with dimensions one-half the receiver group spacing in the in-line direction and nominal line spacing in the cross-line direction. Traces that fall within the cell make up a common cell gather. Due to cable feathering, not all these traces are from the same shot line. The same data processing techniques in 2-D marine apply to 3-D marine data.

APPLICATIONS OF 3-D SEISMIC DATA

Enhanced Oil Recovery

A number of techniques will allow mapping of EOR processes. One of the most familiar techniques uses 3-D seismic, borehole seismic, and microseismic. Defining the EOR front depends on the change of density of the reservoir rock as a result of the process. Commonly used techniques that can induce substantial density changes in the rock include steam flooding and in-situ combustion. The particular seismic method used depends on the goal to be accomplished. For example, a complete description of the developing front can be obtained by using 3-D seismic mapping, while limited knowledge can be obtained by using conventional 2-D seismic along the line.

Several considerations are involved in field application of 3-D seismic mapping in enhanced oil-recovery projects using steam injection.

Working with this technique is expensive in terms of both fieldwork and data reduction. The process may have to be shut in during tests. A large number of stations and high-frequency sources are required to obtain the necessary resolution. The large amount of data and special processing require longer time.

The surveys do, however, provide a clearer and more accurate solution, since time-slice maps and sections can be obtained in any direction through the 3-D data sets. From this technique one can monitor the changes and map the swept zone.

With this information an engineer can optimize production through better control, foresee developing problems, and be able to take the proper corrective action in a timely fashion.

Figure 10–12 illustrates the 3-D seismic survey as it was used to map the steam flood project at Street Ranch Test. The recording pattern consisted of four seismic survey lines, each of which ran through the center injection well of an inverted five-spot pattern.

In order to facilitate the seismic survey, the pilot was shut in for 3 days. The formation was 460 m (1,500 feet) in depth. The pay zone was 16 m (50 feet) thick, although steam was injected only in the upper 8 m (25 feet) of the zone.

A VSP survey was conducted to help determine the reflection characteristics of the target formation. A high-frequency energy source was used to obtain high resolution of the EOR swept front.

In Figure 10–12b there are four slices taken at lines 3, 1, 4 and 2, which were taken around the target formation. The changes in reflection seismic data were due to changes in reservoir impedance caused by an increase in gas saturation.

The seismic sections show definite changes in wavelet shape over the center of the injection well and extend partially to the production wells. In the zone where steam was injected, a second peak developed on top of the peak of the pay.

The most useful approach is to make a series of surveys at time intervals that would allow the development of the EOR process to be established. This requires surveys at three-month to six-month intervals.

One of the problems of the resolution can be amplified by the heterogenous nature of the reservoir geology, especially with the surface techniques of conventional 3-D seismic survey.

Caution must be exercised in establishing the geometry of the observation survey layout. It is possible to miss critical developments by improper placement of source and receiver.

In addition, if the response measured, such as wave characteristics, is not related in a linear manner to the EOR process, it is possible to misinterpret the data. It is advisable to make laboratory measurements that will aid in the interpretation of the field measurements.

FIGURE 10–12. 3-D seismic mapping of steam flood—Street Ranch Pilot Test (copyright © 1983, Society of Petroleum Engineers, from Britton et al., "The Street Ranch Pilot Test of Fracture-Assisted Steamflood Technology," *Journal of Petroleum Technology* (March 1983))

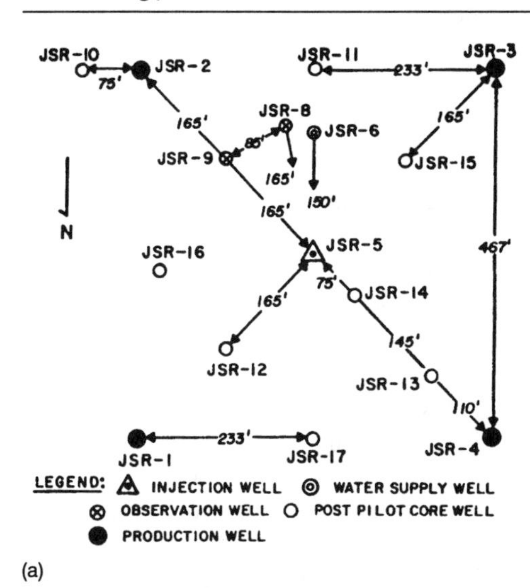

(a)

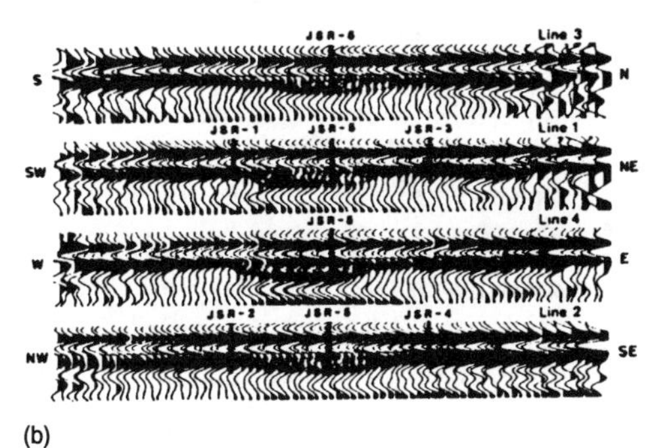

(b)

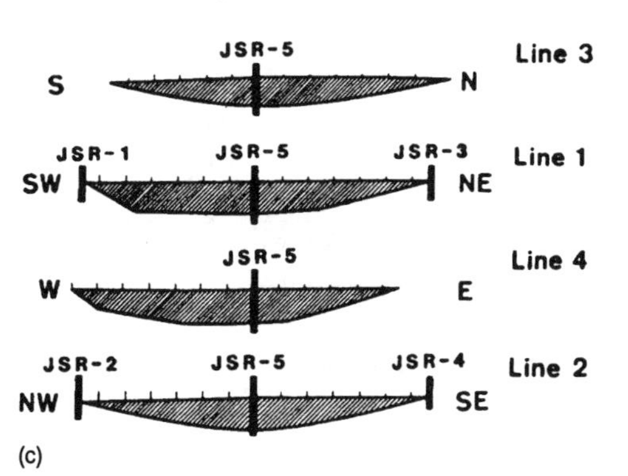

(c)

IN ORDER TO FACILITATE THE SEISMIC SURVEY, THE PILOT WAS SHUT IN FOR 3 DAYS. THE RECORDING PATTERN CONSISTED OF FOUR LINES, EACH RAN THROUGH THE INJECTION WELL OF AN INVERTED FIVE SPOT PATTERN. THE FORMATION WAS AT 460 M (1,500 FT) DEPTH 16 M (50 FT) PAY ZONE. STEAM WAS INJECTED IN THE UPPER 8 M (26 FT) PORTION OF THE ZONE. VSP WAS RECORED TO HELP LOCATE THE TARGET FORMATION REFLECTION CHARACTERISTICS. HIGH FREQUENCY SURFACE SOURCE WAS NEEDED TO OBTAIN GOOD RESOLUTION OF THE FEATURE OF EOR FRONT.

THE CHANGES IN REFLECTION SEISMIC DATA WERE PROBABLY DUE TO CHANGES IN RESERVOIR IMPEDANCE CAUSED BY AN INCREASE IN GAS SATURATION. THE SEISMIC SECTIONS SHOW DEFINITE CHANGE IN THE WAVELET SHAPE OVER THE CENTER OF THE INJECTION WELL AND EXTEND PARTIALLY OUT TO THE PRODUCTION WELLS. IN THE ZONE WHERE STEAM WAS INJECTED A SECOND PEAK DEVELOPED ON THE TOP OF THE PEAK OF PAY.

HIGH-RESOLUTION 3-D IMAGING

A high-resolution seismic survey was conducted at Amoco's Gregoise Lake In-situ Steam Pilot (GLISP) located in northeastern Alberta, Canada. This field is in a channel sand overlying an unconformity.

A base survey was run to use as a basis of comparison with future surveys of a similar type, in order to monitor the progress of in-situ heat movement.

1. The Target
   Specialized data acquisition, processing, and display techniques were used to manipulate the 3-D data to image the Devonian unconformity surface at a depth of 240 m (790 feet). The results demonstrate that exceptional lateral and vertical resolution (necessary for accurately mapping fronts) have been achieved.
2. The Test
   The well configuration for the pilot site was to consist of a central injector surrounded by three equidistant producers separated by only 80 meters. Three infill observation wells were to be drilled. Only two wells, H-3 injector and HO-7 observation hole, will be handled in this example. Figure 10–13 shows the location of the wells and the 3-D seismic grid. Note how finely the surface was sampled (CDP grid 4 × 4 m).

FIGURE 10–13. Plat showing well locations and 3-D grid (Pullin & Matthews, 1987, courtesy of Society of Exploration Geophysicists)

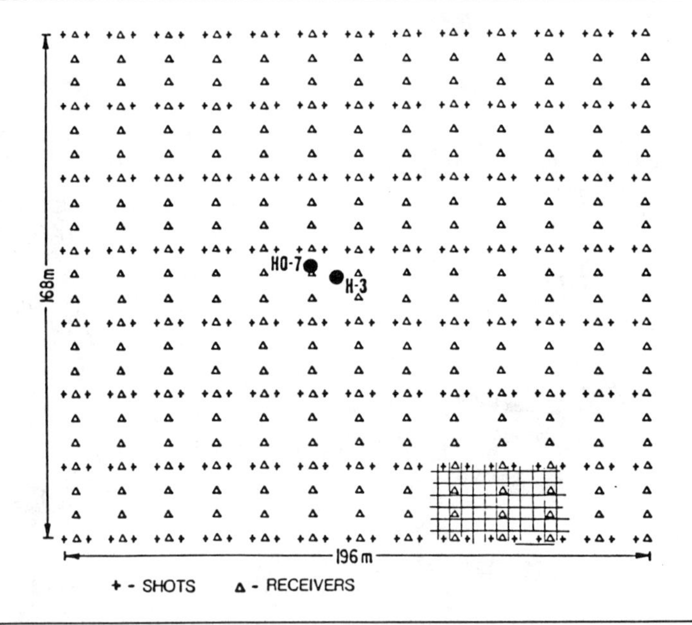

3. Geology
   Figure 10–14 illustrates the geology of the test site. It is characterized by a 50-meter section of discontinuous McMurry tar sands resting on a Devonian unconformity approximately 240 m below surface.

   The first well drilled was H-3, which encountered a 4 m sand aquifer directly above the erosional surface. It was assumed that this wet zone would influence the performance of the steam-based recovery process.

   Later drilling of HO-7 indicated this aquifer was absent, raising fears of a potential problem with the proposed heating method. So it was very important to accurately map the unconformity, not only from the standpoint of EOR system design, but also to prove the resolving power of seismic for such conditions.
4. Noise Test and Field Work
   Before conducting the survey, all data available were studied. It was determined that a field noise test must be conducted to establish the optimum data acquisition parameters. These parameters would produce a data record with high signal-to-noise ratio over a broad band of high frequencies.

   Figure 10–15 shows a display of the field test to select the geophone depth and the depth and size of the dynamite shot. Figure 10–15a shows a comparison between surface-planted geophones and subsur-

FIGURE 10–14. Geology of the street ranch test site (Pullin & Matthews, 1987, courtesy of Society of Exploration Geophysicists)

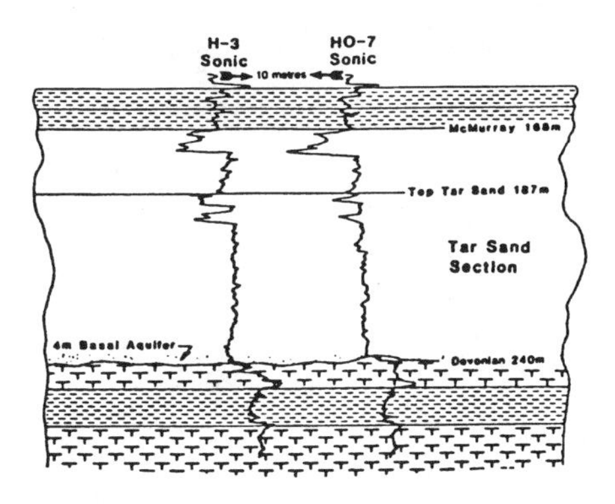

GEOLOGY OF THE GLISP SITE IS CHARACTERIZED BY A 50-METER SECTION OF DISCONTINUOUS MCMURRY TAR SANDS RESTING ON A DEVONIAN UNCONFORMITY APPROXIMATELY 240 M BELOW SURFACE. WELL H-3 ENCOUNTERED A 4-M SAND AQUIFER DIRECTLY ABOVE THE EROSIONAL SURFACE. IT WAS ASSUMED THAT THIS BASAL WET ZONE WOULD INFLUENCE THE PERFORMANCE OF THE STEAM-BASED RECOVERY PROCESS. LATER DRILLING OF HO-7 INDICATED THIS AQUIFER WAS ABSENT, LEADING TO A POTENTIAL PROBLEM WITH THE PROPOSED HEATING METHOD. ACCURATE MAPPING OF THE UNCONFORMITY WAS DESIRABLE FROM THE STANDPOINT OF IN-SITU DESIGN. IT ALSO PROVED THE SEISMIC RESOLVING POWER.

FIGURE 10–15. Results of tests to determine effects of geophone depth and charge size (Pullin & Matthews, 1987, courtesy of Society of Exploration Geophysicists)

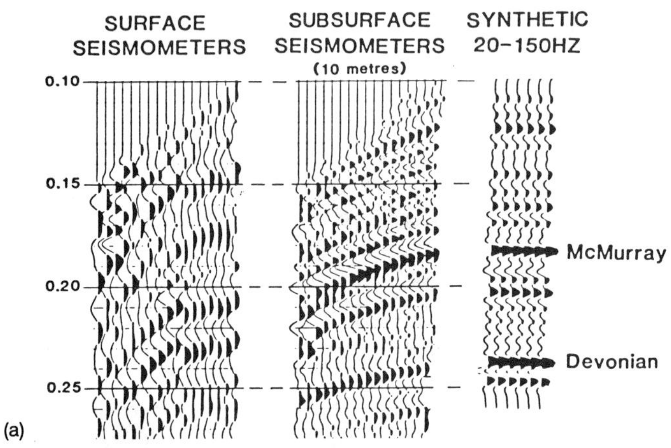

COMPARISON OF H-3 SYNTHETIC WITH SURFACE AND SUBSURFACE FIELD PROFILES GENERATED BY A 50-G SOURCE AT 13-M DEPTH.

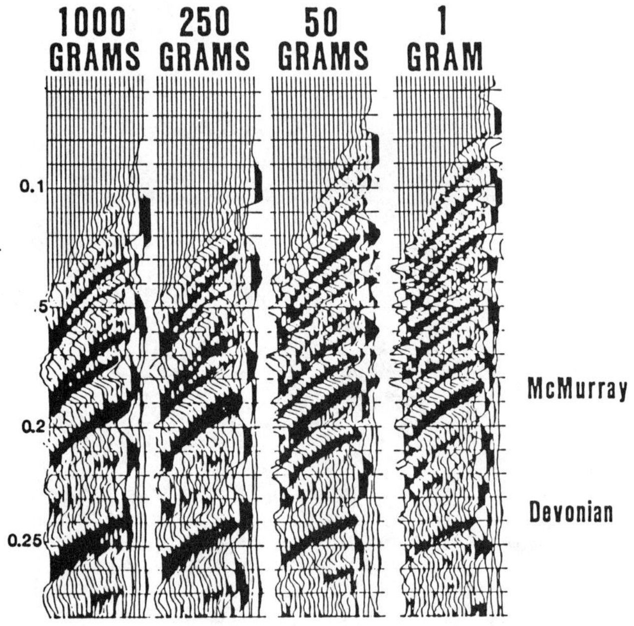

SHOT RECORDS (BURIED SEISMOMETERS) FOR DIFFERENT CHARGE SIZES.

face geophones buried at a depth of 10 meters. At well location H-3, a sonic log was available, and a synthetic seismogram was made to tie lithology to the seismic record. Figure 10–15b shows test results for different charge sizes with buried geophones to select the most economic charge size for the fieldwork.

After investigation of the field tests, a recording geometry was designed consisting of 22 single geophones, buried and cemented at 13 meters below surface and spaced 8 meters apart. Charge size was 18 grams, cemented in a hole 18 meters in depth. The survey area was 196 × 168 meters. CDP cell size was 4 × 4 meters in order to access the shot hole for future monitoring. The geophone holes were cased to total depth with 4-inch PVC tubing.

The geophone leads were protected between surveys by a length of PVC pipe with a protective top cap. A 120-channel recording system was used, with 1 ms sample rate recordings.

3. Processing

The data was processed using sophisticated processing flow, including surface consistent (Q) compensation for attenuation losses, surface consistent deconvolution, and 3-D migration.

The final frequency filter passband chosen for the 3-D data was 25-220 Hz.

Figure 10–16 shows the ability of 3-D migration to improve spatial resolution. Migration has the ability to shrink the Fresnel zone by collapsing the diffraction. The Devonian unconformity reflection on the migrated section demonstrates an apparent channel feature that is not imaged on the unmigrated section. This channel is verified by wells H-3 and HO-7.

Figure 10–17 shows 3-D as a powerful tool to understand microstructure and stratigraphy. An interactive interpretation station can be used to display 3-D output. One effective way to study the erosional surface of an unconformity is with time slices. After the processing is done, it is possible to extract data from this 3-D cube to produce a 2-D vertical profile or a horizontal time slice at any specific time. Figure 10–18 is an example of a time slice. This slice, taken at 224 ms, is shown before and after migration. On the channel overlying the unconformity, the unmigrated slice shows no definite relationship between wells H-3 and HO-7 after 3-D migration. It is clear that HO-7 is off the channel, as documented by drilling.

Other Benefits of 3-D Surveys

3-D migration gives sharper definition than 2-D. It defines fault planes more clearly, gives better definition of deeper faults, better coherency for deeper events, and better signal-to-noise ratio, which help in making better interpretation. At the top left of Figure 10–19 is a 3-D migrated sec-

FIGURE 10–16. Post-stack 3-D migration improves spatial resolution (Pullin & Matthews, 1987, courtesy of Society of Exploration Geophysicists)

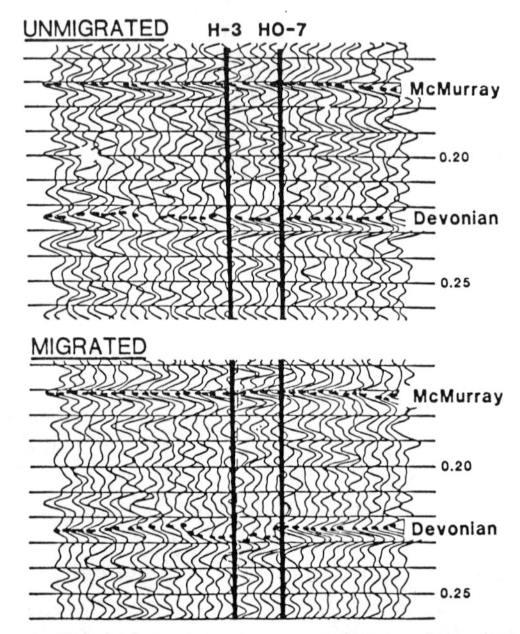

3-D MIGRATION HAS THE ABILITY TO IMPROVE SPATIAL RESOLUTION. MIGRATION HAS THE ABILITY TO CONSIDERABLY SHRINK THE FRESNEL ZONE, BY COLLAPSING THE DIFFRACTIONS. THE DEVONIAN UNCONFORMITY REFLECTION ON MIGRATED SECTION EXHIBITS AN APPARENT CHANNEL FEATURE WHICH IS NOT IMAGED ON THE UNMIGRATED DATA. THE CHANNEL IS VERIFIED BY WELLS H-3 AND HO-7. THIS HIGH RESOLUTION SURVEY WAS CONDUCTED AFTER EXTENSIVE FIELD TESTS WERE DONE TO GET THE BEST PARAMETERS. A VERY SPECIALIZED DATA PROCESSING FLOW WAS USED. THE FINAL FREQUENCY FILTER PASSBAND CHOSEN FOR THE 3-D DATA WAS 25-220 HZ.

tion that shows a better definition of the lateral extent of the closure compared to the 2-D and the final stacked section, which is heavily populated with diffractions. See how the faulting on the migrated time slice is well defined as its shown with the event's terminations.

3-D Land Surveys

3-D statics programs use the leverage available from high multiplicity inherented in the data volume to optimize the near-surface control with better refraction and residual static applied in a surface consistent manner. Good static solutions will allow better stacking velocity analyses; hence, they allow better stacking and coherency and improve the signal-to-noise ratio. Figure 10–20 illustrates a comparison between 2-D and 3-D residual static sections.

FIGURE 10–17. 3-D images are powerful tools for microstructure and stratigraphy (Pullin & Matthews, 1987, courtesy of Society of Exploration Geophysicists)

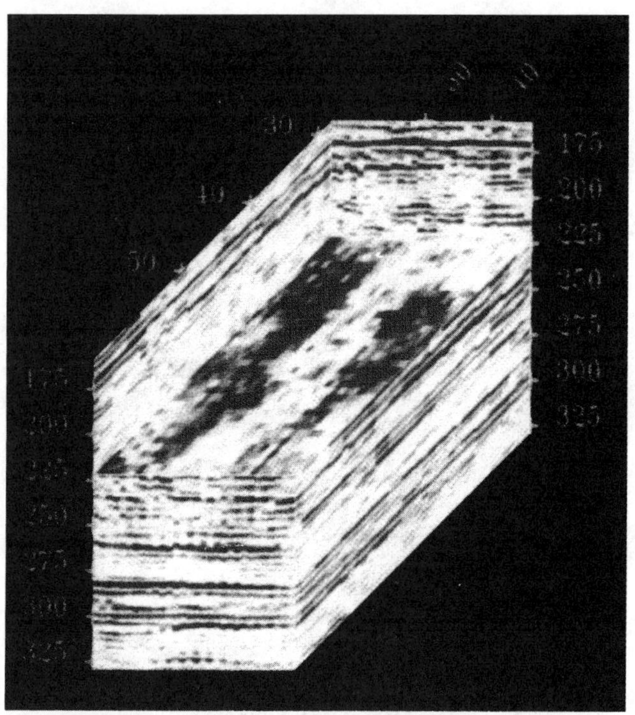

"CHAIR" DISPLAY OF A PORTION OF THE MIGRATED 3-D DATA. THE "SEAT" IS AT THE DEVONIAN.

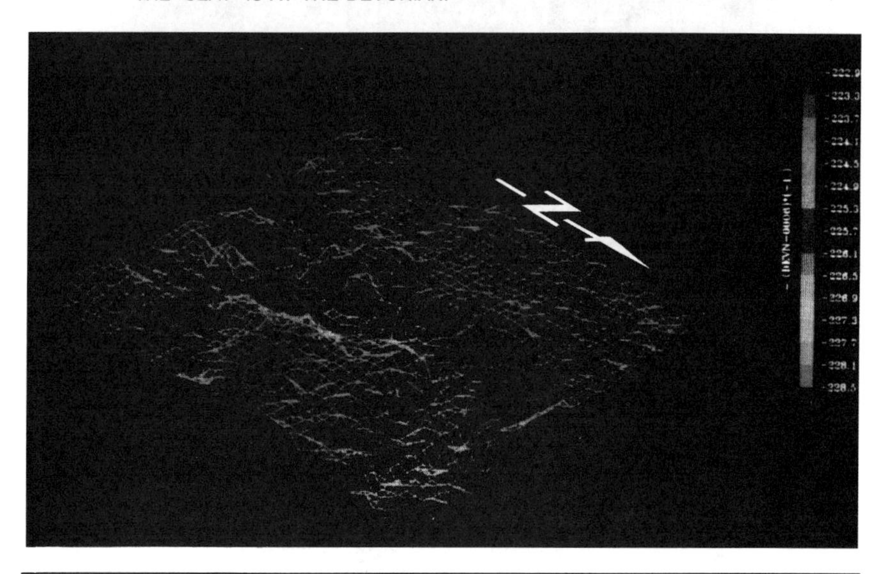

FIGURE 10–18. Time slice before and after migration (Pullin & Matthews, 1987, courtesy of Society of Exploration Geophysicists)

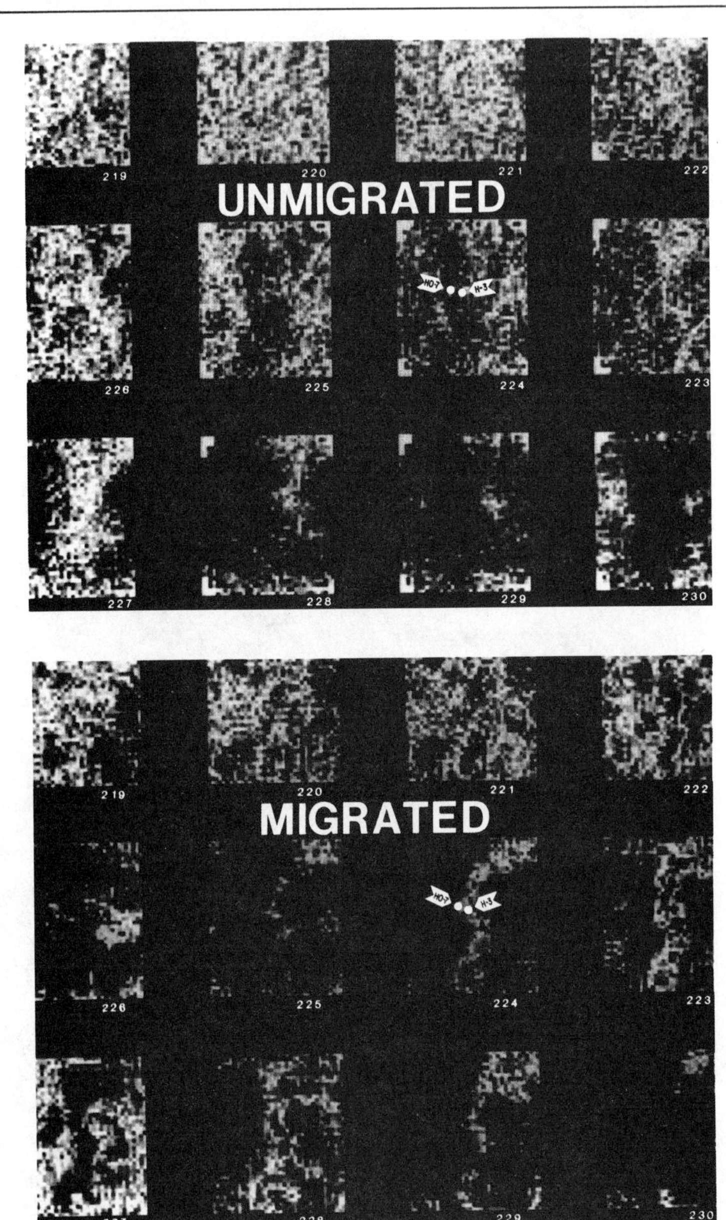

FIGURE 10–19. Unmigrated and migrated 3-D displays (Brown, 1991, reprinted by permission of the American Association of Petroleum Geologists)

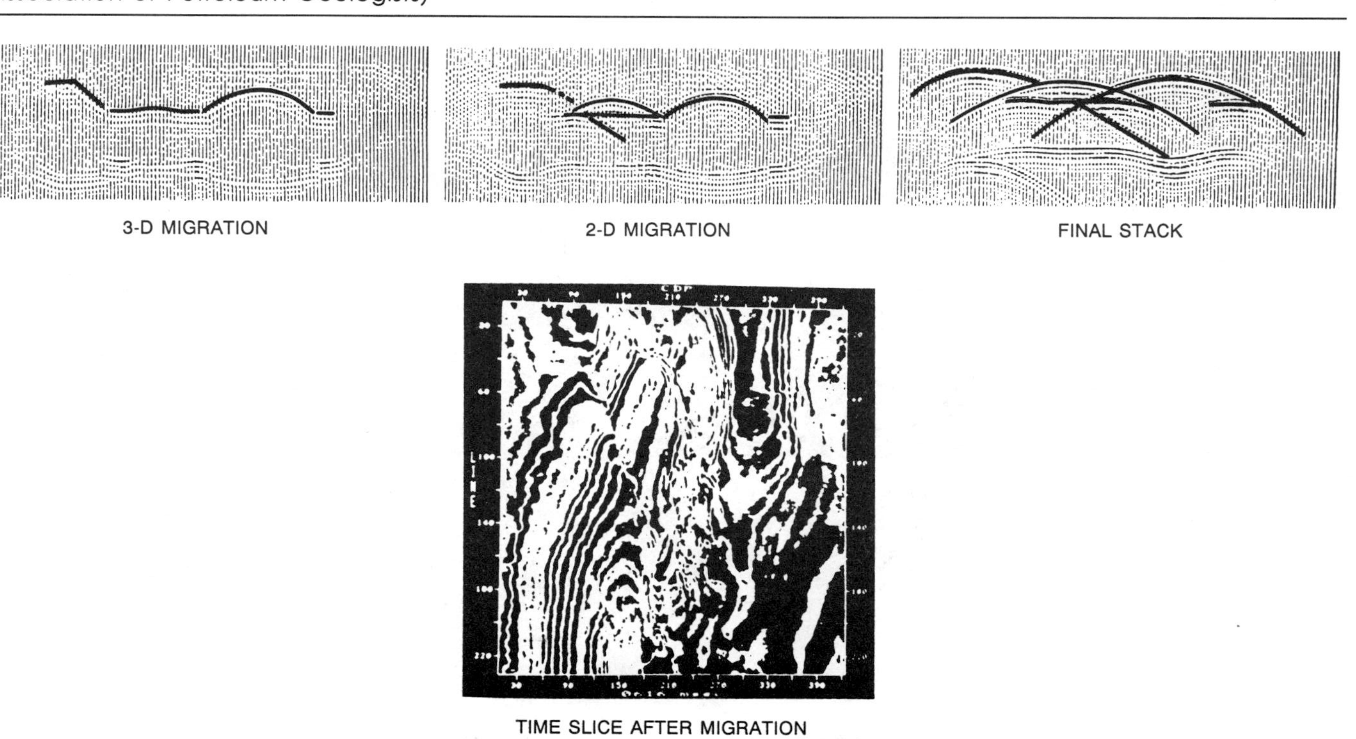

FIGURE 10–20. 3-D land surveys (courtesy Halliburton Geophysical Services)

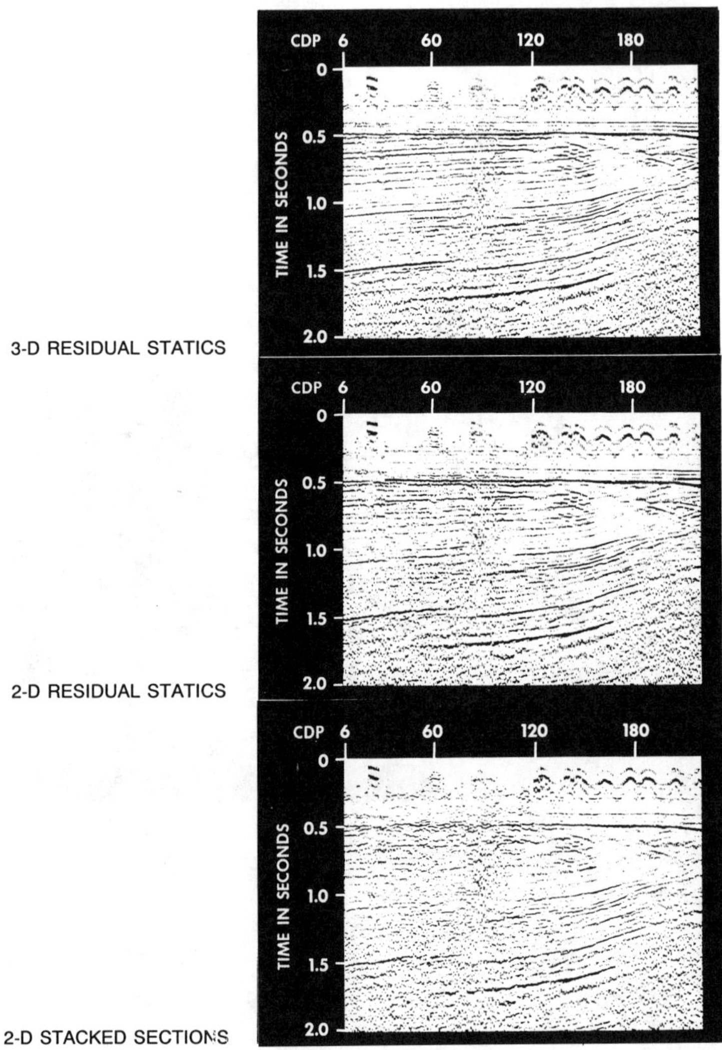

3-D STATICS PROGRAMS USE THE LEVERAGE AVAILABLE FROM HIGH MULTIPLICITY IN THE DATA VOLUME TO OPTIMIZE THE NEAR-SURFACE CONTORL. THE ABOVE SECTIONS SHOW GREAT IMPROVEMENTS ON THE 3-D RESIDUAL STATICS STACK SECTION. THE 3-D STACK SECTION SHOWS BETTER CONTINUITY, COHERENCY AND IMPROVED SIGNAL-TO-NOISE RATIO. THIS IS DUE TO THE USE OF MULTIPLICITY CONTAINED IN THE 3-D DATA VOLUME. EXPECT BETTER STACKING VELOCITIES.

FIGURE 10–21. 3-D marine surveys (Brown, 1991, reprinted by permission of the American Association of Petroleum Geologists)

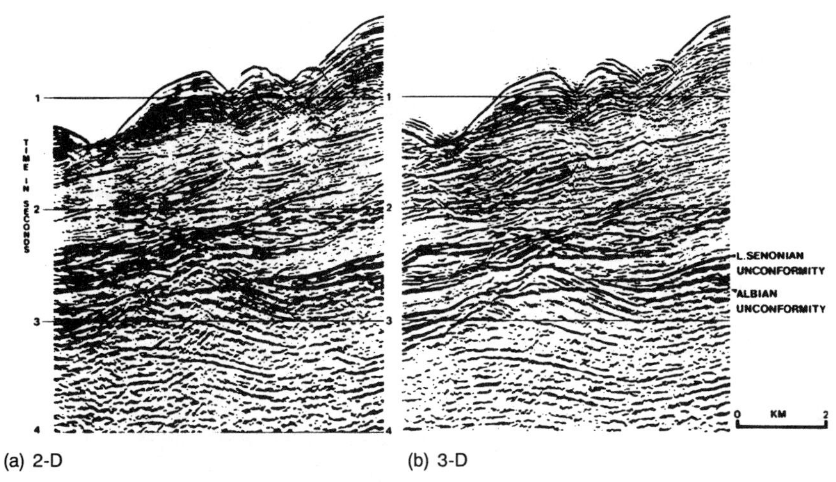

(a) 2-D (b) 3-D

SECTION (b) FROM 3-D MARINE DATA, USING A CLOSELY-SPACED SET OF SUBSURFACE LINES AND DUAL SOURCE AND/OR STREAMER. THE MAIN BENEFITS SHOWN BETWEEN 2-D AND 3-D USING CROSS LINE MIGRATION FROM 3-D IN-LINE SECTIONS ARE:

1. BETTER DEFINITION ON 3-D SECTION NEAR THE EROSIONAL HIGH OF ALBIAN UNCONFORMITY AND FLUID CONTACT (FLAT SPOT)
2. A GREAT IMPROVEMENT IN THE SIGNAL-TO-NOISE RATIO ON THE 3-D MIGRATED SECTION.

3-D MARINE SURVEYS

Figure 10–21 shows the difference between 2-D and 3-D migration of a marine seismic line. Observe the definition on the 3-D section near the erosional high of Albian unconformity and fluid contact (flat spot) zones. It is also clear that 3-D migration has better signal-to-noise ratio.

SUMMARY AND DISCUSSION

3-D seismic data provide more details than 2-D data, because of the dense grid of lines. It helps to develop a more accurate and complete structural and stratigraphic interpretation.

In the past, the technique has been used for development purposes on existing fields. Today, it is also used for exploration in highly complicated structure and stratigraphic areas.

3-D surveys helped find additional reserves by defining the reservoir boundary and providing data for more effective drainage. They also helped in defining salt structures as well as meandering channels.

In the enhanced oil-recovery (EOR) projects, 3-D surveys proved their success, especially in steam flooding. There are many applications for 3-D surveys, and it is widely used because of its cost effectiveness, reasonable turnaround time, and reliable processing software applications.

The cost of a 3-D seismic survey of a particular area depends on its size, geological setting, depth of target and grid configuration.

In the U.S. Gulf Coast, for a medium size 3-D seismic survey (say 3 × 5 miles for a total of 15 square miles) the price range is $15,000–$20,000 per square mile for a typical grid of 110 feet bins, 24 fold.

Data processing cost varies according to the objective of the survey and problems encountered. The average cost is about $1,500 per square mile.

Under normal circumstances, a land field crew can survey one square mile per day. For data processing of a medium-sized survey, six to eight weeks turnaround time is needed for routine processing. It may take more time if problems are encountered.

These cost numbers were quoted by several contractors in the first quarter of 1993.

The price changes from one survey to another. It depends on the availability of the field crews within your time frame.

KEY WORDS

Azimuth
Bins (binning)
Cable feathering
Common cell gather
EOR
Flat spot
Horizontal resolution
Migration aperture
Swath shooting

EXERCISES

1. The radius, R, of the first Fresnel zone is given by:

$$R^2 + Z^2 = \left(Z = \lambda/4\right)^2$$

Where Z is the depth to the reflector and λ is the wavelength. Solving for R gives:

$$R = \left(\lambda Z/2 + \lambda^2/16\right)^{0.5}$$
$$R = \left(\lambda Z/2\right)^{0.5}$$

a. Calculate the radius of the first Fresnel zone for a reflector at depth of 2,000 m, with seismic frequency of 30 Hz and seismic velocity 3,000 m/sec.
b. Using the familiar relationships

$$Z = Vt/2 \text{ and } \lambda = V/f$$

Where t is arrival time, V is the velocity, and f is frequency, The above equation reduces to:

$$R = \left(V/2\right)\left(t/f\right)^{0.5}$$

Calculate the first Fresnel zone radius for a 20 Hz reflection at 2.0 seconds and velocity of 3 km/sec.

BIBLIOGRAPHY

Brown, A. R. "Interpretation of Three-Dimensional Seismic Data. *Memoir* 42 (1988).

Dalley, R. M. et al. "Dip and Azimuth Displays for 3-D Seismic Interpretation." *First Break.* (March 1989): 86–95.

Dunkin, J. W. and F. K. Levin. "Isochrons for a Three-Dimensional System." *Geophysics* 36 (1971): 1099–1137.

Enachescu, M. "Amplitude Interpretation of 3-D Reflection Data." *TLE* 12 (1993): 678–685.

Greaves, R. J. and T. J. Flup. "Three-Dimensional Seismic Monitoring of an Enhanced Oil Recovery Process." *Geophysics* 52 (1987): 1175–1187.

Hilterman, F. J. "Three-Dimensional Seismic Modeling." *Geophysics* 35 (1970): 1020–1037.

Kluesner, D. F. "Champion Field: Role of Three-Dimensional Seismic in Development of a Complex Giant Oilfield." *AAPG Bulletin* 72 (1988): 207.

Nestvold, E. O. "3-D seismic: Is the Promise Fulfilled." *TLE* 11 (1992): 12–19.

Pullin, N., L. Matthews, and K. Hirsche. "Techniques Applied to Obtain Very High Resolution 3-D Seismic Imaging at an Athabasca Tar Sands Thermal Pilot." *TLE* 6 (1987): 10–15.

Reauchle, S. K., T. R. Earr, R. D. Tucker, and M. T. Singleton. "3-D Seismic Data For Field Development: Land Side Field Case Study." *TLE* 10 (1991): 30–35.

Reblin, M. T., G. G. Chapel, S. L. Roche, and C. Keller. "A 3-D Seismic Reflection Survey Over the Dollarhide Field, Andrews County, Texas." *TLE* 10 (1991): 11–16.

Ritchie, W. "Role of the 3D Seismic Technique in Improving Oilfield Economics." *Jour. Petr. Tech.* (July 1986): 777–786.

Robertson, J. D. "Reservoir Management Using 3D Seismic Data." *Jour. Petr. Tech.* (1989): 663–667.

Walton, G. G. "Three-Dimensional Seismic Method." *Geophysics* 37 (1972): 417–430.

CHAPTER 11

TOMOGRAPHY

INTRODUCTION

The word *tomography* is derived from the Greek words *tomos* (section) and *graphy* (drawing). Sectional viewing of the subsurface is the ultimate aim of seismic surveying. Tomography using radiation and magnetic resonance is successfully used in the medical field to provide detailed images of the interior of the body. The same principles are being applied to imaging the interior of the earth using seismic energy.

Geophysical applications of traveltime tomography are divided into two groups (Lines, L. R. and LaFehr, E. D. 1989). In global seismology, tomographic methods using about 2 million earthquake traveltimes have produced a velocity model for the earth's mantle. In near-surface seismology, where the earth's near-surface region is investigated to a depth of a few kilometers, there is increased interest in the application of tomography to exploration geophysics.

Traditionally, seismic data have been gathered by surface sources and receivers. The next logical step in the effort to improve seismic resolution is to place both source and geophone in the subsurface.

We can make an analogy between seismic tomography and medical tomography. In some cases, the raypath geometry describing the propagation of a seismic wavefront from surface source to subsurface receivers is close to the physical arrangement of radiation, source, and receiver used in medical tomography.

Few publications describing seismic tomography field experiments exist. Bois, et al. (1971, 1972) explained how seismic velocities can be estimated along ray paths between boreholes and demonstrated their results with real field data. Weatherby (1936) was one of the early American geophysicists to propose this technique. Many other geophysicists, such as Butler and Weir (1981), have discussed procedures for recording cross-hole seismic data.

During the past few years, geophysicists have applied seismic tomography to image velocity variations of the earth with great success. This enhances the accuracy of depth conversion, depth migration, and other applications that will be discussed later in this section.

Seismic tomography is a method of inversion, which is a procedure for obtaining models that adequately describe the seismic data and observations, and show the effect of rock properties on seismic wave propagation.

Forward modeling begins with a description of the geology and the seismic parameters for data acquisition, and it predicts the data that should be obtained from an actual experiment, using solutions based on some theoretical model of the physical process.

On the other hand, inversion starts with data and data acquisition parameters and tries to infer a model of the subsurface, which would predict the observed data using the given method of forward modeling.

TYPES OF SEISMIC TOMOGRAPHY

Tomography has been used in two modes: reflection and transmission.

Reflection tomography involves the propagation of seismic waves from the surface to a subsurface reflecting marker and back to the surface. For raytracing computations, it is important to define these reflecting markers or boundaries. This makes reflection tomography difficult to model.

It was stated by Gulf Oil Company geophysicists in 1980 that traveltime tomography could be used successfully to estimate velocities from seismic reflection times, which can be used in seismic imaging such as migration and depth conversion. This method has been demonstrated by scientists at Amoco Research Company and at the California Institute of Technology.

In transmission tomography, the seismic energy of interest is that which has traveled *through* the subsurface without reflection. This requires placing the source in a borehole and the receivers at the surface, or vice versa, or the source may be in one borehole and the receivers in another.

In borehole-to-surface tomography, the measurements may consist of first arrivals in a vertical seismic profile (VSP), where a series of seismic energy is injected into the ground and the resulting waves are recorded by a seismometer at various depths in the borehole. Or, the energy is injected at depth in the borehole, using a downhole source, and recorded at the surface (reverse VSP).

In borehole-to-borehole measurements, both sources and receivers are placed in the subsurface inside the holes. Seismic energy "illuminates" the region between boreholes. Figure 11–1 shows types of raypaths between surface and subsurface source and receiver positions.

FIGURE 11–1. Raypaths in tomography

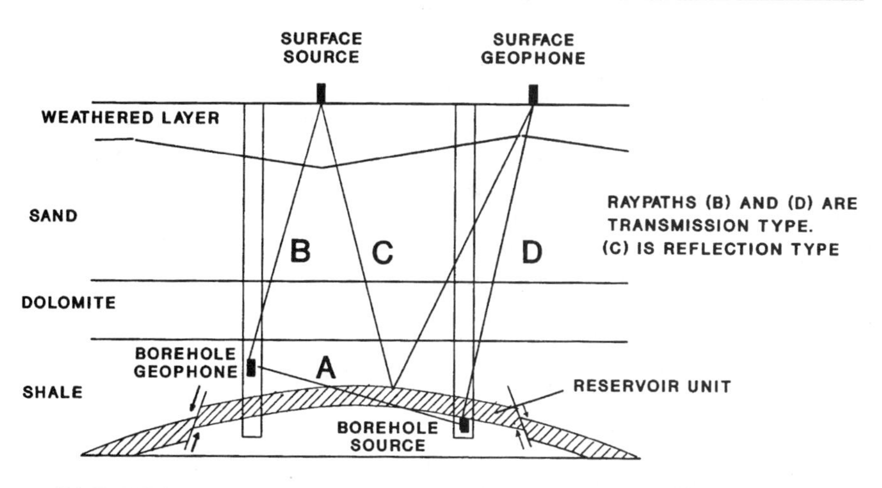

(A) IS THE ONLY RAYPATH WHICH DOES NOT TRAVEL THROUGH THE NEAR-SURFACE WEATHERED LAYER; LITTLE OF THE HIGH-FREQUENCY COMPONENTS ARE LOST DUE TO DISSIPATION AND SCATTERING.

(C) IS THE TYPE INVOLVED IN STANDARD SURFACE REFLECTION SEISMOLOGY. IT DIFFERS FROM OTHER PATHS IN THAT IT IS THE LONGEST, PASSING THROUGH THE WEATHERED LAYER TWICE. HIGH-FREQUENCY COMPONENTS HAVE BEEN ATTENUATED. IT IS THE ONLY SEISMIC REFLECTION INFORMATION FROM THE RESERVOIR.

TRAVELTIME TOMOGRAPHY PROCEDURE

In either reflection or transmission type tomography, the procedure can be done as follows:

- Determination of actual seismic traveltimes.
- Raytrace modeling of energy travel paths.
- The solution of traveltime equations is called "traveltime inversion," because it is involved in producing a velocity model that best fits the observed data.

DETERMINATION OF ACTUAL TRAVELTIME

This process involves seismic data interpretation on an interactive graphics station to pick reliable traveltimes. This step will take the interpreter two or three working days, and it will generate typically 10,000 to 20,000 values.

Traveltimes are normally picked from seismograms such as common source records, common midpoint sort records, or slanted stack records. The digitizing can be done by hand by using a digitizing tablet or more automatically by using a seismic work station.

RAYTRACING

Tomography is based on the approximation that the energy travels from source to receiver along a raypath. Modeling the energy propagation through a medium by raytracing can be done by solving equations related to a velocity model, a set of reflecting boundaries, and receiver pairs (Aki and Richards, 1980). The solution to the two-point boundary for the ray equation is found by shooting upward from the reflecting surface. This can be done by raypath modeling.

TRAVELTIME INVERSION OR TOMOGRAPHIC INVERSION

In seismic tomography, ray bending methods are generally needed. The layered-media model (see Figure 11–2) is normally subdivided into constant-velocity cells.

Lines et al. (1989) illustrated a relationship between traveltimes, distance, and slowness ($1/V$). The traveltime t_i of the ith acoustic ray through the cells may be calculated by expressing it in terms of the sum of the products of distance d_{im} and slowness S_m for these cells within the body.

The traveltime equation for the ith ray can be expressed as:

$$t_i = d_{i,1}S_1 + d_{i,2}S_2 + d_{i,3}S_3 \ldots\ldots d_{i,m}S_m \tag{1}$$

Or in matrix notation as:

$$T = DS, \tag{2}$$

FIGURE 11–2. Traveltime tomography model (modified after Dines and Lytle, 1979)

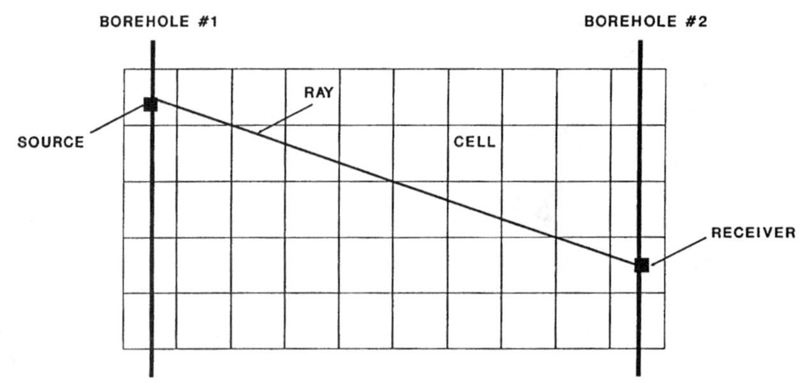

where T is the traveltime vector.
D is an $n \times m$ matrix containing the ray distances.
n is the number of raypaths.
m is the number of slowness cells.
S is the slowness vector.

The above equation is nonlinear, because the distance values of the rays depend on the velocity.

Since a raypath will travel only a small part of the total number of cells, the system traveltime equation will be sparse (i.e., more than 99% of the matrix elements of D are zeros). The solution of large sparse system of linear equations can be achieved.

Traveltime inversion can be done quickly on the computer, using the *row action* method, in which the processor solves a single equation of the system $T = DS$. A least-squares solution of the system eventually results by iteration.

Once the actual traveltimes have been picked and rays have been traced, a large system of traveltime equations is solved for the unknown slowness equation.

TRANSMISSION TOMOGRAPHY

As stated before, transmission tomography can be either borehole-to-borehole or borehole-to-surface. Figure 11–3 illustrates the geometry of each type of transmission tomography.

FIGURE 11–3. Transmission tomography geometry (after McMechan, 1983)

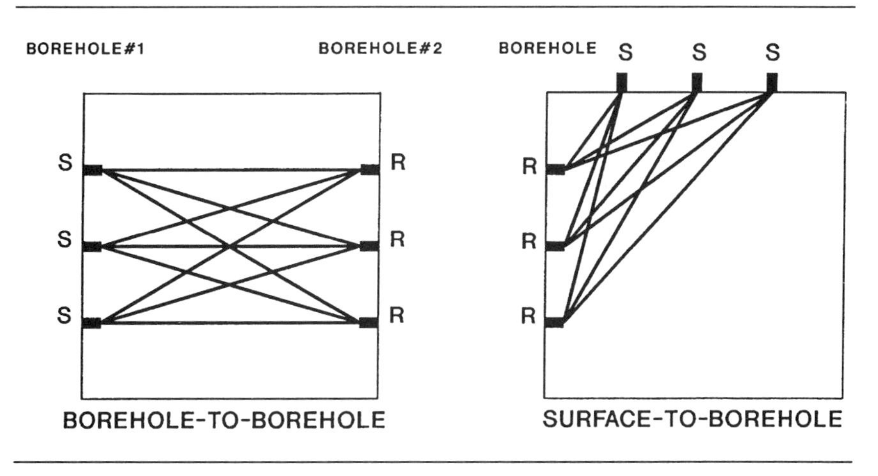

In the borehole-to-borehole type, the sources are placed below the surface in one borehole, and the receivers are placed in another hole. Since the high-frequency component of the seismic frequency band is largely attenuated by the near-surface weathered formations, one would expect this arrangement to result in higher frequencies being recorded.

The surface-to-borehole geometry is typical for a VSP survey.

Models on both reflection traveltime and transmission tomography will be reviewed before discussing real data examples.

EXAMPLES OF TOMOGRAPHIC MODELS

REFLECTION TRAVELTIME TOMOGRAPHY MODEL

Bishop et al. (1985) demonstrated the geometry of reflection tomography using the finite difference approach, which is shown in Figure 11–4.

1. The subsurface model is divided into cells with constant slowness.
2. Curved rays are traced from the source to the reflecting horizon and back to the surface.
3. The reflecting horizons are known.

Amoco Research scientists (Bording et al., 1987) utilized an iterative procedure to image a velocity field. The iterative migration and reflection tomography flow chart is illustrated in Figure 11–5.

FIGURE 11–4. Reflection tomography geometry (after Bishop et al., 1985. Courtesy SEG)

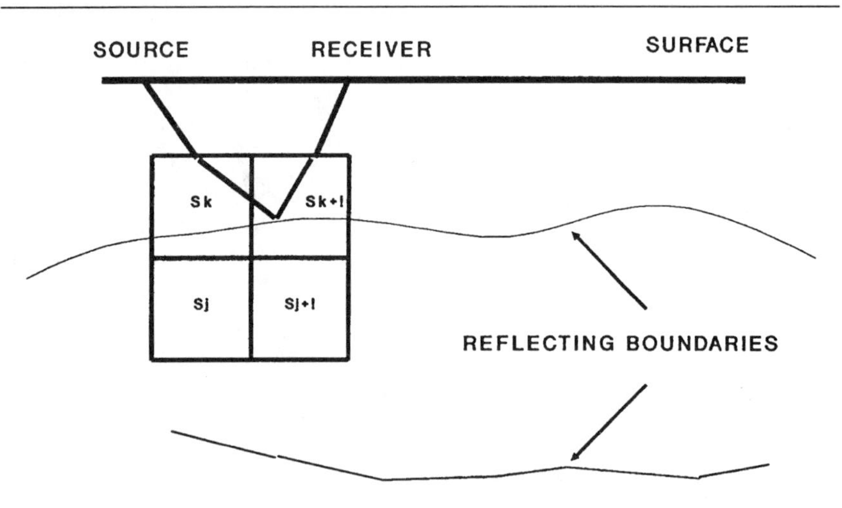

FIGURE 11–5. Iterative reflection and migration tomography (after Bording et al., 1987)

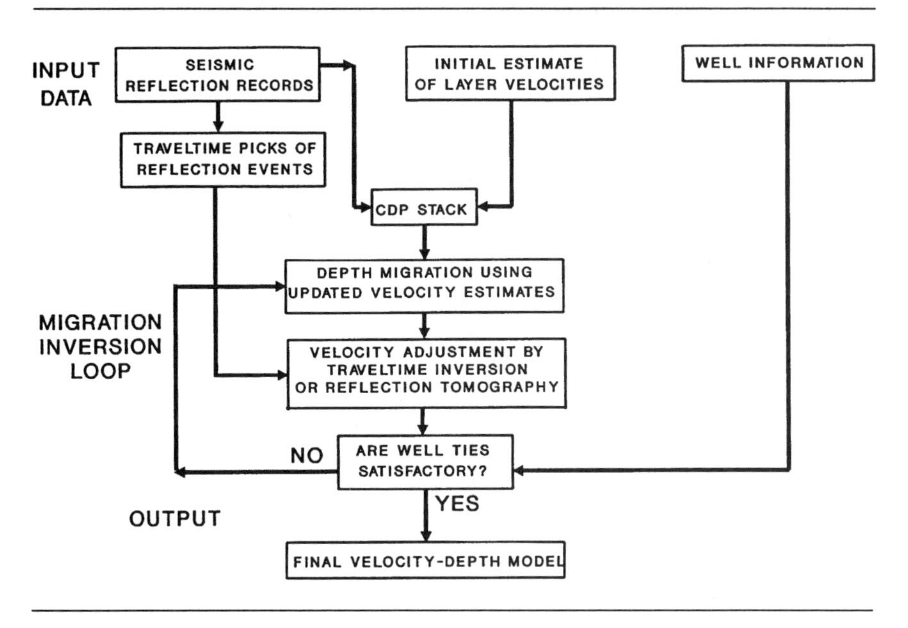

The above procedure combines the ability of tomography to extract the subsurface velocity field with the ability of migration to image the subsurface interfaces. This is called *iterative tomographic migration.*

ITERATIVE TOMOGRAPHIC MIGRATION PROCEDURE

This approach was implemented on the geological model shown in Figure 11–6a, which shows the velocities of the layers.

INITIAL MIGRATION AND TRAVELTIME PICKING

The initial velocity model consists of flat, constant-velocity layers whose velocities were chosen to be correct on the left side of the model section. A CDP stack generated from this velocity is shown in Figure 11–6b.

Using the velocity model, a depth-migrated section was generated (see Figure 11–7a). Note that migration provided a good reconstruction of the left-hand side of the dome, but it did not image the salt dome or deeper markers on the right-hand side of the model.

The main objective was to produce an accurate velocity tomogram to improve the flat-layer migration results by using these refined velocities to migrate the data. To accomplish this, a set of well-spaced common

FIGURE 11–6. Iterative tomographic migration (Copyright © 1987, Blackwell Scientific Publications, Ltd., from Bording et al., "Applications of seismic travel—time tomography," *Geophysics Journal International*, vol. 90, 1987.)

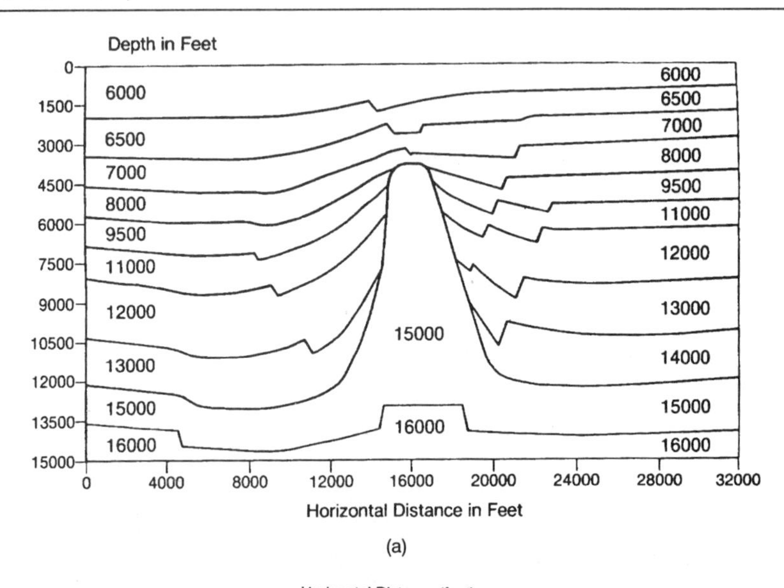

(a)

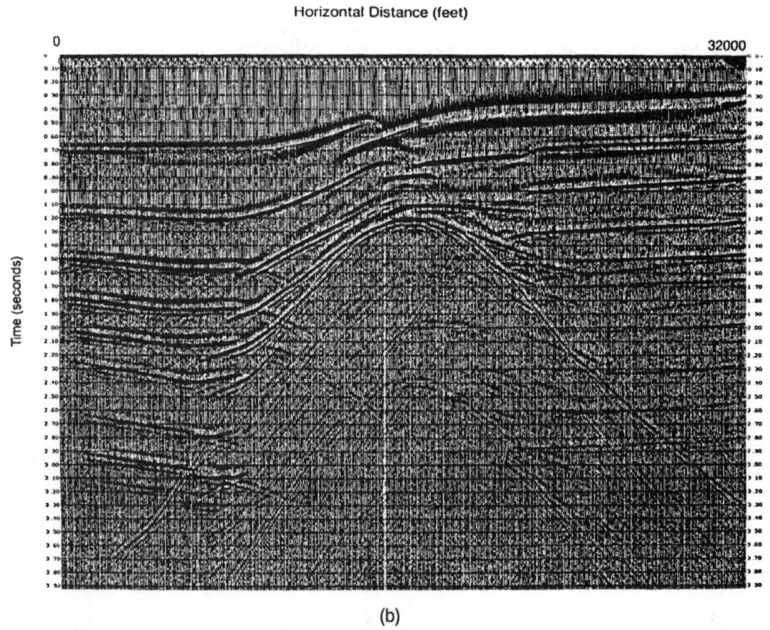

(b)

(a) GEOLOGICAL MODEL USED IN REFLECTION TOMOGRAPHY EXAMPLE. THE LAYER VELOCITIES ARE SHOWN.
(b) CDP STACK USING FLAT-LAYER VELOCITY MODEL.

FIGURE 11–7. Iterative tomographic migration (Copyright © 1987, Blackwell Scientific Publications, Ltd., from Bording et al., "Applications of seismic travel—time tomography," *Geophysics Journal International*, vol. 90, 1987.)

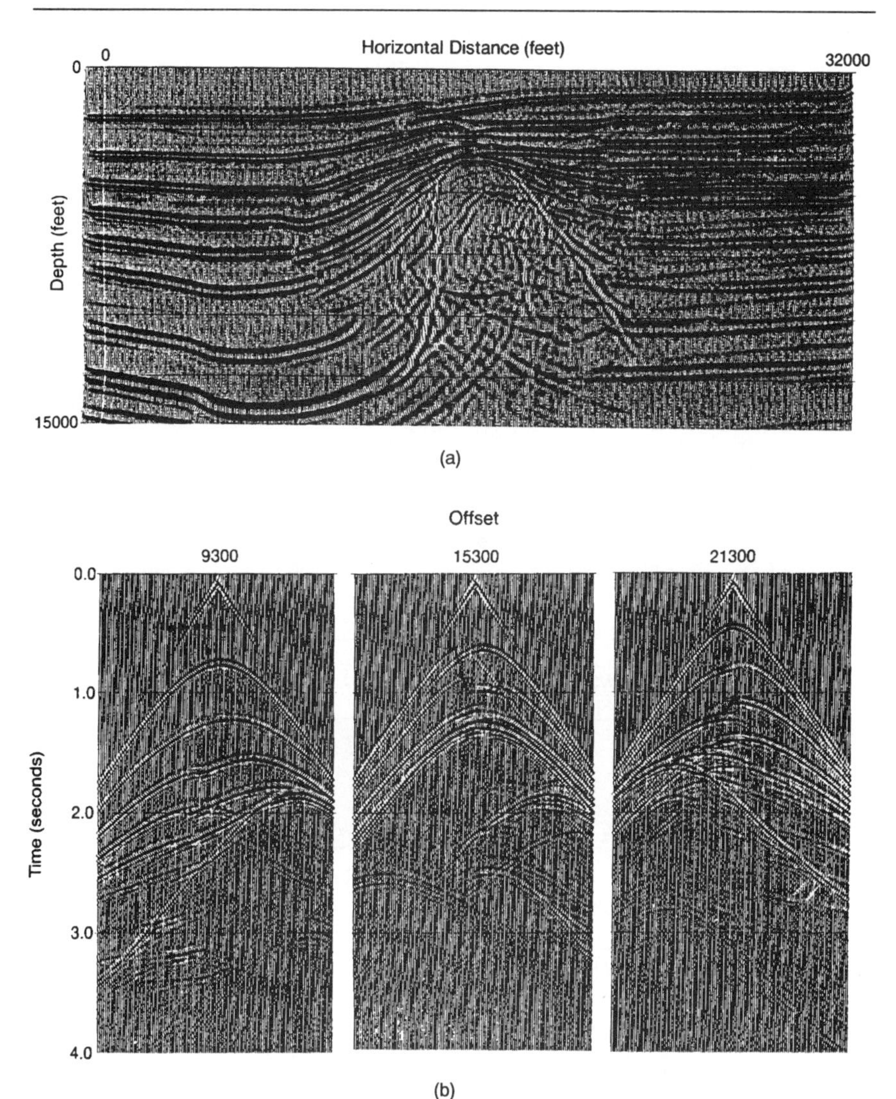

(a) INITIAL DEPTH MIGRATION USING FLAT-LAYER VELOCITIES.

(b) FINITE-DIFFERENCE COMMON SOURCE GATHERS OF OFFSETS OF 9,300, 15,300, AND 21,300 FEET. TRAVELTIME PICKS WERE MADE ON PEAKS JUST BELOW HORIZONS.

shot records (see Figure 11–7b) was chosen. Then the traveltimes of ten events within each of nine records were picked. A total of 4,468 traveltimes were available.

RAY TRACING FOR THE MODEL

The model composed of 40 × 80 cells with dimensions of 400 × 400 feet. Figure 11–8 shows the rays traced on three different reflectors.

1. Tomographic Inversion
   Following the estimation of the traveltimes (from data) and the computation of the traveltimes from the model, a modified version of Eq. (2), $\Delta T = D \bullet \Delta S$ is solved for the slowness deviation ΔS, where ΔT is

FIGURE 11–8. Iterative tomographic migration (Copyright © 1987, Blackwell Scientific Publications, Ltd., from Bording et al., "Applications of seismic travel—time tomography," *Geophysics Journal International,* vol. 90, 1987.)

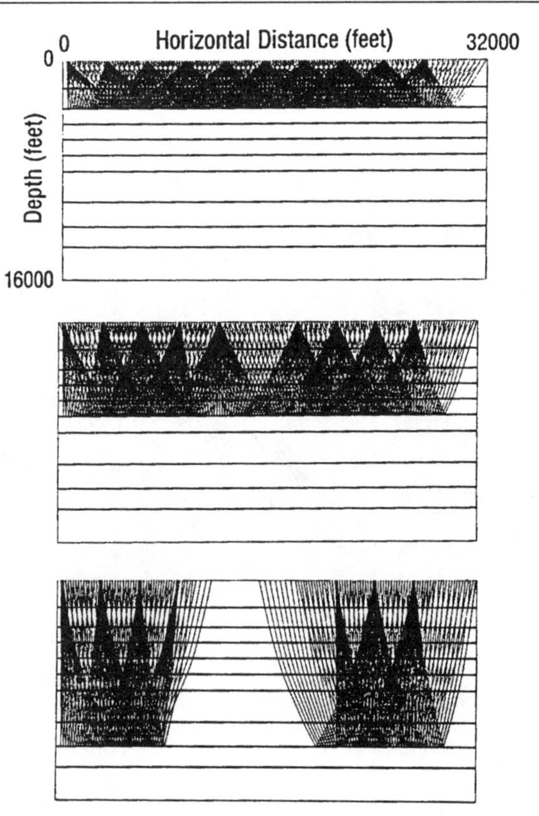

SAMPLE CURVED-RAY ILLUMINATION OF FLAT LAYER INITIAL APPROXIMATION.

the difference between data and initial model response. Only cells that are illuminated contribute significantly to the inversion.

The D matrix is large (4,468 × 3,200) and sparse (about 1% have nonzero values); therefore, sparse iterative solutions are employed.

The slowness deviations are then added to the original slowness model to give an updated slowness solution. As a final step, a filter is usually applied to the output velocity model in order to smooth the solution.

This procedure can be repeated until no variation in ΔS occurs; this means we have reached the optimum solution and the tomographic process is completed with satisfactory results; we are now ready for the final migration step.

2. Final Migration

The final stack is computed using the final velocity determined from the depth model. This new stack is shown in Figure 11–9.

The new CDP stacked section is then depth migrated with tomographically derived velocities (see Figure 11–10). One can see a much improved image of the salt dome flanks compared to Figure 11–7a.

CROSS-BOREHOLE TOMOGRAPHY MODEL

For reflection seismology, significant high-frequency attenuation can occur near the surface due to the weathered layers. Such absorption prob-

FIGURE 11–9. Iterative tomographic migration (Copyright © 1987, Blackwell Scientific Publications, Ltd., from Bording et al., "Applications of seismic travel—time tomography," *Geophysics Journal International*, vol. 90, 1987.)

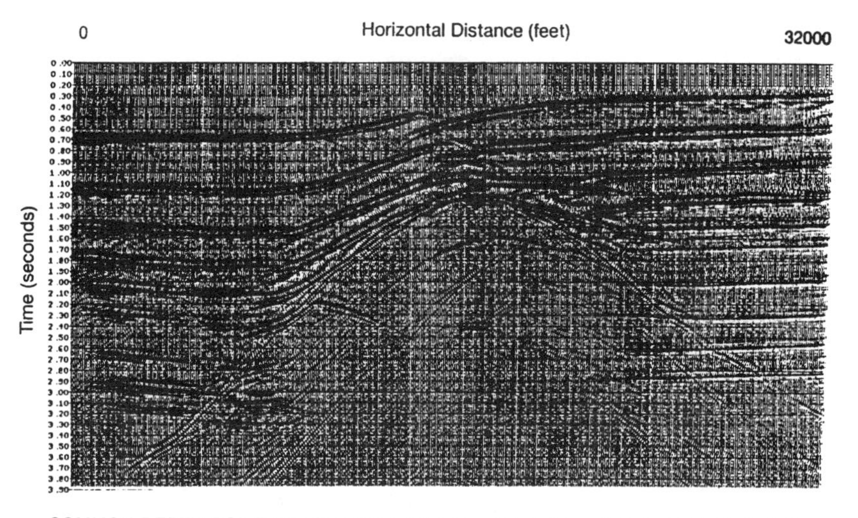

FIGURE 11–10. Iterative tomographic migration (Copyright © 1987, Blackwell Scientific Publications, Ltd., from Bording et al., "Applications of seismic travel—time tomography," *Geophysics Journal International*, vol. 90, 1987.)

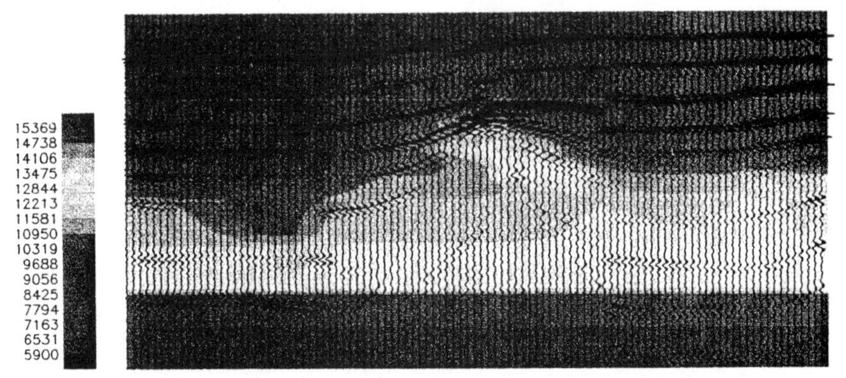

DEPTH MIGRATION USING TOMOGRAPHICALLY DERIVED VELOCITIES OVERLAIN BY COMPUTED TOMOGRAM OF VELOCITY RANGE FROM 6,000 FT/SEC TO 16,000 FT/SEC.

lems are reduced in cross-borehole seismic methods by placing both source and receivers below the depth of weathering (Dines and Lytle 1979; Ivanson 1986).

A seismic section or map is more easily understood if it is in depth domain rather than in time domain, but this requires knowledge of the velocities. Velocity control at the borehole is available from sonic logs, and these velocities can be interpolated. However, a preferable approach involves a cross-borehole survey in which seismic traveltime measurements allow the velocity to be estimated between boreholes.

In addition to velocity information for depth conversion and depth migration, tomography can delineate reservoir boundaries and can be used in enhanced oil-recovery projects.

Transmission tomography can be used in modeling the near-surface formation in an area. Lines and Lafehr (1989) proposed an approach based on a data set from Amoco's Denver Region, which had been used to model the near-surface formation. This study may help in getting better record quality in future field data acquisition.

In their classic paper on seismic inversion problems, Wiggins et al. (1986) dealt with residual statics analysis as a general linear inverse problem. In their analysis, they assumed that the data traces had already been adjusted for the "elevation statics," effects attributable to variations in shot and receiver elevations and those weathering variations which are determined by uphole times. They further assumed that the data had been dynamically corrected for normal moveout corrections (NMO) using stacking velocities.

These assumptions were made for two reasons:

1. Large apparent lateral variations in the velocities are due to the near-surface formation rather than to velocity changes at depth.
2. Lateral averaging of residual NMO tends to stabilize the statics solution.

By linear inversion and iteration, a good solution of statics is obtained.

Applying this solution to the original field data, better velocity analysis can be derived and applied for better stack coherency and better signal-to-noise ratio.

This method is sensitive to the multiplicity. As you know, the higher the multiplicity, the larger the number of the statistics and the closer the statics solution. Minimum fold stack is three fold.

Figure 11–11 illustrates this method before and after applying the statics derived from tomography. The initial model was known geologically; the marker at 2.0 sec is almost flat.

CHOICE OF ERROR CRITERION

To accomplish inversion, it is necessary to have a forward model and its response. In the case of a geological model, one must know its geophysical response—whether it is gravity, magnetic, or seismic signature. An optimization algorithm is needed to allow adjustment of the parameters of the forward model.

Treitel (1989) discussed the effects of the choice of the forward model and the selection of the optimization algorithm on the quality of the inversion. This has become a very important question. How can we minimize the difference between what is observed and what is modeled? Should one use the least-squares method? The least absolute deviation? Or mini-max?

Treitel proposed conducting a sensitivity analysis as an important step to the solution, because unless it is subjected to sensitivity analysis it is not necessarily a satisfactory solution. For example, we have a set of parameters that have been extracted using an inversion procedure, and we find that by changing a particular parameter by 10, 15, and 20 percent, nothing happens. Clearly, the model is not sensitive to this parameter.

The *initial* guess of the model is so important that a close guess can get a good inversion.

To illustrate this point, Figure 11–12 shows seismograms for model inversion after 1 iteration and 10 iterations. This is a prestack model; the left panel (a) shows a model of common source gather for the four-layer case. One can see that the first-guess model (b) has no match with the model. Of course, it gets better as the number of iterations increases (c).

FIGURE 11–11. Residual statics tomography (courtesy of Western Geophysical)

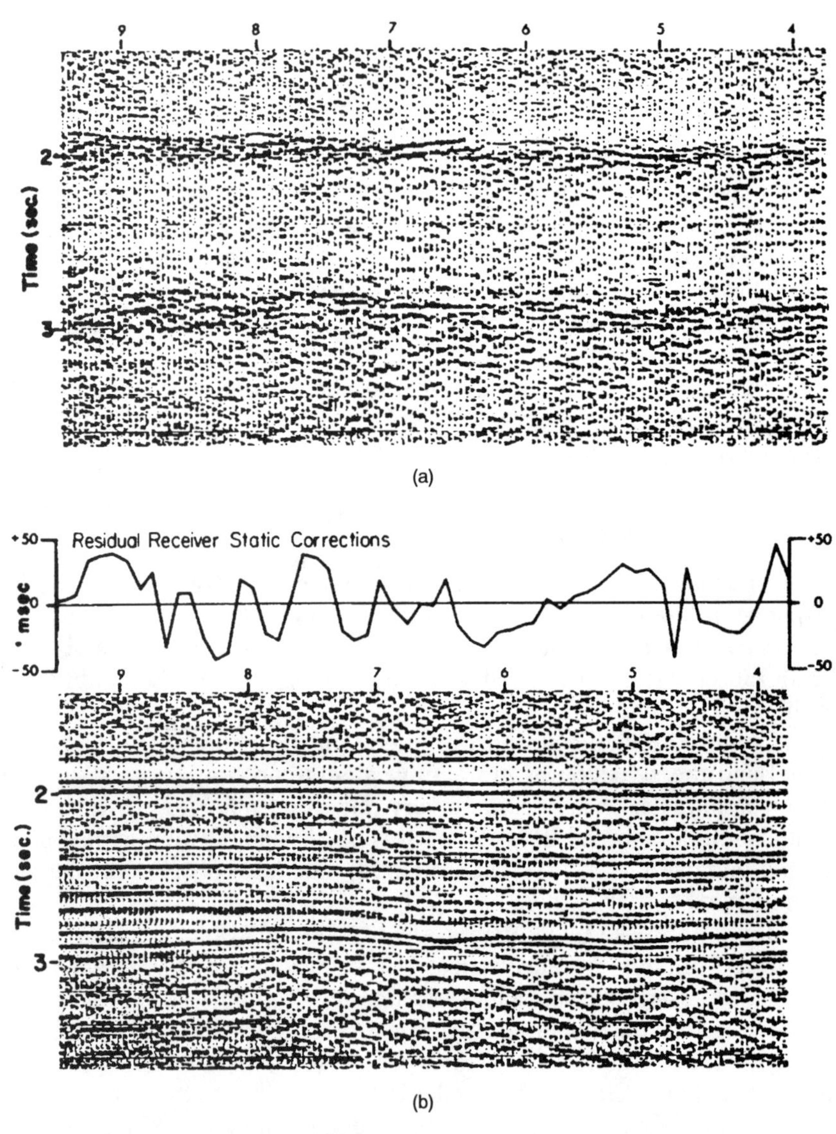

(a) TWELVE-FOLD STACK SECTION OF 48 TRACE RECORDS. ELEVATION-DERIVED STATICS WERE APPLIED. ONE CABLE LENGTH SPANS FOUR INTERVALS.

(b) SAME SECTION AS (a) AFTER SURFACE-CONSISTENT STATIC CORRECTIONS WERE APPLIED. RESIDUAL STATICS EXCEEDS 50 MS IN SOME PORTIONS OF THE LINE.

FIGURE 11–12. Seismograms for model inverson (courtesy of the Society of Exploration Geophysicists, adapted from Treitel, 1989)

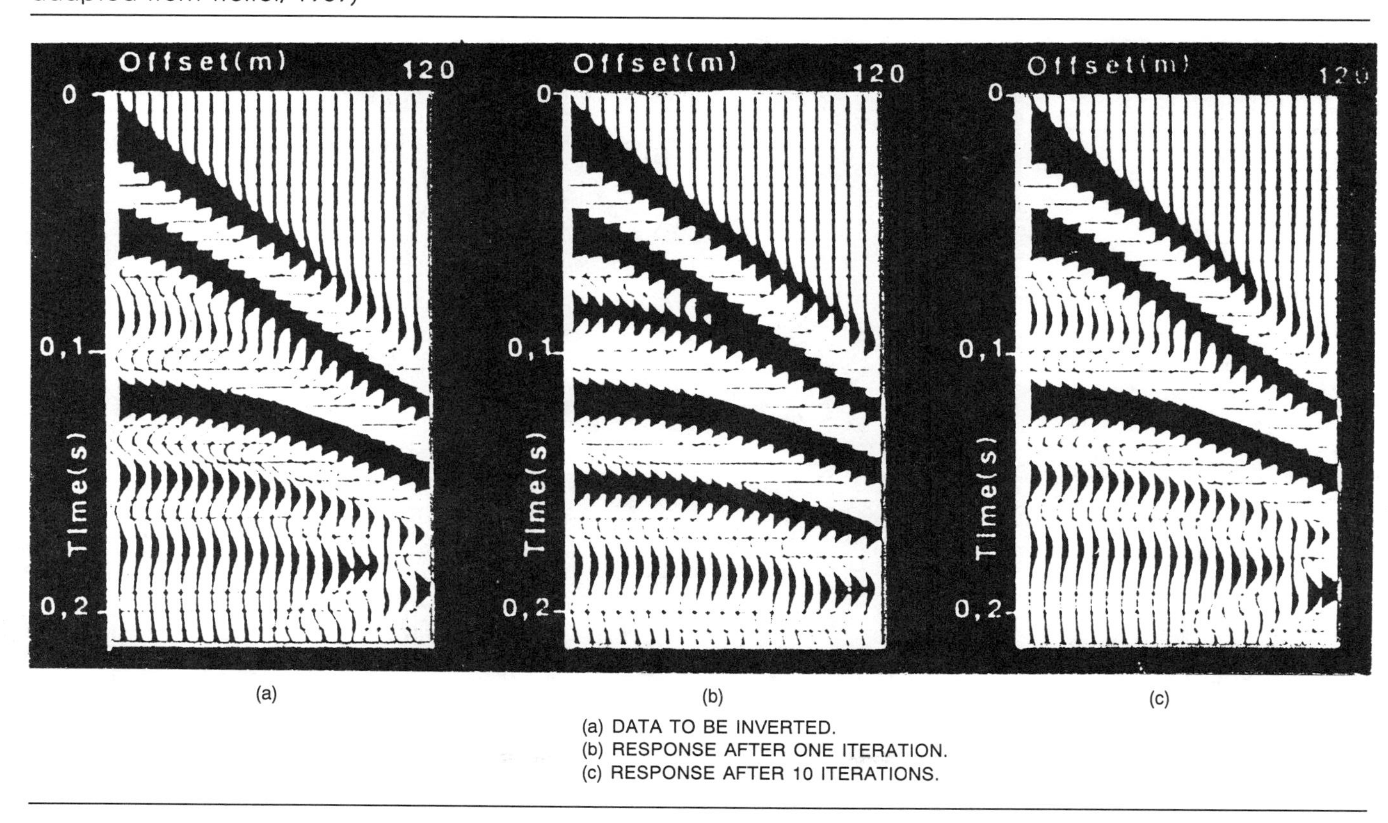

(a) DATA TO BE INVERTED.
(b) RESPONSE AFTER ONE ITERATION.
(c) RESPONSE AFTER 10 ITERATIONS.

A good guess and a disasterous guess are shown in Figures 11–13 and 11–14. It shows the progress of the velocity inversion of the model in Figure 11–12.

In Figure 11–13 we started with a good guess for the velocity model, and at 125 iterations we obtained a good match between the model and observed velocity.

In Figure 11–14 we started with a bad initial guess of the velocity model, and we used the same number of iterations. The result was a bad velocity model.

SEISMIC TOMOGRAPHY AND RESERVOIR PROPERTIES

To improve recovery from existing reservoirs, it is important to obtain much better descriptions of their internal structure and how fluid flows through them. Some of the information we can use are reservoir properties derived from cores, cuttings, well logs, and well testing. But this data is sparse; we need to improve the sampling of the subsurface to be able to get a better description of the reservoir. Because we cannot drill enough wells to get the required data in order to accomplish this, some sort of remote-sensed data must be acquired.

FIGURE 11–13. Progress of velocity inversion—good initial guess of model (courtesy of the Society of Exploration Geophysicists, adapted from Treitel, 1989)

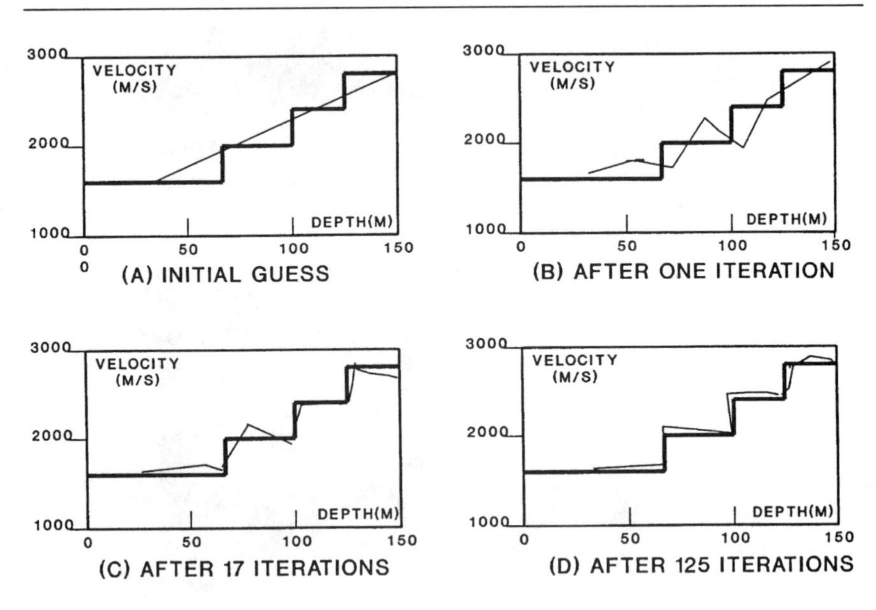

FIGURE 11–14. Progress of velocity inversion—bad initial guess (courtesy of the Society of Exploration Geophysicists, adapted from Treitel, 1989)

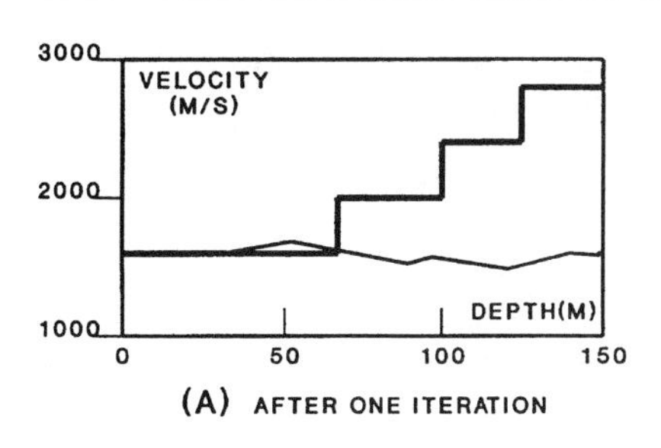

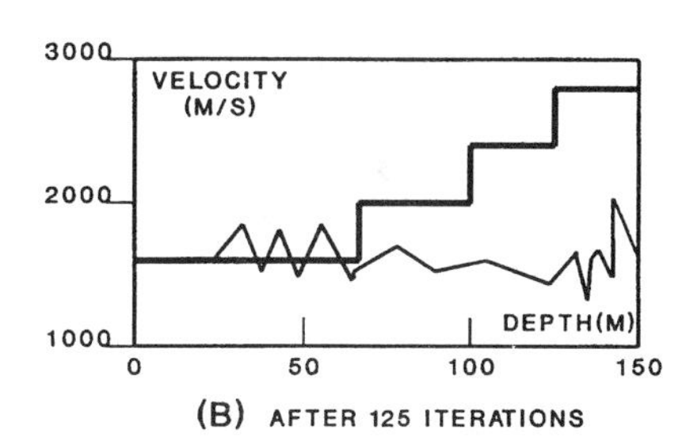

For the decade of the 1990s, we think that seismic borehole measurements and tomography will play an important role in understanding more and more about reservoir properties.

Today, 3-D surveys and extensive VSP surveys may sample the reservoir better. We think and believe in the following statements:

- Reservoir geophysics, especially reservoir seismic, will become vital and will play an active role in the next ten years or so in understanding more about the reservoir characteristics.
- Borehole seismic surveys will be the key in relating wave propagation to the reservoir structure and fluid flow behavior to the seismic velocity response.
- We will have better methods to acquire and process the needed seismic response that will be able to revolutionize reservoir engineering, oil recovery, and enhanced oil-recovery methods.

The measurements we now have may be able to describe the reservoir only within a few feet from the wellbore. Since the reservoir rock is heterogeneous, we need to obtain these properties in their spatial and temporal variations.

The reservoir properties we need to obtain, both laterally and vertically, include:

- Mineralogy.
- Rock properties—porosity, permeability, compressibility, and saturations.
- Fluid properties—chemistry, viscosity, compressibility, and wettability.
- Environmental factors—temperature, stress, and pore pressure.

We should stress the fact that without a more complete understanding of the relationships between reservoir properties, fluid flow, and seismic wave propagation, the application of existing methods and future developments of hardware to acquire more measurements and software to expedite the results will be hindered. The geophysical and engineering communities should be working together to understand each other's terminology, to appreciate the complexity of the methods geophysicists are developing, and to understand more about their applications and limitations.

RECENT DEVELOPMENTS

DOWNHOLE SOURCES

A consortium of major oil companies has been organized to conduct research and development of a seismic source for generating seismic energy in the borehole. Some tool development work has been done by various other organizations. Several different tools have been tested. These include three impulse sources—an air gun, a modified core gun, and a sparker. The remaining two are a hydraulic vibrator and a pneumatic vibrator.

Some cement damage was induced by both the air gun and the core gun. Very little change of the cement bond log was observed after using the vibrators.

The hydraulic vibrator generates more energy than the impulse sources. Moreover, impulse sources generate tube waves and other types of body waves that tend to mask out any arrivals later than the direct P-wave arrival. Hydraulic vibrators can generate a very wide range of frequencies on the order of 5 to 720 Hz.

N.P. Paulson of Chevron has suggested that the downhole seismic source for cross-hole imaging and reverse VSP should have the following characteristics:

- Nondestructive.
- Large bandwidth.
- Powerful enough for reverse VSP.
- Repetitive.
- Depth-independent seismic output.
- Short cycle time.
- High temperature capability.
- Cased- and open-hole operation.
- Both P & S waves transmitted.
- Usable in wells between 5"–12".
- Reliable.

The best test results to date were obtained from a clamped hydraulic vibrator from which P, S, and SV waves were obtained. This tool is still under research for improvement.

RESERVOIR PROPERTIES

Reservoirs are much more heterogeneous than anyone likes to think. As time goes on, more and more reservoirs are reclassified as severely heterogenous. Therefore, the first task and application of tomography is to describe the reservoir heterogeneity. The second application is to monitor the recovery of hydrocarbon in place. Recovery plans may be reevaluated and modified as production proceeds.

Much laboratory work has been done by many reservoir geophysicists to relate acoustic velocity and seismic response to the petrophysical properties of the reservoir rock.

It may be possible to go from seismic measurements, in this case dispersion, to reservoir properties, which in this case is the hydraulic permeability. This is good news to the reservoir engineer and geologist, since through seismic tomography permeability can be mapped both vertically and laterally.

Figure 11–15 illustrates the relationship between porosity and clay content for 75 shaly sandstones and the effect on P and S velocities.

The effect of temperature on velocity is shown by Figure 11–16.

The effects of saturation and pressure are demonstrated in Figure 11–17.

Figure 11–18 shows the relationships between compressional and shear velocities for various minerals and as a function of depth for selected Gulf Coast formations.

APPLICATIONS IN PETROLEUM ENGINEERING

Because of the increasing economic importance of oil recovery, the growing complexity of finding new oil fields, and the growing realization of the heterogeneity of the reservoirs, a major shift to using seismic methods is taking place. It is very important to understand the relationship between the seismic measurements of the reservoir and related reservoir properties.

One of the most important applications of tomography is relating reservoir seismic velocity in the description of the reservoirs and in the monitoring of their recovery. Some of these applications are:

- Porosity and permeability mapping.
- Fracture detection.
- Anomalous pore pressure detection.
- Tracking thermal fronts and mapping flooded zone.
- Monitoring gas cap movement.
- Water flooding.
- Monitoring CO_2 flood process.

FIGURE 11–15. Effect of porosity and clay content on compressional velocity Vp (after Han, 1986)

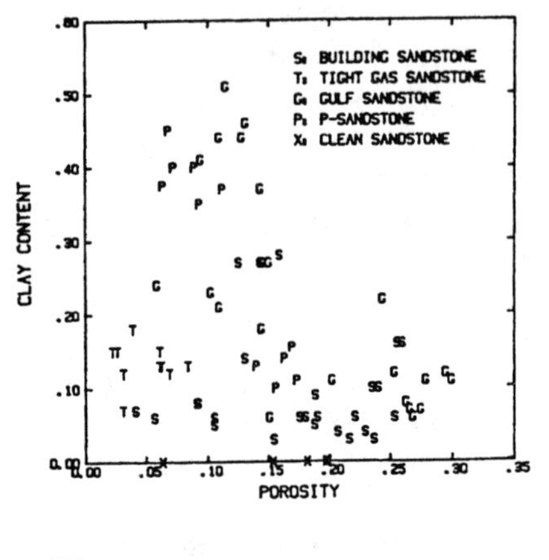

(a) RANGE OF CLAY CONTENT AND POROSITY FOR 75 SHALY SANDSTONES POROSITY RANGE FROM 2–30 PERCENT AND CLAY CONTENT RANGES FROM 0–50 PERCENT. THE DATA INDICATE THAT SANDSTONES WITH HIGH CLAY CONTENT TEND TO HAVE LOW POROSITIES.

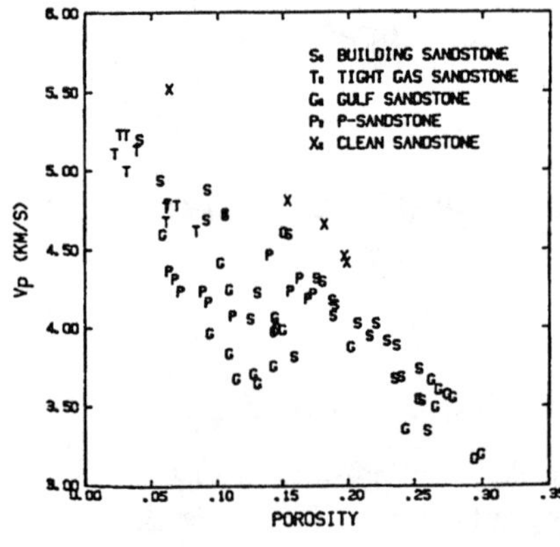

(b) MEASURED COMPRESSIONAL VELOCITIES VERSUS POROSITY FOR 75 SAMPLES AT Pc = 40 MPa AND Pp = 1.0 MPa.

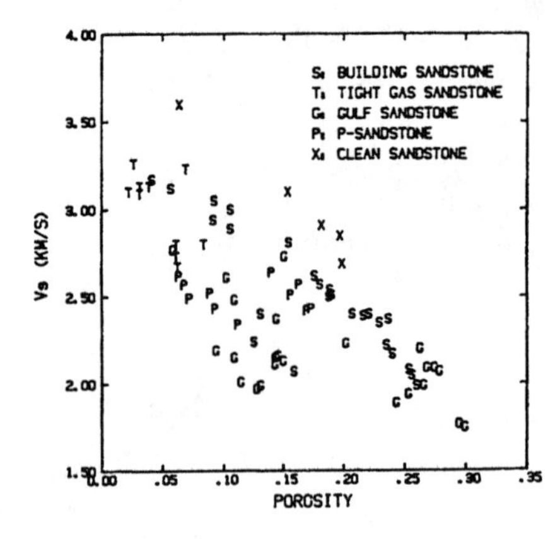

(c) MEASURED SHEAR VELOCITIES VERSUS POROSITY UNDER THE SAME CONDITIONS.

FIGURE 11–16. Effect of temperature on velocity (after Nur, 1989, courtesy SEG)

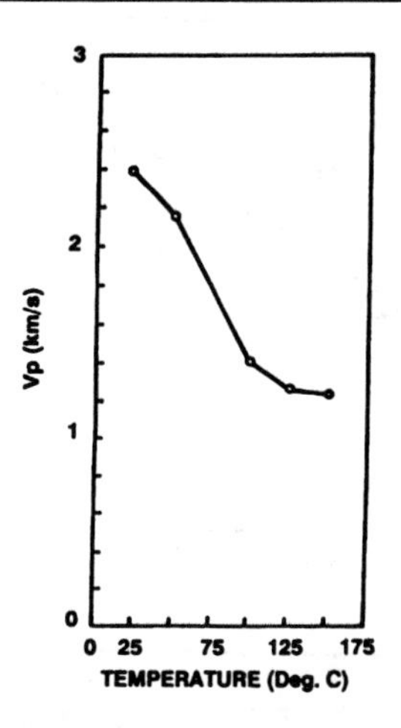

FIGURE 11–17. Effect of saturation and pressure on velocity (after King, 1965, courtesy SEG)

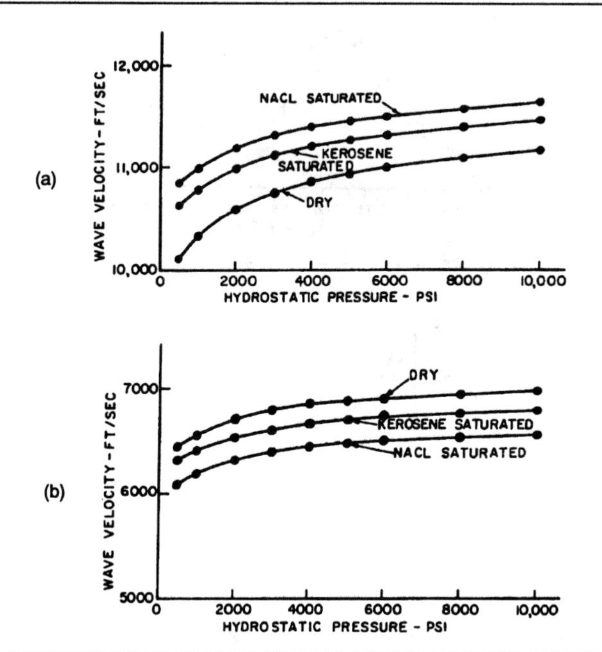

(a) COMPRESSIONAL WAVE GROUP VELOCITIES OF BOISE SANDSTONE.
(b) SHEAR WAVE GROUP VELOCITIES OF BOISE SANDSTONE.

FIGURE 11–18. Vp/Vs (after Castagna et al., 1984, courtesy SEG)

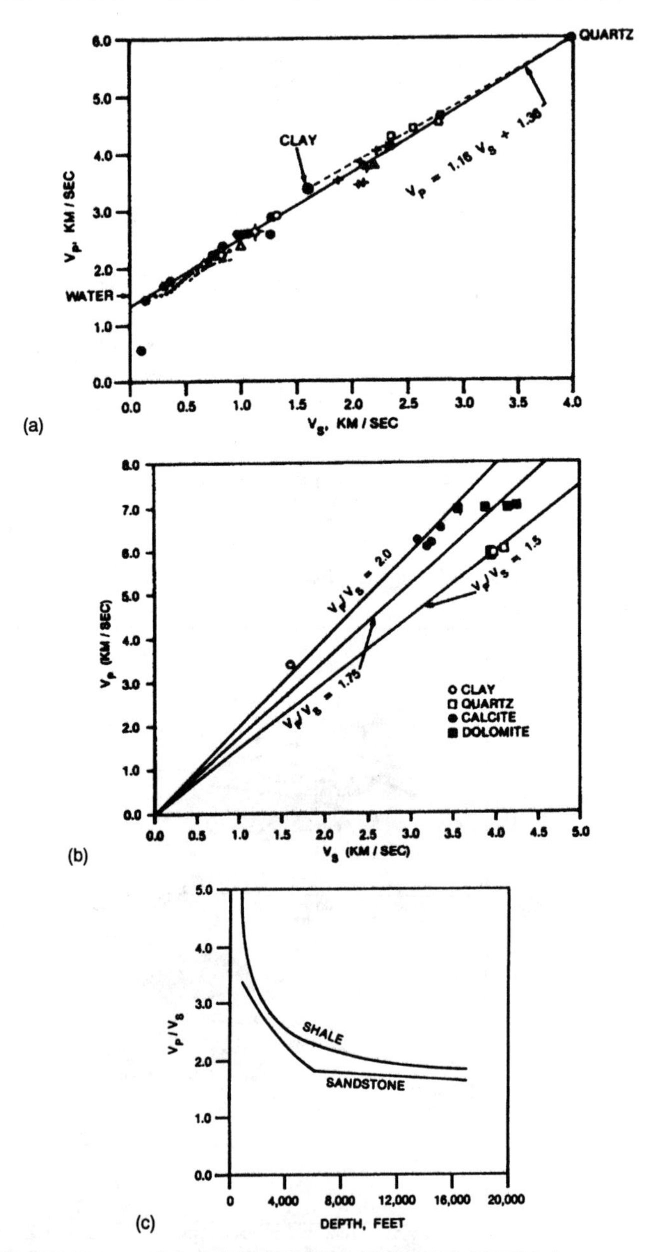

(a) COMPRESSIONAL AND SHEAR VELOCITIES FOR SOME MINERALS.
(b) COMPRESSIONAL AND SHEAR WAVE VELOCITIES FOR MUDROCKS FROM IN-SITU SONIC AND FIELD SEISMIC MEASUREMENTS.
(c) Vp/Vs COMPUTED AS A FUNCTION OF DEPTH FOR SELECTED GULF COAST SHALES AND WATER-SATURATED SANDS.

SUMMARY AND DISCUSSION

Tomography has been used in two methods—reflection and transmission. Travel time tomography is used successfully to estimate velocities from seismic reflection times that can be used in seismic imaging, such as depth conversion and migration.

Transmission tomography can be either borehole-to-borehole or borehole-to-surface measurements. In the borehole-to-borehole type, the source is placed in one borehole and the receiver in the other borehole. Research and development are under way to refine a borehole source that can generate P waves, S waves, Sv (vertical component of the shear wave) waves, and a borehole receiver that can respond to these waves. The source should be safe, repetitive, flexible, able to sustain the borehole conditions, and have broad frequency bandwidth.

The energy source can be lowered into one hole and the receivers into the surrounding wells. Velocities can be measured and mapped between wells. These velocities can be related to the reservoir petrophysical properties. One can map porosity, permeability, fluid content between wells, and account for vertical and horizontal changes in these parameters.

This is good news for the engineer and geologist, because they will be able to understand more about the heterogeneity of the reservoir rock and more accurately describe the reservoir characteristics.

Tomography coupled with borehole measurements will be the key to success in improving our hydrocarbon recovery methods and enhanced oil-recovery projects. The subject matter is very involved, and it needs the integration of a lot of information; namely, geophysical, geological and engineering. It needs the efforts and effective communications of all disciplines involved.

KEY WORDS

Least squares
Matrix
Ray tracing
Slowness
Sparse system

BIBLIOGRAPHY

Aki, K., and P. G. Richards. "Quantitative Seismology." *W. H. Freeman and Co.*, 1980.

Beydoun, W. B., J. Delvaux, M. Mendes, G. Noual,, and A. Tarantola. "Practical aspects of an elastic migration/inversion of crosshole data for reservoir characterization: a Paris Basin example." *Geophysics* 54 (1989): 1587–1595.

Bois, P., M. La Porte, M. Lavergne, and G. Thomas. "Well to Well Seismic Measurements." *Geophysics* 37 (1972): 471–480.

Bording, R. P., A. Gersztenkorn, L. Lines, J. Scales, and S. Treitel. "Applications of Seismic Traveltime Tomography." *Geophys. J., Roy. Astr. Soc.* 90 (1987): 285–303.

Bregman, N. D., R. C. Bailey, and C. H. Chapman. "Cross-Hole Seismic Tomography." *Geophysics* 54 (1989): 200–215.

Castagna, J. P., M. L. Batzle, and R. L. Eastwood. "Relationships Between Compressional-Wave and Shear-Wave Velocities in Clastic Silicate Rocks." *Geophysics* 50 (1985): 571–581.

Dines, K. A. and R. J. Lytle, "Computerized Geophysical Tomography." *Proc. IEEE* 67 (1979): 1065–1073.

Dyer, B. C. and M. H. Worthington. "Seismic Reflection Tomography: a Case Study." *First Break* 6 (1988): 354–366.

Han, D., A. Nur, and D. Morgan. "Effects of Porosity and Clay Content on Wave Velocities in Sandstones." *Geophysics* 51 (1986): 2093–2107.

Ivansson, S. "Seismic Borehole Tomography—Theory and Computational Methods." *Proc. IEEE* 74 (1986): 328–338.

Ivansson, S. "A Study of Methods for Tomographic Velocity Estimation in the Presence of Low-Velocity Zones." *Geophysics* 56 (1985): 969–988.

Justice, J. H., A. A. Vassiliou, S. Singh, J. D. Logel, P. A. Hausen, B. R. Hall, P. R. Hutt, and J. J. Solanki. "Acoustic Tomography for Monitoring Enhanced Oil Recovery." *The Leading Edge* 8 (1989): 12–19.

King, M. S. "Wave Velocities in Rocks as a Function of Changes in Overburden Pressure and Pore Fluid Saturations." *Geophysics* 31 (1966): 50–73.

Krohn, C. "Cross-Well Continuity Logging Using Guided Seismic Waves." *TLE* 11 (1992): 39–45.

Lines, L. R. and E. D. LaFehr. "Tomographic Modeling of a Cross-Borehole Data Set." *Geophysics* 54 (1989): 1249–1257.

Lines, L. R. "Applications of Tomography to Borehole and Reflection Seismology." *TLE* 10 (1991): 11–17.

Lytle, R. J. and M. R. Portnoff. "Detecting High-Contrast Seismic Anomalies Using Cross-Borehole Probing." *IEEE Transactions Geosci: Remote Sensing* 22 (1984): 93–98.

Marcides, C. G., E. R. Kanasewich, and S. Bharatha. "Multiborehole Seismic Imaging in Steam Injection Heavy Oil Recovery Projects." *Geophysics* 53 (1988): 65–75.

McMechan, G. A. "Seismic Tomography in Boreholes." *Geophys. J., Roy. Astr. Soc.* 74 (1983): 601–612.

Myron, J. R., L. R. Lines, and R. P. Bording, "Computers in Seismic Tomography." *Computers in Physics* (1987): 26–31.

Nur, A. "Four-Dimensional Seismology and (True) Direct Detection of Hydrocarbons: the Petrophysical Basis." *The Leading Edge* 8 (1989): 30–36.

Treitel, S. "Quo vadit inversio?" *The Leading Edge* 8 (1989): 38–42.

Wiggens, R. A., K. L. Larner, and R. D. Wisecup. "Residual Statics Analysis as a General Linear Inverse Problem." *Geophysics* 41 (1976): 922–938.

Wong, J., N. Bregnan, G. West, and P. Hurely. "Cross-Hole Seismic Scanning and Tomography." *The Leading Edge* 6 (1987): 36–41.

Worthington, M. "An Introduction to Geophysical Tomography." *First Break* 2 (1984): 20–27.

Zhu, X., P.D. Sixta, and B. G. Angstman. "Tomostatics: Turning-Ray Tomography + Static Corrections." *TLE* 11 (1992): 15–23.

CHAPTER 12

MAPS AND MAPPING

1. SUBSURFACE STRUCTURAL CONTOUR MAPS

INTRODUCTION

A topographic map of the earth's surface shows such features as hills, slopes, river valleys, and so forth as shown in Figure 12–1. The spacing of the contour lines is a measure of the steepness of the slope; the closer the spacing, the steeper the slope.

A subsurface structural map shows by contour lines the topography of a given horizon or formation as it would appear if all the rocks above were stripped away. Structural contour maps reveal the slope of the formation, structural relief of the formation, its dip, and any faulting and folding.

SEISMIC CONTOUR MAP DATUMS

When a subsurface map will be constructed from seismic data, a reference datum must be selected before the processing begins. The datum may be sea level or any other depth above or below sea level.

More often than not, another datum above sea level is selected in order to image a shallow marker on the seismic cross-section, which may have a great impact on the interpretation of the zone of interest.

These lines may also be processed using different datums to make structural contour maps (time or depth). All the lines should be corrected to the same datum or to sea level. This correction is twice the difference in elevation between the new datum and the old datum divided by the correction velocity (sometimes called "elevation velocity" or "replacement velocity").

FIGURE 12–1. Topographic map

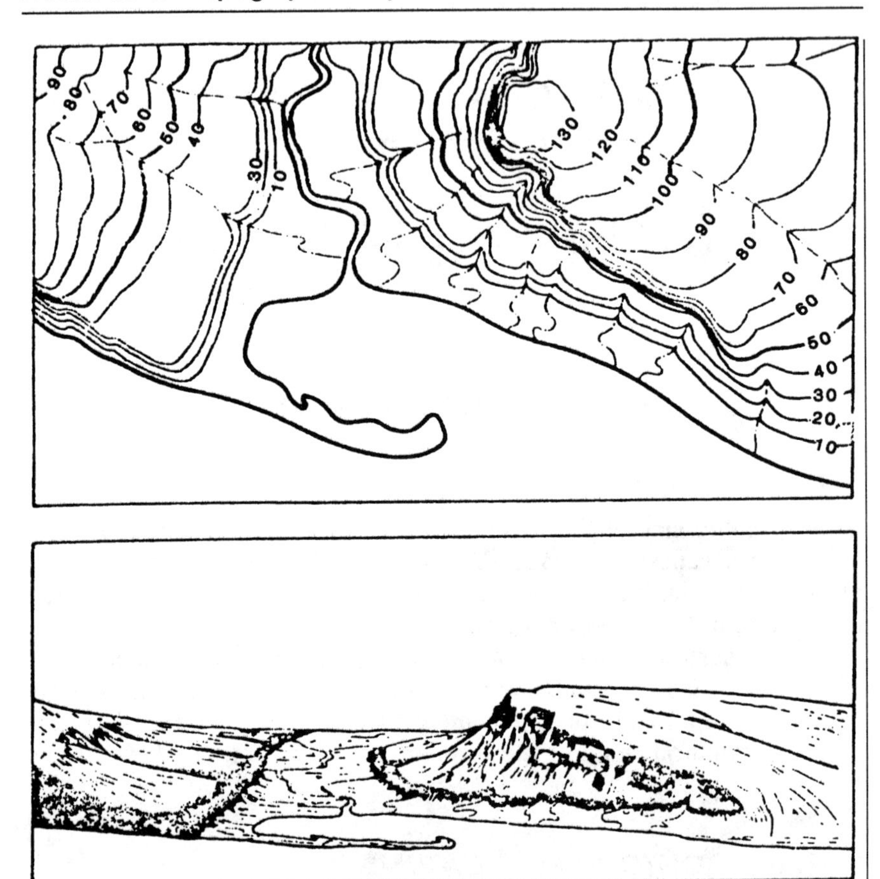

One of the most frequently encountered problems is a mis-tie of two intersecting lines. This can occur when working with seismic lines shot by different sources, receivers, or instrumentation in the same prospect area.

An example of this is shown on Figure 12–2, which is a two-way time map of a shallow seismic marker. At the intersections of the NW-SE line 5 with the NE-SW lines, one can observe that the time values on line 5 are smaller than on the intersecting lines. The mis-tie is about 20 ms, and this value will be added to the time picks on line 5 in order to make a smooth map. This technique of adjusting the values is not always as simple as this example because the mis-tie may not be constant over the entire line.

FIGURE 12–2. Line intersection mis-ties

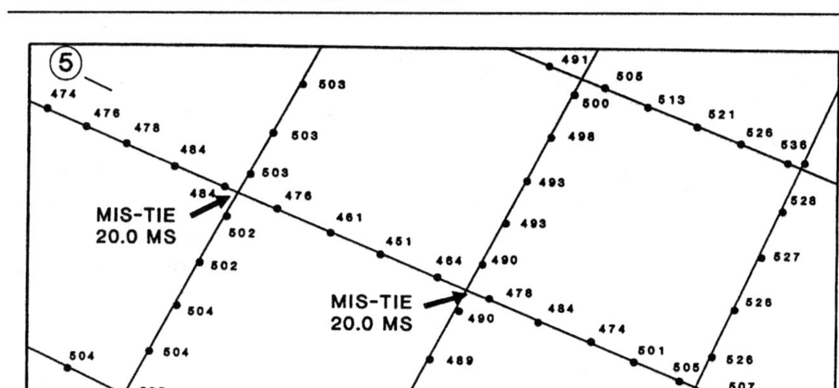

CONTOURING TECHNIQUES

There are three general contouring techniques:

- Mechanical spacing.
- Uniform spacing.
- Interpretative contouring.

MECHANICAL SPACING

Mechanical spacing may be done using a multipoint divider or computer-contouring program in which the contour lines are drawn in a mathematical relation to the data point.

In areas of few data points, this technique may be very misleading and some modifications of the computer-generated contour maps are usually required.

UNIFORM SPACING

Uniform spacing mapping produces pretty drawings; however, it is not logical contouring and does not reflect the geology of the subsurface.

INTERPRETIVE CONTOURING

This is the technique used in constructing a structure contour map. A knowledge of the general character and form of the structures in the regions being mapped greatly aids in correctly interpreting the subsurface where the well control is sparse.

An attempt should be made to contour the data so that the structure pattern bears out regional trends or tendencies. In this technique, contours are usually drawn parallel to each.

Figure 12–3 illustrates the three general types of contouring techniques.

CONTOUR DRAWING

There are a number of general rules and some basic techniques to follow in constructing a structure contour map. These are summarized in the following points.

- Study the elevations (depth or time), noting particularly the highest and the lowest points.
- A contour line must pass *between* points whose numerical values are higher and lower, respectively, than that of the contour line, as shown in Figure 12–4.

FIGURE 12–3. Contouring techniques

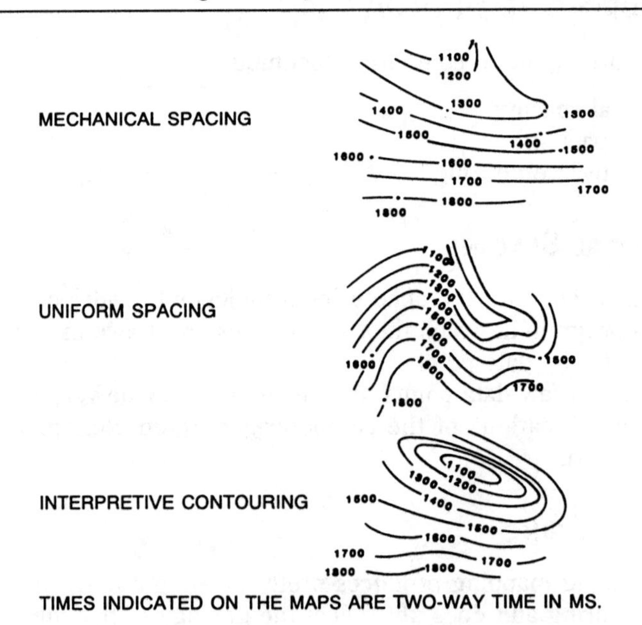

TIMES INDICATED ON THE MAPS ARE TWO-WAY TIME IN MS.

FIGURE 12–4. General rules for contouring

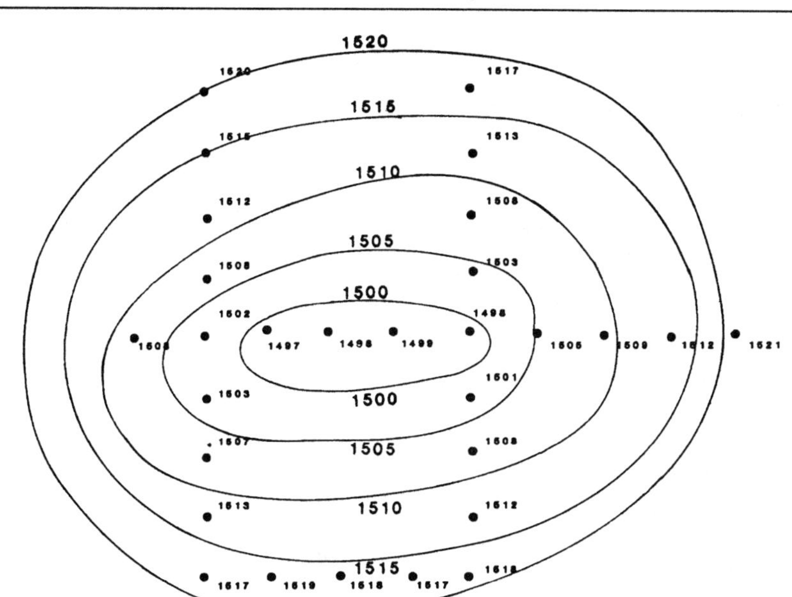

NUMBERS INDICATED ON THE MAP ARE TWO-WAY TIME IN MS.
CONTOUR INTERVAL: 5 MS

- A contour line never crosses over itself or another contour, except in case of overturned folds or reverse faults.
- Where structure slope reverses direction (as in a ridge or valley), the highest or the lowest contour must be repeated. A single contour may not mark the axis of reversal (see Figure 12–5).
- A contour line may not merge with contours of different values or with different contours of the same value. When a steeply sloping surface is projected upon the map, the contours sometimes appear to merge (see Figure 12–6).
- Contours should close, or end at the edge of the map, if there is enough data.
- The contour map may be read more easily if every fifth contour line is drawn heavier than the others. Number the contours so that the figures are evenly distributed over the map.
- Contour lines can be drawn for any preselected contour interval, and the same interval should be used over an entire prospect area.
- Always draw contours lightly with pencil and ink over them later.
- In a stream or a valley, the V-shaped contour points upstream.
- Use dashed lines for contours when control points are lacking.

FIGURE 12–5. Contouring lows and highs

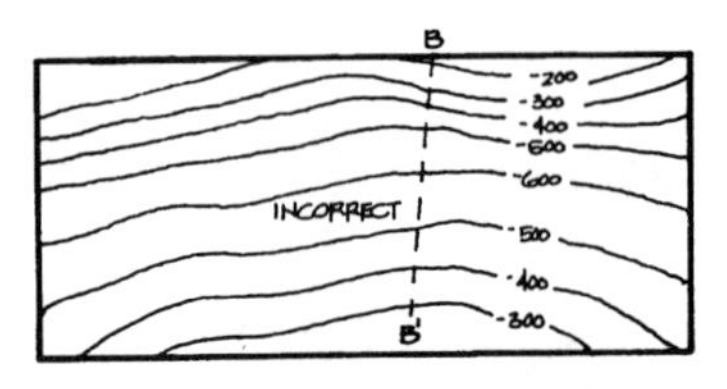

THIS MAP IS DRAWN WITH THE BOTTOM OF THE LOW AT CONTINUOUS-600 FT. THIS IS MECHANICALLY POSSIBLE, BUT VERY IMPROBABLE.

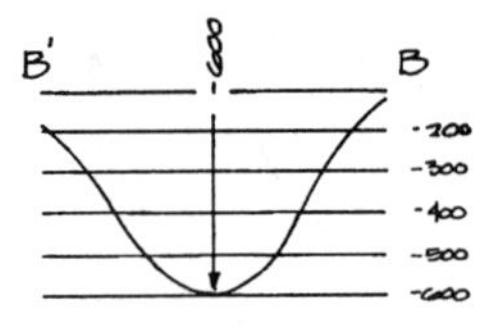

THE BOTTOM OF THE LOW MAY OCCASIONALLY TOUCH THE -600 FOOT LEVEL, BUT CERTAINLY NOT CONTINUOUSLY

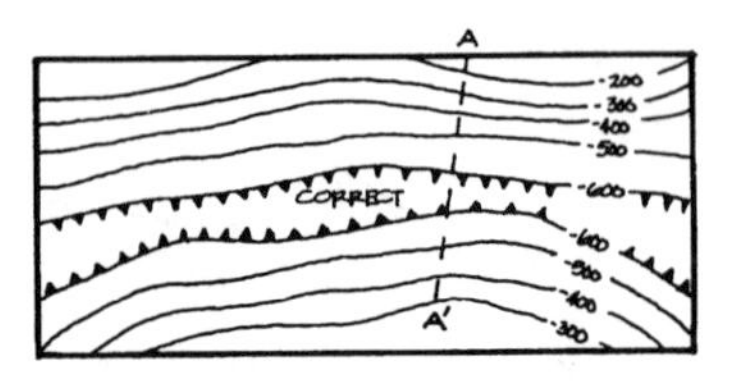

WHEN THE BOTTOM OF THE LOW IS CONTINUOUSLY BELOW-600 FOOT LEVEL, A REPEAT OF THE -600 FOOT CONTOUR IS NECESSARY.

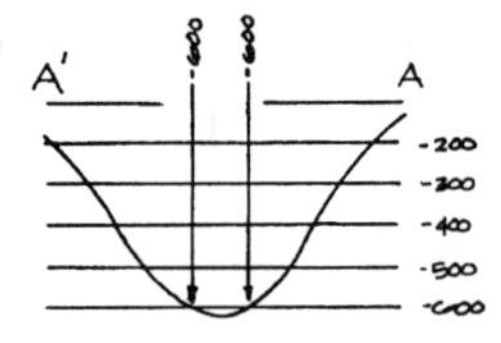

ONE MAY ASSUME THE BOTTOM OF THE LOW IS ACTUALLY BELOW THE -600 FOOT LEVEL.

FIGURE 12–6. Contouring steeply sloping surfaces

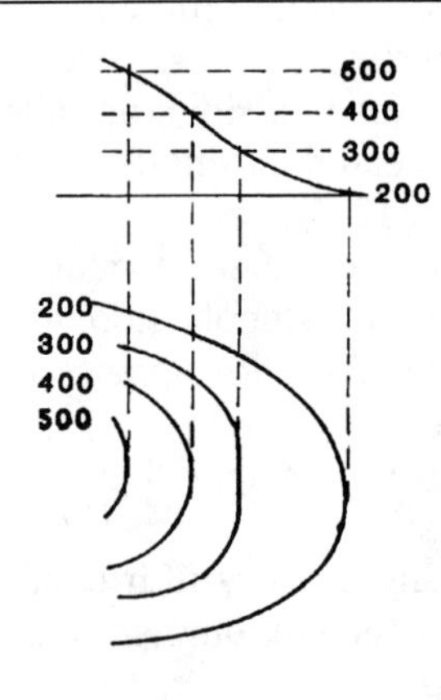

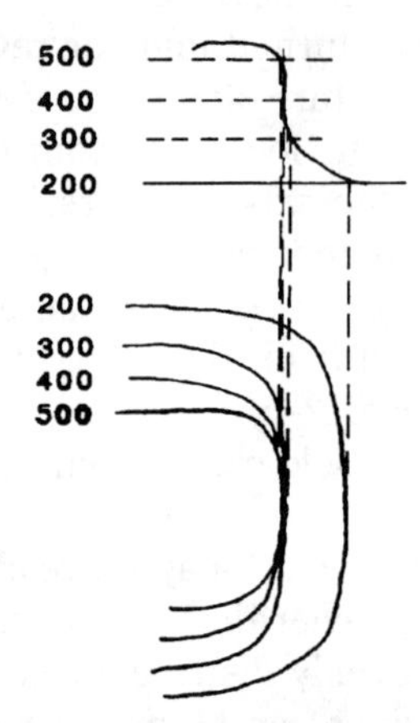

CONTOUR LINES MAY NOT MERGE WITH CONTOURS OF DIFFERENT VALUES OR THE SAME VALUE. WHEN A STEEPLY SLOPING SURFACE IS PROJECTED UPON A MAP, THE CONTOURS APPEAR TO BE MERGING.

- Contour by hand several lines (rather than a single line) at a time. The contours should be smooth and have as equal spacing as possible.
- Keep things as simple as possible.

The resulting contour map is only a first approximation of the true subsurface structure.

STRUCTURE CONTOUR MAP FROM SEISMIC DATA

- After the seismic loop is tied by applying necessary corrections to the marker to be mapped, values of the picks are spotted on the map (see Figure 12–7).
- The data is contoured using the same general rules as discussed before.
- The contour interval must be chosen to show the desired feature clearly. Symbols for faults and folds are the same as those used in subsurface contouring (see Figure 12–8).
- Use a light pencil and eraser. Do not hesitate to use the eraser until you have created a geologically sound structure map. Ink the map later.

2. ISOTIME AND ISOPACH MAPS

INTRODUCTION

An *isopach* contour is an imaginary line connecting points of equal thickness. An isopach map shows variations in the stratigraphic thickness of formations. Drawing an isopach map requires the construction of maps on two horizon or key beds, one on the top and the other on the bottom of the stratigraphic unit of interest. *Isochron*, or *isotime*, maps display contours of equal time or time difference. An *isochore* map is one that shows by contours the drilled thickness of a formation regardless of its true thickness. Isochores are no longer used.

Isopach maps have many uses in regional geology. They are used to locate buried structures and to indicate the direction of probable wedging, as well as help to restore the original depositional edge of the stratigraphic unit.

ISOCHRON MAPS

An isochron map is a time-interval map that shows the time difference between two markers. One can think of it as showing paleostructure; that is, the structure of the lower horizon at the time the upper horizon was deposited.

FIGURE 12–7. Tying seismic lines around a closed loop (courtesy of Seismograph Service Corporation)

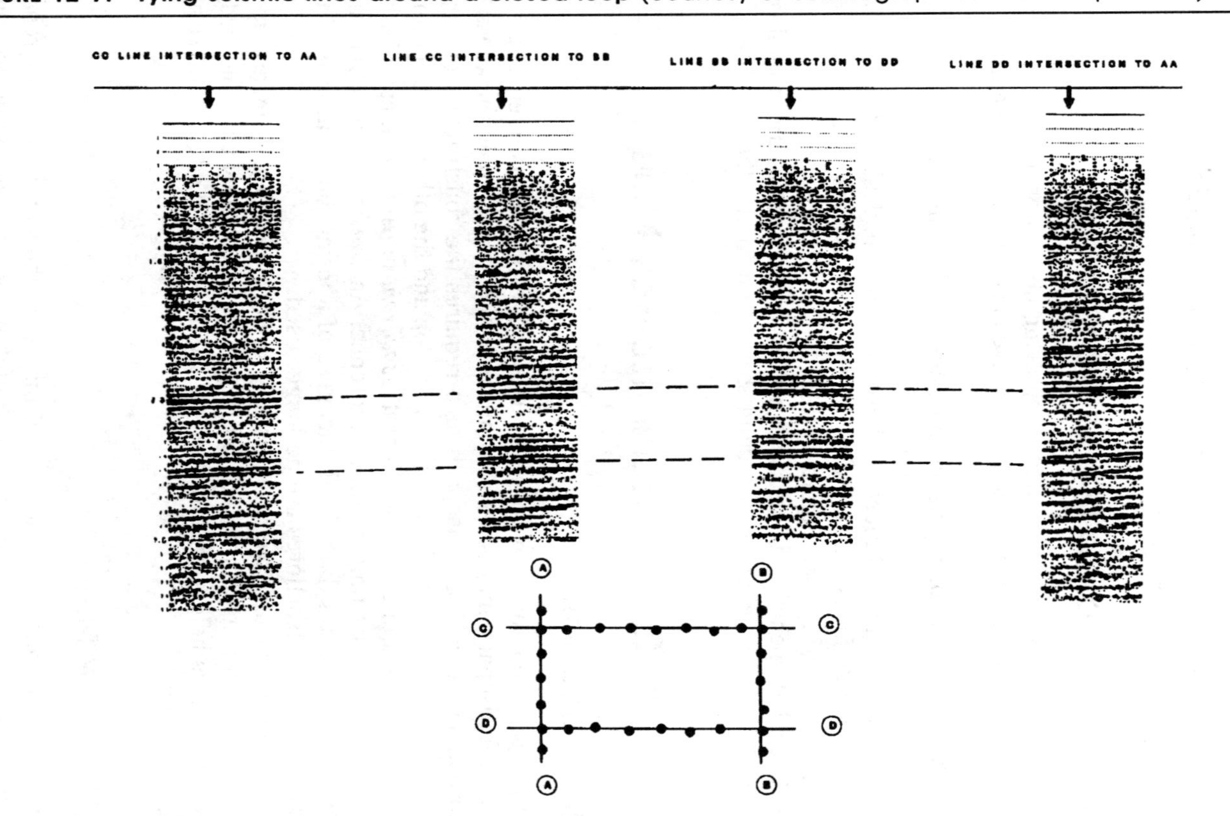

FIGURE 12–8. Seismic mapping scheme with standard symbols

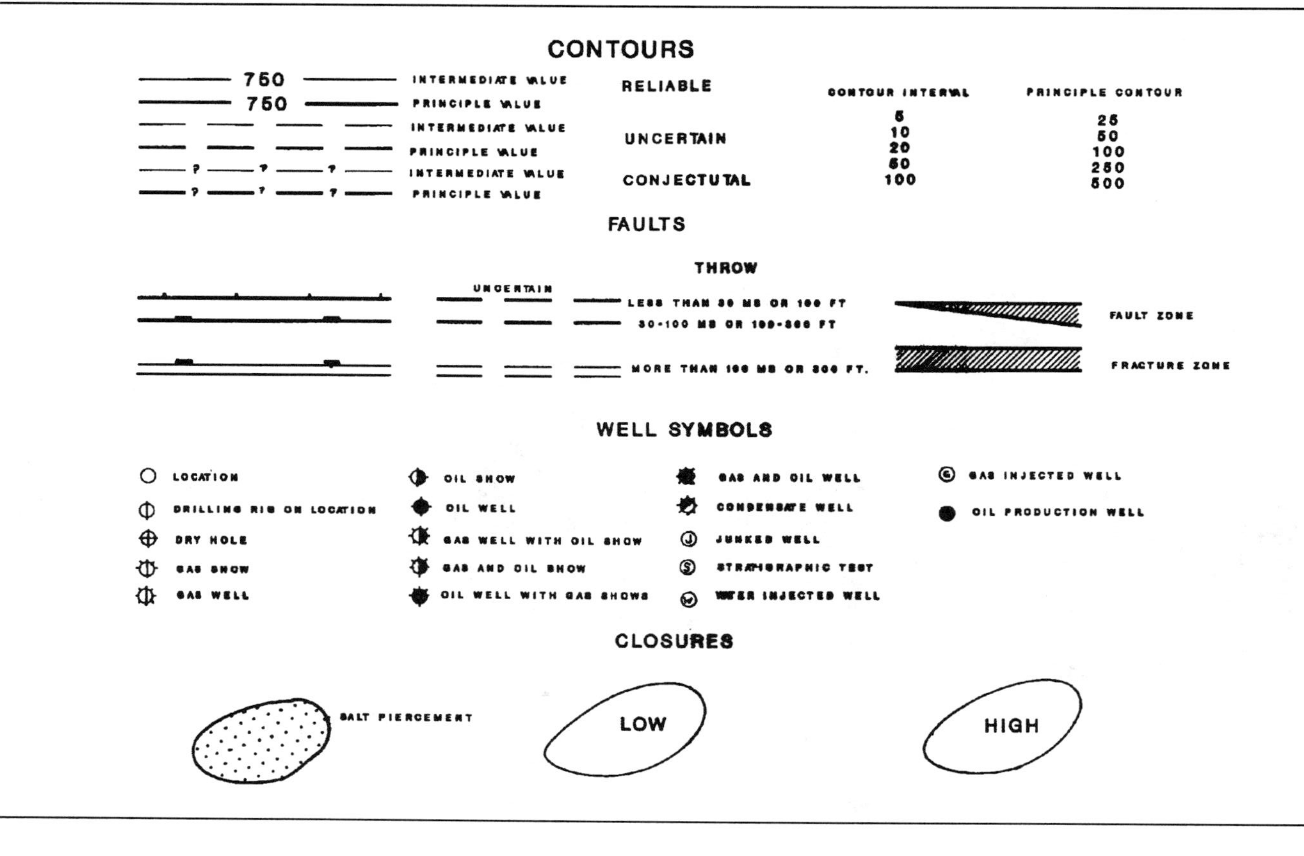

There are two ways to make a time interval map. One is by subtracting the times at shot points. The other, which is quicker but less accurate, is to make a structural contour map on each horizon and overlay one on the other, matching locations carefully. Then subtract wherever a contour on one horizon crosses a contour on the other map. This gives a new point, which is contoured using the previously discussed methods. Figure 12–9 illustrates this kind of seismic map.

SEISMIC ISOPACH MAPS

An isopach map can be prepared from seismic data if the velocities are known for the shallow and deep markers. The time picks are converted to depth by simply multiplying one-way time (one-half the time picked from the seismic section) by the velocity of the marker.

FIGURE 12–9. A typical two-way time interval map

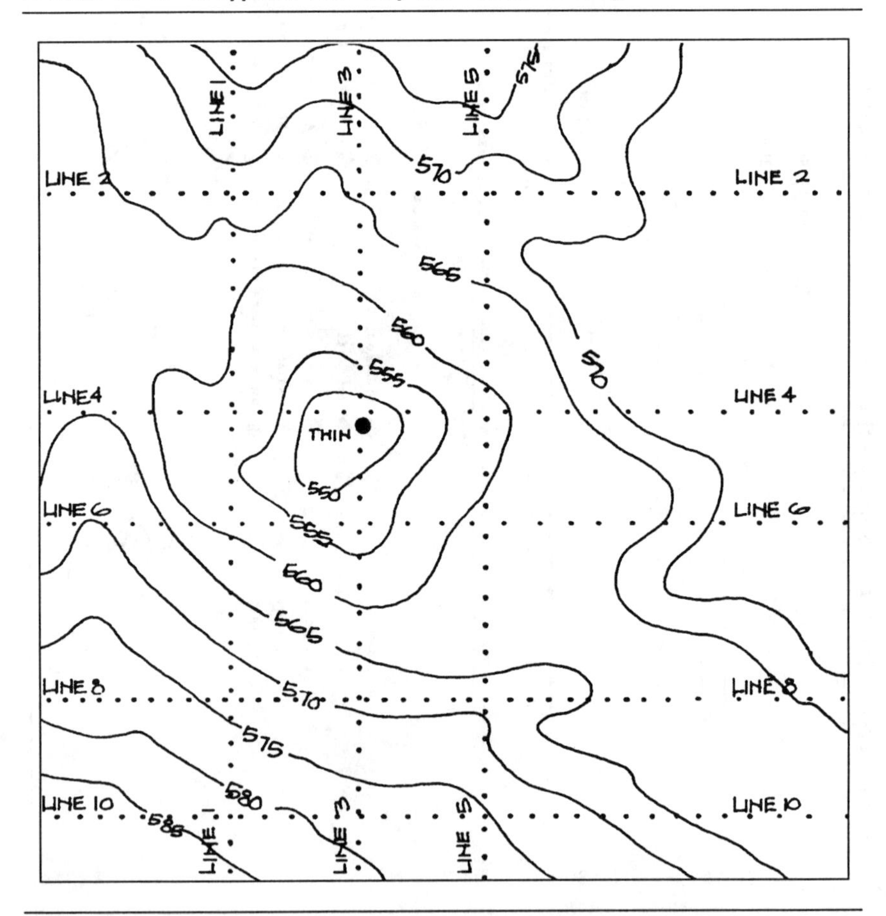

The shallow horizon-depth values are subtracted from the deep horizon values and spotted on the map. The values are then contoured. The velocity is assumed to be constant.

VELOCITY GRADIENT MAPS

Velocity measurements made with check shots at wells are accurate, but the wells are usually far apart and the velocities can change in a short distance due to changes in lithology, structure, and other factors.

Velocity determined from seismic data, even from the best stacking velocity, is not very accurate. Accuracy decreases with depth; and if one uses these velocity values to convert to depth, there will be some problems when tying to the wells. This is probably enough reason to make a seismic reflection time map; otherwise, there would be miscommunication with engineers, geologists, and managers who want to know the depth.

It is usually better to map in time rather than in depth, but there will be occasions when, in spite of all the problems, you do need to map in depth. In the development stage of a field, depth maps are more important than time maps.

Such problems as velocity pull-up or velocity pull-down make the time structure incorrect below the feature that causes the velocity effect. Some areas have sudden changes in velocities, and in such areas the apparent structure on a time map is quite different from the structure measured in depth.

For a velocity that varies in the area, one can make a velocity map to the formation at each well or at some points where fairly good velocity information is available and make them fit consistently (Figure 12–10). Knowing the horizon tops from several wells close by the seismic data, one can get velocity values at this well. Contour these values and make a velocity gradient map. Read the velocities at shot points and write these below the time picks. You are now ready to make a depth map for this horizon.

Following the same procedure as above, one can make a velocity gradient map for another horizon.

An isopach map can be drawn from a structural contour map derived from time picks and a velocity gradient map. This is done by calculating the depth differentials, plotting them on the map, and contouring them.

3. INTERACTIVE INTERPRETATION VIA COMPUTERS

Advances in computer systems technology are rapidly changing in hardware configuration and software sophistication. Current hardware systems are very impressive, very fast, and friendly to use.

FIGURE 12–10. Average velocity map

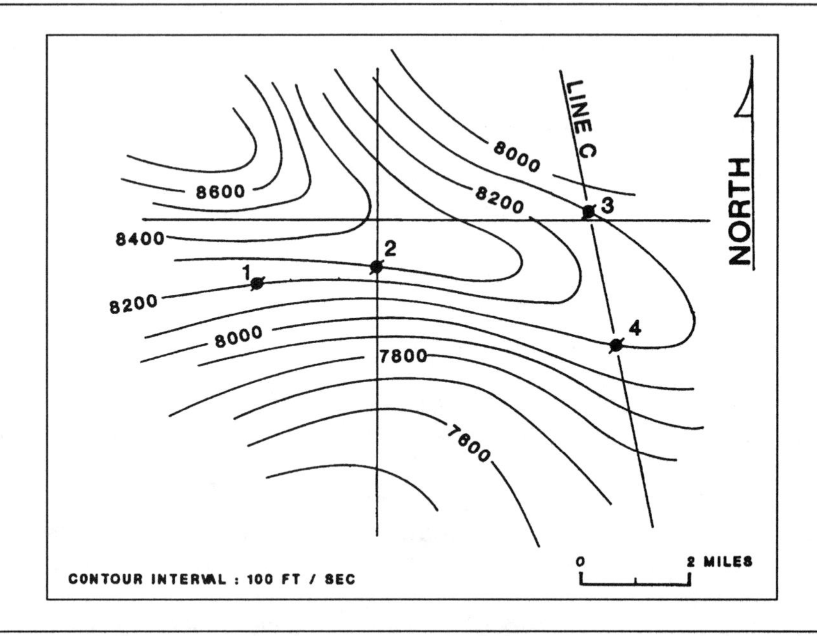

The computer industry continues to build smaller and faster equipment. Last year's mainframe computer is today's desktop personal computer, which may in a few months be a hand-held computer.

Software systems are equally impressive and can be utilized in seismic analysis, reservoir modeling, reservoir simulation, and data management.

Desktop systems can provide a great visual output that can monitor complex reservoir configurations in three dimensions. These programs are user friendly and guide the user through the most complicated application.

As the production cost increases, it becomes critical to understand more about the reservoir characterization.

Advances in computer graphics software help geoscientists to better understand the reservoir configuration by using 3-D imaging software. The software available for interpretation is flexible enough to let the geoscientist experiment with a number of possibilities on the screen to help interpret a structural or stratigraphic problem.

Seismic surveys create a vast amount of data, and the processing, sorting, and retrieving of these data can be time consuming and can impact the efficiency and profitability.

A wide range of systems are available to the explorationist and geoscientist that can help in exploration data management and make it possible for any geoscience professional to view, select, and request any item from the exploration data base.

Users can view the selected items at any point of the operation such as well logs, seismic lines, base maps.

Work stations have proved their effectiveness in 3-D reconstruction of seismic data.

As we discussed in Chapter 10, horizontal slices and 2-D seismic sections are combined to create a 3-D volume. Three-dimensional computer graphics have been integrated with measurement, analysis, interpretation and other reservoir properties to help the interpreter better understand more about the reservoir.

In addition to the fast analysis of the seismic data, super computers can run tomograms for determining average velocity variation for depth conversion and migration.

Super computers run tomograms that can show the variation of some petrophysical properties, such as porosity, permeability, and fluid content between wells.

The systems available vary in price from tens of thousands to hundreds of thousands of dollars, depending upon the purpose and the application of the system.

Interactive interpretation helps cut down the turnaround time in data processing and interpretation and enables oil companies to manage a vast amount of data. They can save a lot of time and effort and meet their deadlines and commitments for drilling, lease sale, or monitoring secondary recovery project.

For readers who would like to increase their knowledge about interactive computer systems, we recommend J. A. Coffeen's book, *Seismic on Screen*, published by PennWell.

BIBLIOGRAPHY

Anstey, N. A. *Seismic Interpretation: The Physical Aspects*, Boston: IHDRC, 1977.

Anstey, N. A. *Seismic Exploration for Sandstone Reservoirs*, Boston: IHRDC, 1978.

Coffeen, J. A. *Interpreting Seismic Data*. Tulsa, OK: PennWell Publishing Company, 1984.

Coffeen, J. A. *Seismic On Screen*. Tulsa, OK: PennWell Publishing Company, 1990.

Levorsen, A. I. *Geology of Petroleum*. San Francisco: Freeman & Co., 1958.

Pettijohn, F, J. *Sedimentary Rocks*, New York: Harper & Row, 1948.

Pirson, S. J. *Geologic Well log Analysis*, Houston, Texas: GPC, 1970.

Sheriff, R. E. *A First Course In Geophysical Exploration and Interpretation*. Boston: IHDRC, 1978.

Wharton, Jay B., Jr. "Isopachous Maps of Sand Reservoirs." *Bulletin of the AAPG* 32, no. 7 (1948): 1331–1339.

Chapter 13

Case Histories and Applications

This chapter presents some case histories that show successful application of new technologies in the field of reservoir seismology. In addition, there are some reprints of technical papers on the applications of these technologies.

Is AVO the Seismic Signature of Lithology? A Case History of Ship Shoal-South Addition*

by Fred Hilterman
Geophysical Development Corporation
Houston, Texas

Many times we've heard, "I've tried amplitude-versus-offset (AVO) and it doesn't work in my area." Before casting AVO aside, let's look at the process one more time. AVO analysis is actually a two-step process. The first step relates the seismic response (CDP gather) to the rock's velocity, density and Poisson's ratio; the second step relates these rock properties to the lithologies (sand, shale, etc.).

With regard to the first step, R.T. Shuey introduced a classic expression in 1985 for the AVO seismic response in terms of the rock properties. An approximation of Shuey's reflection coefficient (*RC*) equation is

$$RC(\theta) = NI\cos^2(\theta) + 2.25\Delta\sigma\sin^2(\theta),$$

where NI is the normal incidence reflection coefficent and $\Delta\sigma$ is the difference of Poisson's ratio between the lower and upper media. In fact, Shuey's equation or one similar to it is routinely implemented to extract the seismic rock properties NI and $\Delta\sigma$ from CDP gathers.

Now the interpreter (in the second step of the AVO process) must predict the lithologies from the rock properties which are encoded in NI and $\Delta\sigma$. To accomplish this, the rock properties of the expected lithologies need to be cataloged, and the lithologic prediction tested by forward modeling.

Recent publications have suggested many inversion schemes to accomplish the first step of the AVO process, but limited attention has been given to the second step. In order to answer the question, "Is AVO the seismic signature of lithology?," a project was undertaken in which seismic rock properties, in a small geographical area, were cataloged vis-à-vis lithology, and subsequent AVO modeling was tested for predictive accuracy.

STUDY AREA

Figure 1 illustrates a major Plio-Pleistocene depocenter which is notorious for its poor match of conventional one-dimensional (1-D) synthetics to

*Reprinted, by permission, from *Geophysics: the Leading Edge of Exploration*, June 1990, courtesy of The Society of Exploration Geophysicists and the authors.

FIGURE 1. Study area in Ship Shoal-South Addition, offshore Louisiana. Ninety-seven wells are indicated along with a seismic line through the well tie.

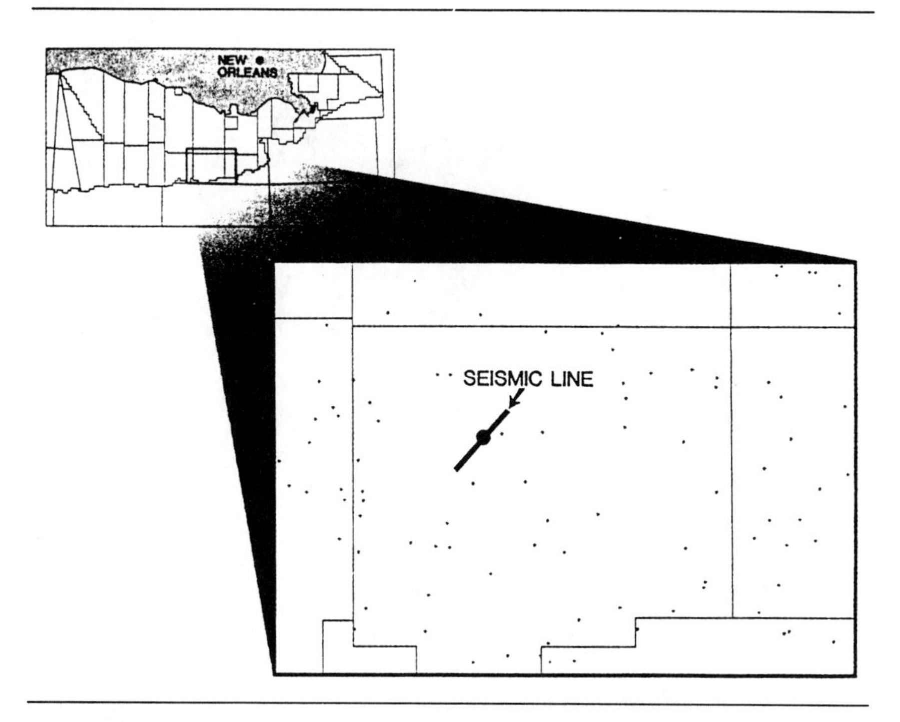

stack sections. The major lithologies in the study area are sand/shale sequences which have low acoustic impedances. To assist in building the rock properties catalog, 97 wells distributed throughout the area were selected.

In addition, AVO analyses were done on the seismic line indicated in Figure 1 and shown in Figures 2 and 3. Selection of the line was not arbitrary, but based on several criteria which included:

- No complex geology or near-surface problems
- CDP gathers with AVO anomalies
- A well location within 200 ft which included a full suite of logs (from 200 ft to TD) that contained gas-sand, clean water-sand, and shaly-sand packages

The reflection amplitude preservation (RAP) section equivalent to the shaded area in Figure 2 is shown in Figure 3 along with a CDP gather inserted at the well location. The mute zone, which represents incident angles over 60 degrees, covers twice the angle range that would normally be included in a conventional stack. There are two AVO anomalies evi-

FIGURE 2. Migration-before-stack seismic line.

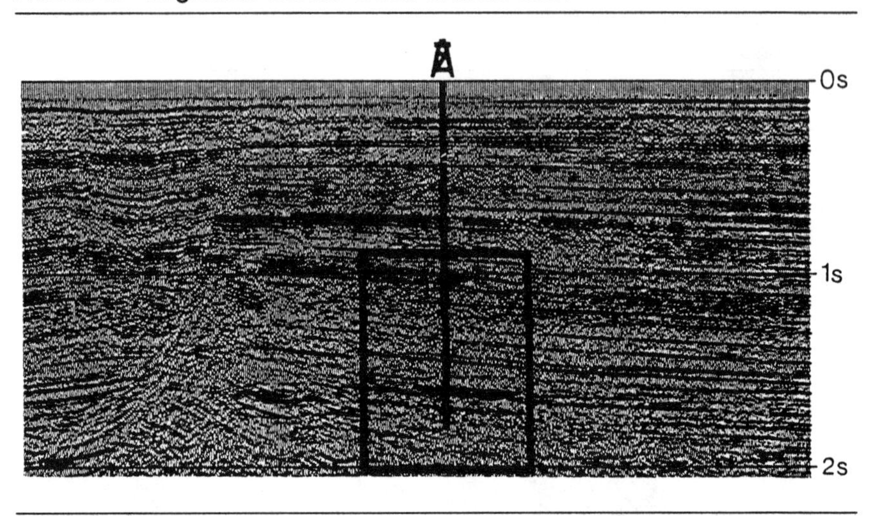

FIGURE 3. CDP gather inserted at well location.

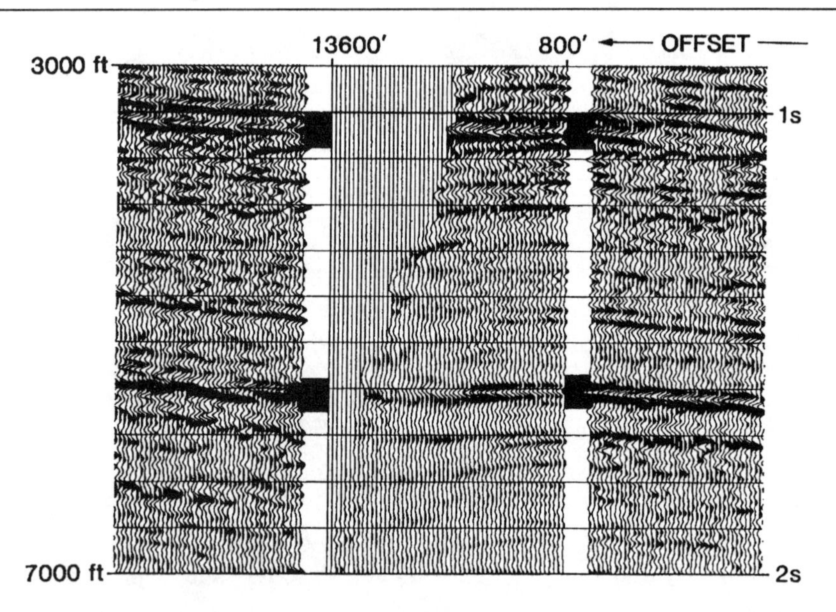

dent in Figure 3—one at 1.050 s (3000 ft) and the other at 1.600 s (5500 ft). One of the AVO anomalies is a gas sand while the other is a clean water-saturated sand. Using AVO analysis, how does one discern fluid type and determine the appropriate thicknesses?

If either modeling or inversion is chosen as the vehicle for step 1 in the AVO analysis, we still need to relate the elastic rock properties to the expected lithologies. With this in mind, a "classical" solution for cataloging the "average" rock properties to lithology was undertaken in order to provide the necessary information for step 2.

ROCK PROPERTIES

The first step in cataloging the rock properties involved building three series (A,B,C) of velocity histograms covering 500 ft depth intervals from all 97 wells. Figure 4 shows the 16 histograms which span the range 0–8000 ft for each series. The lithologic constraints are listed in Figure 5.

FIGURE 4. Velocity calibration of 97 wells.

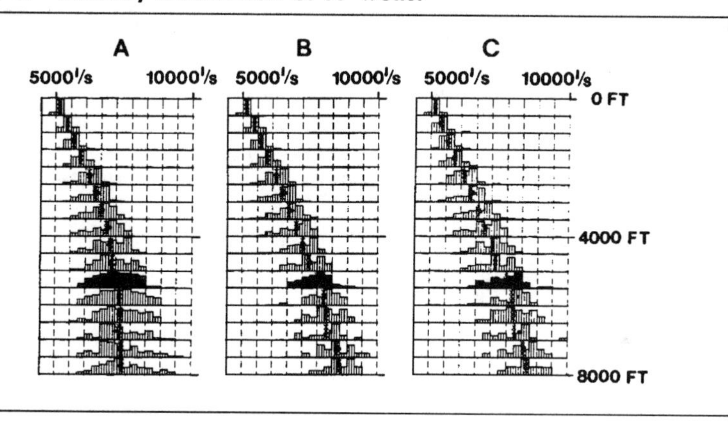

FIGURE 5. Analyses of 5000–5500 ft histograms from Figure 4.

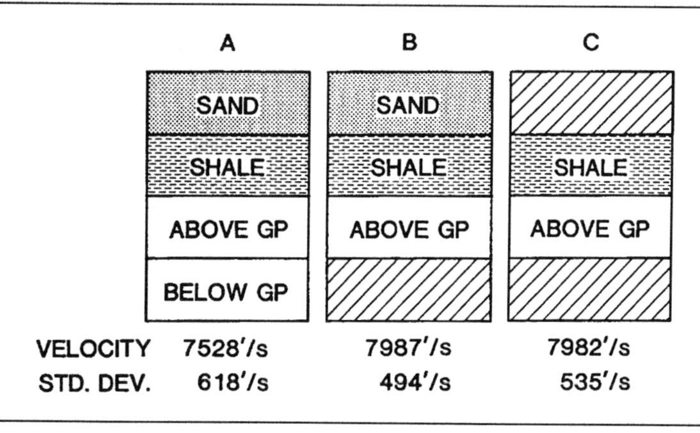

The A-series was wide open. Basically, all sand/shale lithologies that were not "washed out" or hydrocarbon saturated were included. As indicated in Figure 5, the 5000–5500 depth interval of the A-series has an average velocity of 7528 ft/s with a standard deviation of 618 ft/s.

Following typical statistical analysis, when lithologies are subdivided further and a lower standard deviation results from this subdivision, then the classification is considered to be improving. Now examine the B-series where the lithologies having abnormal pore pressure (geopressure) were omitted. As is indicated in Figure 5, the standard deviation did decrease and, as expected, the average velocity increased. However, when only shales above geopressure were examined (C-series), the average velocity hardly charged, but the standard deviation increased. This could be interpreted to mean that sand and shale lithologies above geopressure have essentially the same average velocity. Additional histogram series were completed for both velocity and density. Comparisons of the average trends are shown in Figures 6 and 7. Notice in these results that the average sand velocity is always greater than the shale's, while the average shale density is always greater than the sand's. (Shaded zones separate the sand from the shale trends.) These average trends indicate that a single expression would not be sufficient to approximate density from velocity for both sand and shale, as is often done by the expression $\rho = .23 V^{.25}$.

FIGURE 6. Average velocity trends. Sand values are indicated by triangles; shales are circles.

FIGURE 7. Average density trends.

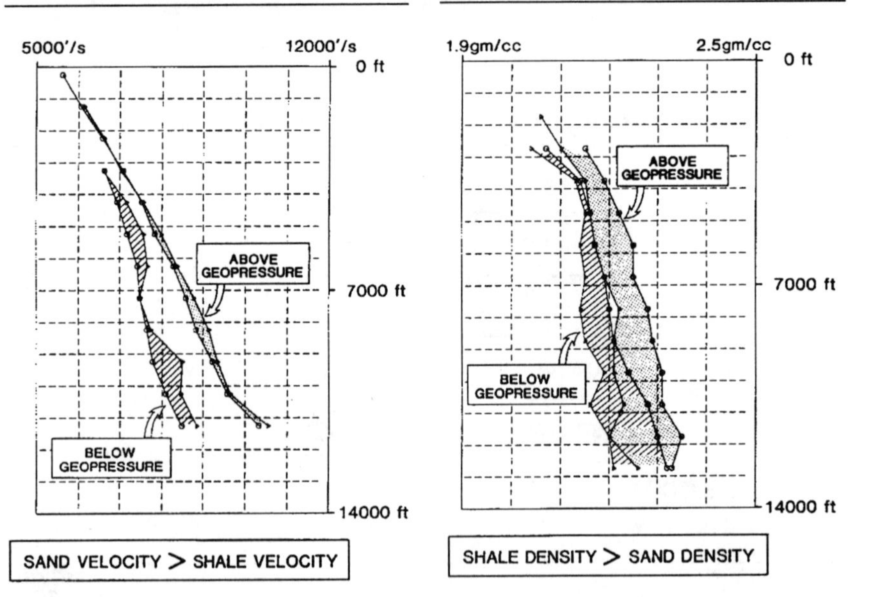

Continuing in a classical approach, the trend curves for velocity and density were fitted to six different analytic expressions. Such expressions are always desirable for classification because rock-property variations between lithologies can then be easily documented. For all expressions tried, a linear function with depth was the best fit for velocity and an exponential function with depth for density (Figure 8).

Having established analytic expressions for the rock properties in this well, the next step was to set up a model for one of the AVO anomalies depicted in Figure 3 (Figures 9 and 10). The shale compressional velocity and density expressions are the same as in Figure 8. An appropriate method for estimating the shear velocity of shale is the "mud line" equation as described by J.P. Castagna et al. in 1985. Sand compressional velocity and density were trend fitted and the sand shear velocity was

FIGURE 8. Shale properties above geopressure.

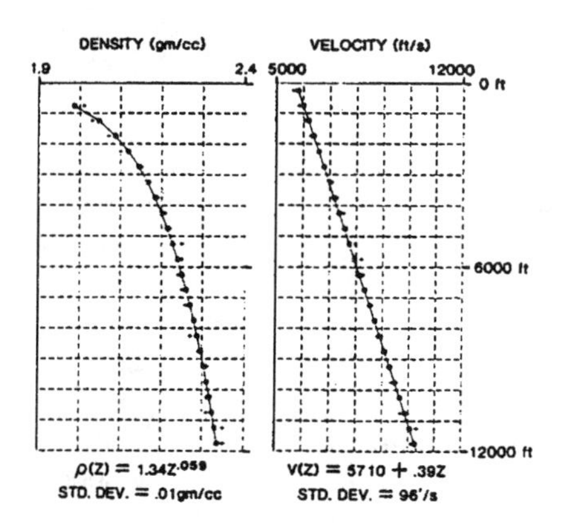

FIGURE 9. Analytic expression of rock properties above geopressure.

SHALE:

$Vp(Z) = 5710'/s + .39Z$

$\rho(Z) = 1.34\ Z^{.059}$

$Vs(Z) = 1077'/s + .34Z$ (MUD LINE)

SAND:

$Vp(Z) = 5617'/s + .44Z$

$\rho(Z) = 1.61\ Z^{.034}$

$Vs(Z) =$ GASSMAN (Vp, ρ, ϕ, etc.)

FIGURE 10. AVO model rock properties for depth of 3000 ft.

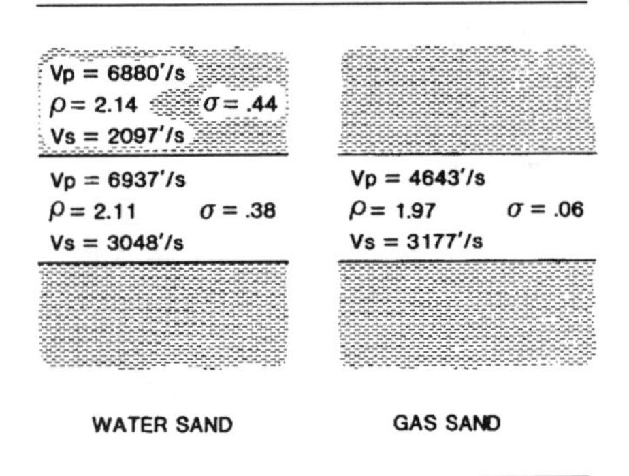

FIGURE 11. AVO synthetics based on properties in Figure 10. A 40 ft water sand is shown in the upper right with the lower right is a 6 ft gas sand.

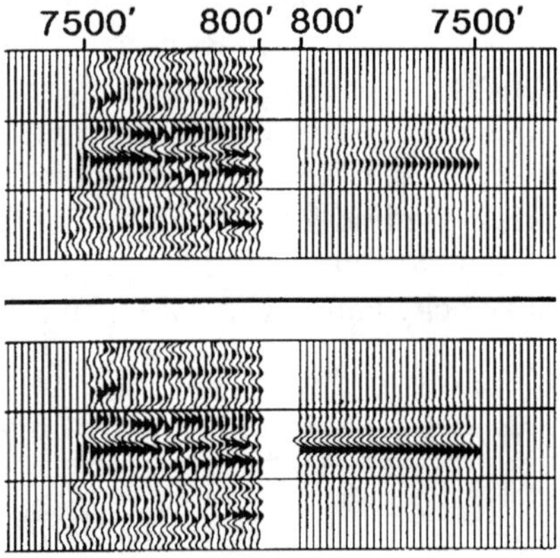

7500′ 800′ 800′ 7500′

estimated by Gassman's equation. With these equations, the rock properties to be expected for the 3000 ft AVO anomaly were developed. The thicknesses of the water sand and gas sand were allowed to vary until a best "eyeball" fit was obtained (Figure 11). Based on this subjective criterion, the gas-sand model was thought to be a better fit than the water-sand model.

The paper would conclude at this point if a gas sand been found at 3000 ft in the well logs. To our chagrin, the 3000 ft AVO anomaly was a clean water sand. Now the question is, "What went wrong with this classical approach of cataloging the rock properties to lithology?"

SINGLE WELL HISTOGRAMS

In the classification scheme, rather than examining a multitude of wells together, an alternative approach is to examine histograms from single wells. These histograms will illustrate rock property trends and variations within a well. Another well near the seismic line was selected for illustrating this principle. The histogram series on the right in Figure 12 was developed with one-foot samples in 100 ft depth intervals and then the frequency values were contoured. Only those depth samples which are blackened in the left portion of Figure 12 were classified as shale and only these depth zones were included in the contoured acoustic impedance histogram. Both high and low impedance trends are indicated, implying at least two (if not three) different shale types exist in this well. Which

FIGURE 12. Contoured histograms based on acoustic impedance function from one well.

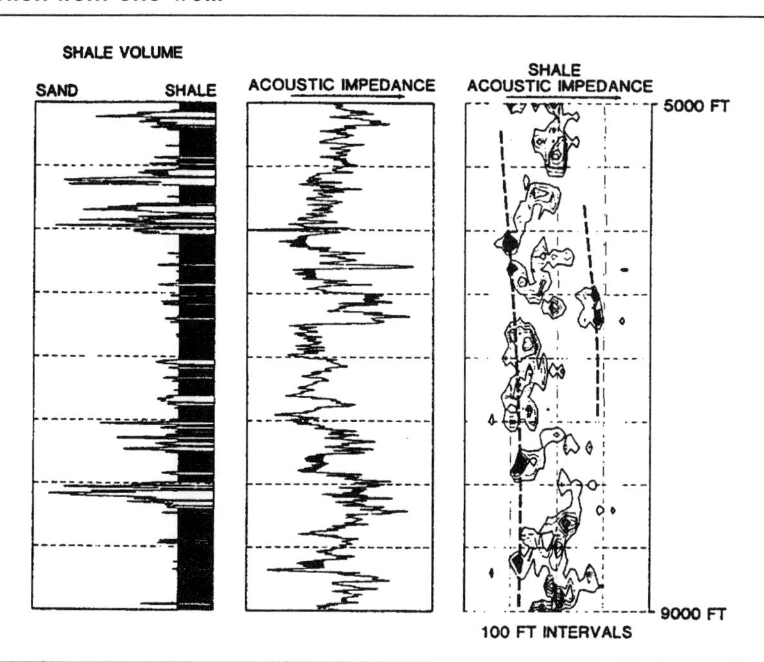

shale type should we have used in our previous modeling—type 1, 2, or 3? There isn't a single "average" rock property for shale in this well!

In fact, the results displayed in Figure 13 are even more alarming. Here a series of histograms from a single well in the northern portion of the study area are compared to a series of histograms from all 97 wells (trend A, Figure 4). The variation in the shale velocity for this one well is almost as great as the velocity variation found in all wells, irrespective of the lithology. Off-hand this would indicate a failure to classify rock properties. A common ground can be found, however, when the variation of shale acoustic impedances are viewed in subareas, such as within major fault blocks. Examine the similarity between three wells located in the southern portion of the study area to three wells located in the northern portion (Figures 14 and 15). In Figure 14 (southern well histograms), one shale type exists to the top of the "hard" geopressure (approximately 8000 ft). This is a second pressurized zone which appears to be isolated from the shallow geopressure zone which has a top at approximately 2500 ft. The high impedance at the top of the deeper pressured zone is common in wells located in the southern portion of the study area.

The shale impedances in Figure 15 (northern well histograms) vary up to 25 percent within 200 ft intervals. Consequently, an accurate description of the depositional environment is necessary because if the

FIGURE 13. Within-well versus within-region velocity analysis.

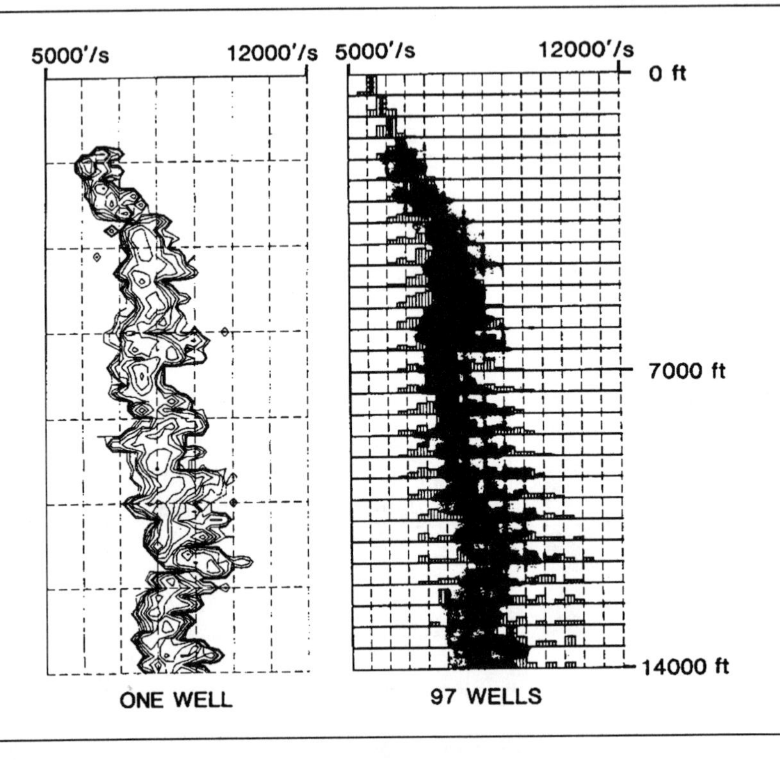

FIGURE 14. Shale acoustic impedance functions in southern portion of study area.

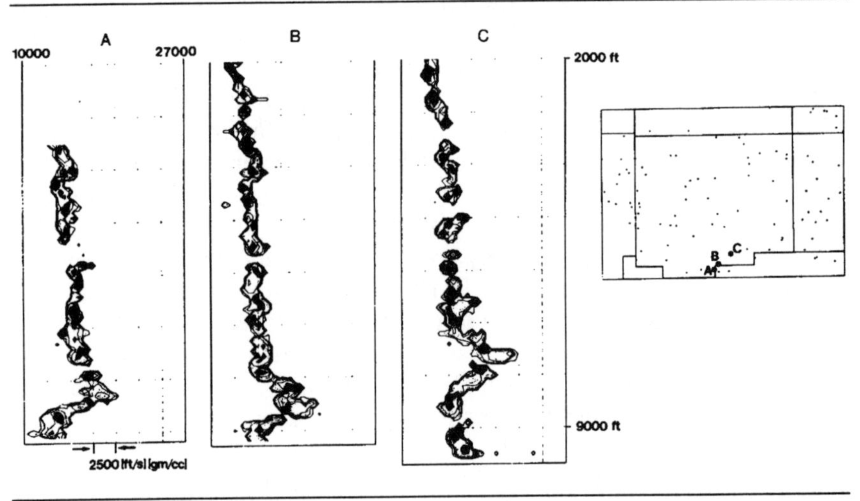

FIGURE 15. Shale acoustic impedance functions in northern portion of study area.

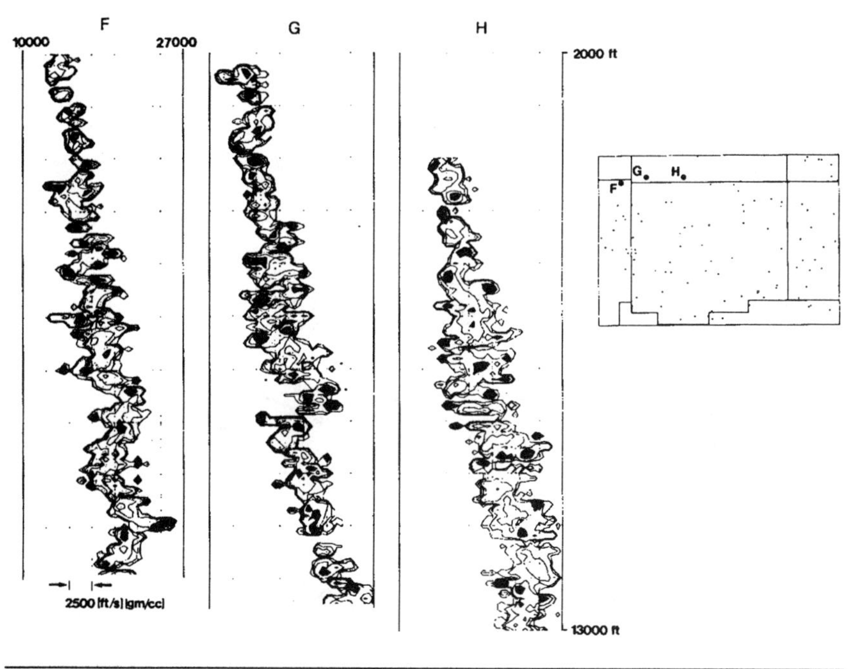

proper shale type is not selected to model AVO anomalies, subsequent lithologic and thickness determinations are useless.

PETROPHYSICAL ANALYSIS

After observing the large variation in shale acoustic impedances for different shale types, detailed volumetric analyses were conducted. Of the 97 wells, 47 had sufficient neutron-density curves for a volumetric analysis. The petrophysical flow and a typical volumetric analysis are given in Figures 16 and 17. The volumetric analyses required five quality control steps before a final result was obtained. The major problem that had to be resolved was the establishment of "baseline" curves that classified the various rock types encountered. In fact, a cluster analysis on the 47 wells identified five major shale types in this study area that must be treated separately if proper rock properties for the shale were to be modeled. In short, an average property must be associated with a rock type, and broad classification such as sand, shale, limestone, and dolomite is inadequate for AVO analysis.

FIGURE 16. Flow diagram for volumetric analysis.

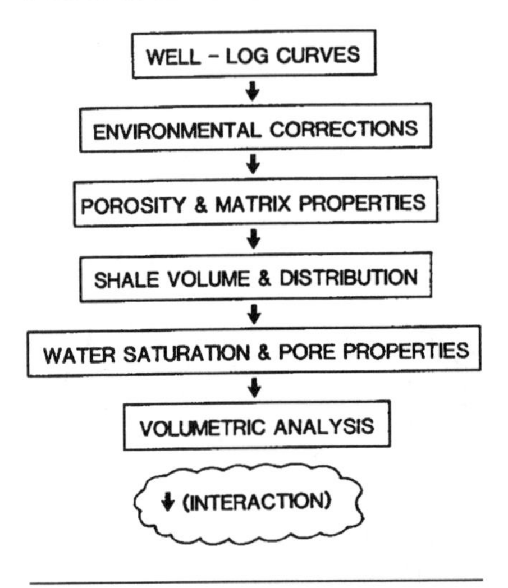

FIGURE 17. Typical volumetric analysis.

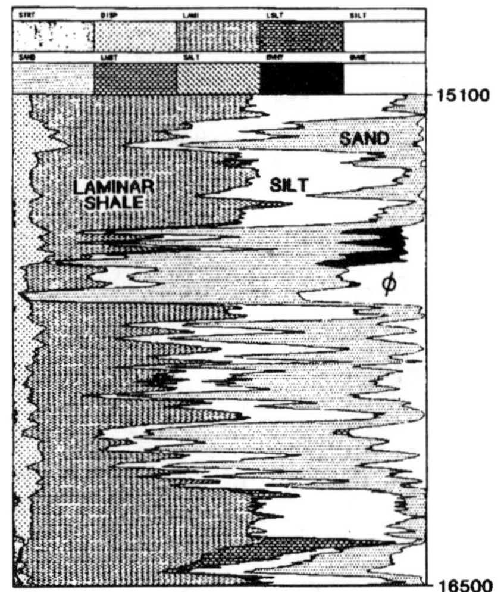

Regression analysis on the five shale types to determine if they are independent with respect to the volumetric properties did not appear to be the most practical solution to aid the intepreter. Rather, it was felt that when analyzing an AVO response, less error in rock type selection would occur if the interpretation "takes off" from the nearest well in the major fault block or geologic sequence.

The next culprit to be examined was the sand lithology. As reported by D. Han et al. in 1986 and Castagna, both compressional and shear velocity have a dependence not only on porosity but on clay content. Thus, the volumetric analysis (Figure 17) was developed to split the clay content from the silt content in the total volume of shale. Also, baseline curves were established to estimate the lime content in both sand and shale.

In order to tie these volumetric rock properties to the CDP gather at the well, an estimate of the shear velocity was necessary. Unfortunately, applying the "mudline" equation to shales and Gassman's equation to sands proved to be too awkward. No definite shale volume cutoff for discriminating shale from sand could be defined. Instead, an empirical approach which allows Gassman's equation to be applied to both sands and shales when the dryrock Poisson's ratio is based on clay content (Figure 18) was developed.

This empirical relationship was applied to tie well data, and crossplots of the resulting Poisson's ratio versus the input compressional ve-

FIGURE 18. Empirical dry-rock Poisson's ratio for Gassman's equation.

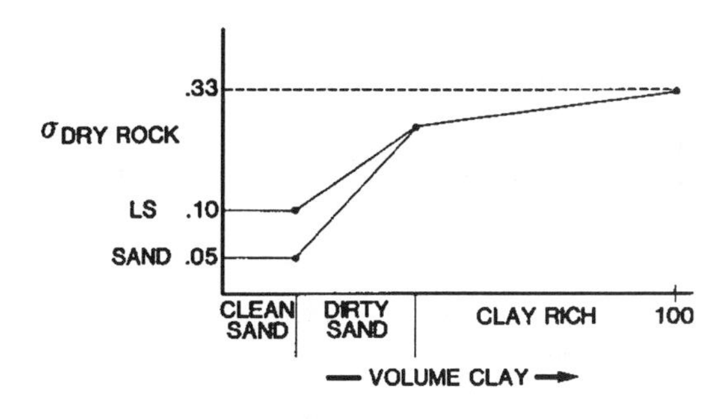

FIGURE 19. Compressional velocity versus Poisson's ratio for *clean* sand and shale. Sand values are indicted by triangles; shales are circles.

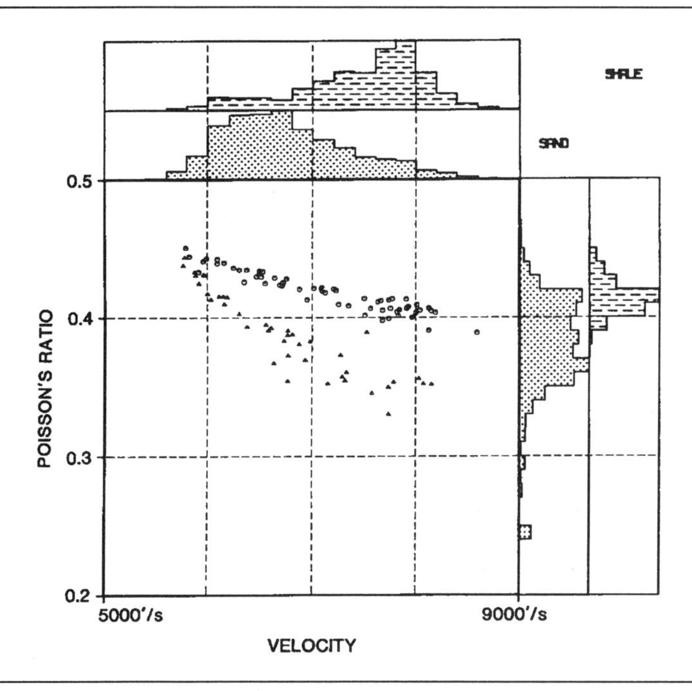

locity are shown in Figures 19 and 20. Clean sands are distinctly separated from shales by Poisson's ratio while dirty sands are not. Of course, these results follow from the specification of the dry-rock Poisson's ratio (Figure 18). Note that as compressional velocity approaches 5000 ft/s, sand and shale Poisson's ratio converge as expected.

FIGURE 20. Compressional velocity versus Poisson's ratio for *dirty* sand and shale.

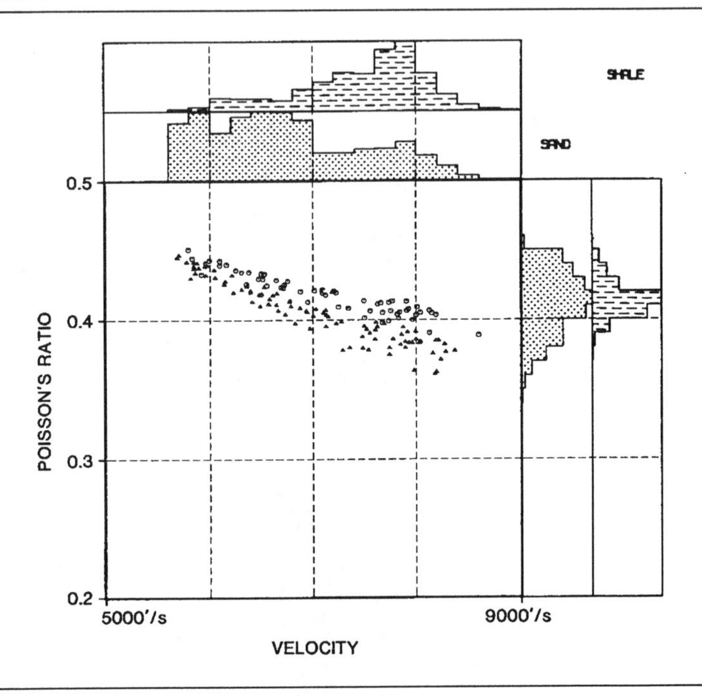

The velocity trends as reported in Figure 6, which averaged different rock types of shale, run contrary to the trend for this tie well. The upper histogram of Figure 19 indicates that sand has a velocity mode which is 1200 ft/s *lower* than shale.

SEISMIC COMPARISON

With the shear velocity estimated as described above, an AVO synthetic with all elastic wavefields was generated via GDC's SOLID program. The AVO synthetic and the CDP gather at the well location are shown in Figure 21. The match was not artificially forced. That is, no edits of the sonic or density logs or time shifting of the logs were allowed to facilitate a better match with the field seismic. If the logs were adjusted arbitrarily for the sake of a better match, then the integrity of the subsequent interpretation would be questionable.

In Figure 21, the CDP gather was matched to the volumetric analysis by correlating the AVO character in the synthetic CDP and assigning the corresponding lithology. Where the 1-D synthetic was somewhat

FIGURE 21. Migration-before-stack CDP gather at well location compared to total elastic AVO synthetic derived from well-log rock properties.

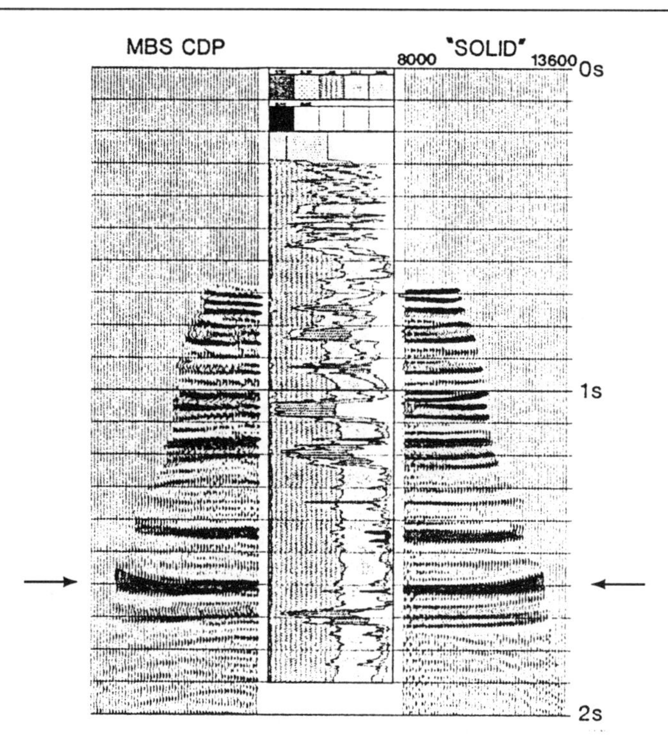

ambiguous as to the tie with the seismic section, the AVO character match was not. The biggest discrepancy, a gas sand at 1.600 s (as indicated by arrows) shows brighter on the field data than on the synthetic for a good reason. The sonic log indicated only 5 ft of pay while the resistivity logs indicated 10 ft of pay. In this zone, the sonic log was definitely questionable but no editing was performed for the sake of this example.

With the CDP gather at the well tied to the volumetric analysis via the AVO synthetic, it was no problem then to tie the seismic line as shown in Figure 22. There are several points worth noting. First, the continuous reflectors at 0.710 s, 1.000 s, 1.160 s, and 1.450 s are not associated with major sand packages as one might expect. The shallow AVO response (1.020 s) is a clean sand package with limited lateral extent. The thick shaly-sand package around 1.700 s is chaotic and doesn't show itself as a distinct continuous reflector. In fact, the sand formations that have gradational shale content have only a slight stack response.

FIGURE 22. Migration-before-stack seismic with well-log volumetric analysis.

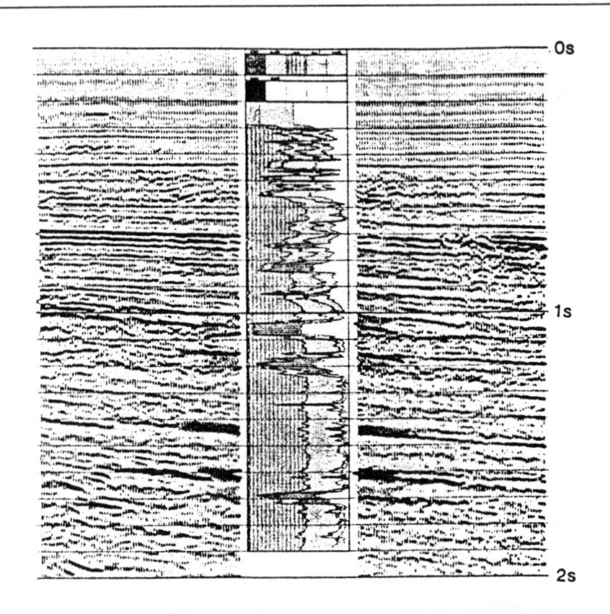

AVO WELL LOG MODELING

Having matched the well-log AVO synthetic to the seismic line, the next step would be to verify one's interpretation at a nearby prospect with AVO modeling. One problem that arises if several models are to be tested is that the total elastic AVO solution requires a sizable amount of computation to properly define the lithology. As only the largest supercomputers can provide results in a reasonable amount of time, an alternate approach was sought. Fortunately, in some areas (especially where the velocity gradient is small), ray-trace modeling matches the total elastic modeling (Figure 23) and thus is a good alternate for AVO modeling.

If ray tracing is an adequate method of modeling in an area, then Shuey's approximation can be implemented (Figure 24) to illustrate what portion of the AVO response can be associated with Poisson's ratio. The sand package at 1.200 s has been gas-charged in this model to illustrate the AVO response caused by Poisson's ratio variations. The Poisson's ratio contribution to the AVO for the shaly gas-sand is almost identical to that of the clean water-sand above it. Thus, the degree of shaliness is necessary to know if an adequate AVO match is to be obtained. To further identify the AVO responses from this well, histograms of Poisson's

FIGURE 23. Total elastic wave solution versus ray-tracing solution.

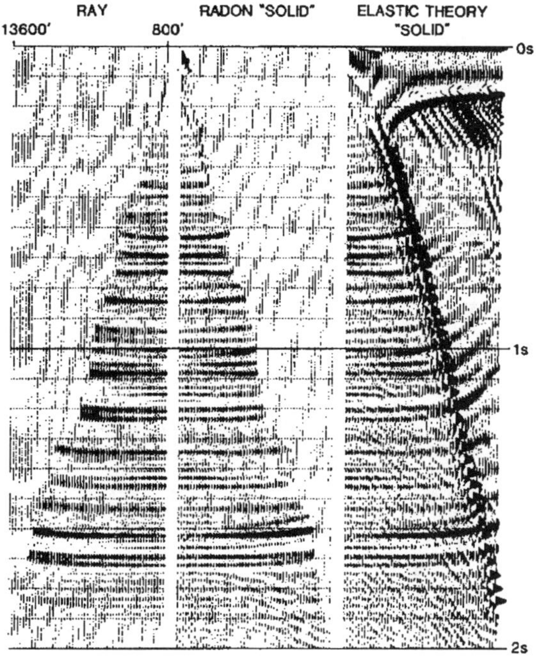

RAY
13600'
800'
RADON "SOLID"
ELASTIC THEORY
"SOLID"
0s
1s
2s

FIGURE 24. Lithologic signatures. Overplot on left-side panel represents 10-degree increments in incident angles.

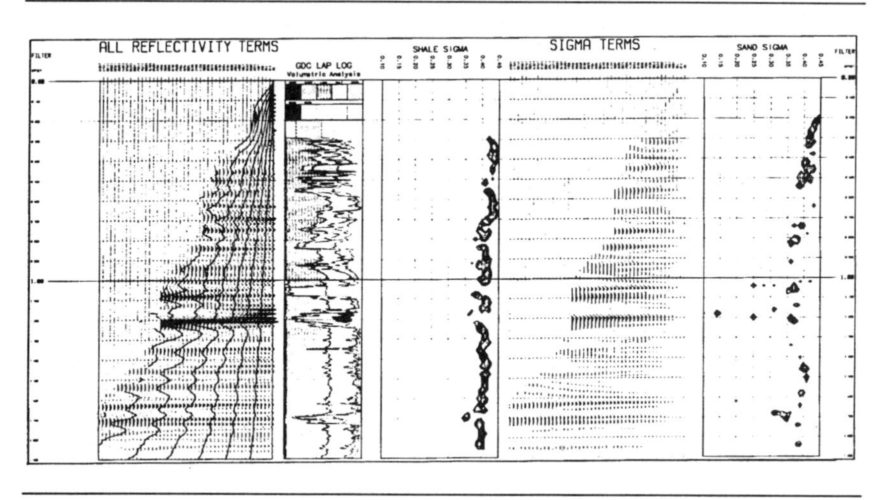

ratio have been displayed to indicate that the AVO responses are associated primarily with sand packages in this well.

The necessity to know shale types and sand shaliness in this area suggests a philosophy for the initial modeling of AVO responses (Table 1). Reinforcing this philosophy of in situ versus blocked models are the synthetics from this well shown in Figure 25. In particular. the top of the gas sand in the "in situ gas" model does not have an *NI* bright spot because of the shale gradient at the top of the sand package. This shaliness also accounts for the convergence of the reflectors with increasing offset. Contrast this with the large amplitude AVO response at the interfaces for

TABLE 1. AVO philosophy

- Tie existing well control
- Petrophysical analysis (one-foot sample)
- AVO models

AVO models	
In situ Gas substitute in situ Clean water sand Gas substitute clean sand	vary thickness
Other lithologies	

FIGURE 25. In situ shaly sands versus clean sands.

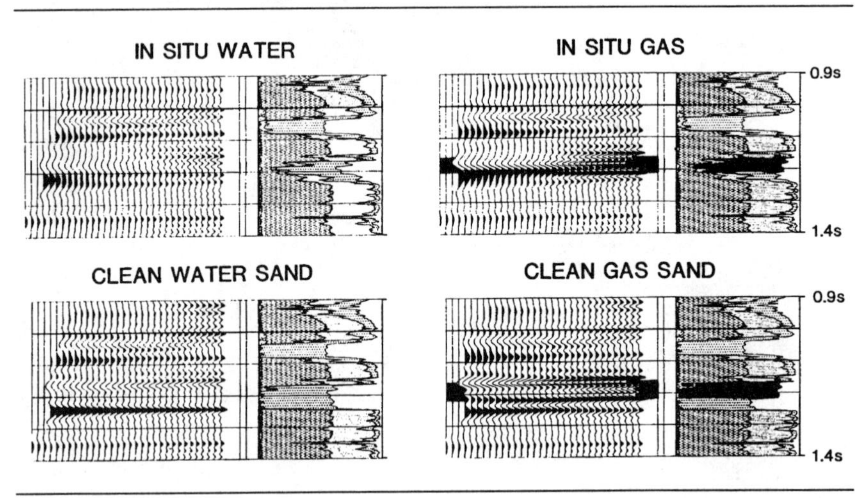

the blocked sand in the "clean water sand" model and the continuity of the AVO response with offset. Obviously, sharp boundaries resulting from coarse blocking produce much stronger and distinct AVO effects on the synthetics than are observed on field data.

CONCLUSIONS

After calibrating the rock types in this study area, a few points were observed:

- There is no unique velocity/density trend for shale; we classified five different shale types in this area.
- The AVO response for a sand package depends not only on the degree of shaliness but also the shale distribution within the sand package.
- AVO synthetics are alternate methods to 1-D synthetics for correlating seismic data to the well-log lithology.
- To reduce risk and the amount of models necessary to interpret a prospect, tie an existing well within the prospect's major fault block or equivalent geologic setting.

In summary, there are two steps in an AVO analysis: (1) extraction of the rock properties from the CDP gather and (2) relating lithology to the extracted rock properties. In this area, the second step was the weak link; the relationship between the rock properties and the expected lithologies was initially oversimplified and accurate models could not be produced until the detailed in situ lithology was considered.

SUGGESTIONS FOR FURTHER READING

Shuey's equation can be found in *A simplification of the Zoeppritz equations* (GEOPHYSICS 1985). The "mud line" equation is derived in *Relationships between compressional-wave and shear-wave velocities in clastic silicate rocks* by J.P. Castagna, M.L. Batzle and R.L. Eastwood (GEOPHYSICS 1985). The dependence of seismic velocity on clay content is examined in *Effects of porosity and clay content on wave velocities in sandstones* by D. Han, A. Nur, and D. Morgan (GEOPHYSICS 1986).

ACKNOWLEDGMENTS

A study such as the one outlined above is definitely a team effort and many colleagues were major contributors. Scott Burns coordinated the seismic processing modeling and overall analysis; John Puffer developed the petrophysics theory; Mark Wilson conducted the log analysis; Luh Liang directed the programming; Folke Engelmark cataloged the rock properties; Peg Guthrie conducted the petrophysical AVO modeling; and

Richard Verm integrated the processing analysis and modeling of petrophysics and seismic into a single system. Finally we appreciate the excellent seismic data made available from JEBCO's recent well-tie surveys. RAP is a trademark of Western Geophysical and SOLID is a trademark of Geophysical Development Corporation.

ABOUT THE AUTHOR

Fred Hilterman earned a degree in geophysical engineering (1963) and a doctorate in geophysics (1970) from Colorado School of Mines. He has worked for Mobil Oil and as a professor of geology at the University of Houston where he was also principal investigator of the Seismic Acoustic Laboratory. Hilterman has been vice-president of Geophysical Development Corporation since 1981. He received SEG's Best Paper Award in 1970, CSM's Diest Gold Medal in 1971, served as SEG vice-president in 1982–83, received the Society's Kauffman Gold Medal Award in 1984, and was SEG Distinguished Lecturer in 1986.

How 3D Seismic-CAEX Combination Affected Development of N. Frisco City Field in Alabama*

by **Mark Stephenson, John Cox, and Pamela Jones-Fuentes**
Paramount Petroleum Co.
Houston, Texas

By applying the latest in 3D seismic and computer aided exploration and production (CAEX) technology, small and midsize independents are changing the methods by which fields are discovered and profitably developed.

The combination of 3D and CAEX has, in many cases, altered oil field economics.

Nuevo Energy Co.'s North Frisco City development—located in the updip Jurassic Haynesville trend of Southwest Alabama—offers a case in point. The 3D technology employed at North Frisco City produced an accurate, detailed picture of the subsurface. Ultimately it more than doubled the drilling success rate over that of a nearby, closely related field in which 3D was not used.

Further, all the needed seismic information at North Frisco City was present from the start of development, eliminating the need to permit, shoot, and process more lines.

This accelerated the field's development cycle and cash flow, in turn improving such economic factors as return on assets and return on investment—a significant advantage for Paramount Petroleum Co. and the other field partners, Nuevo, Howell Petroleum Corp., Rimco, GEDD Inc., Shore Oil Co., and the field operator, Torch Operating Co.

Under a management agreement, Torch provides technical and financial services to Nuevo and Paramount.

As a result, Paramount—an independent until acquired by Nuevo in February 1992—has purchased a second CAEX workstation and is using 3D technology elsewhere in the play to develop another field and explore a prospect. Various advantages obtained with the use of 3D technology at North Frisco City field are discussed in the following case study.

---

*Reprinted, by permission, from *Oil and Gas Journal,* October 1992, courtesy PennWell Publishing.

COMPLEX GEOLOGY

Founders of Paramount launched the Haynesville play in 1987 with the accidental discovery of Frisco City field in Monroe County, Ala. (Figures 1, 2).

An initial well was drilled there to test Jurassic Smackover, the area's dominant play. An unexpected, highly prolific interval, the Frisco City sand, was found in the overlying Lower Haynesville formation.

Well information gained from development of the Frisco City field led to a then-maverick hypothesis regarding the area's lithology and subsurface structure.

The new model—since confirmed by seismic studies and well data at North Frisco City and elsewhere—proposed that the Jurassic age Norphlet, Smackover, and Buckner sequences pinch out on the flank of the Paleozoic basement structure. This igneous and metamorphic basement rock forms the core of the anticlinal structure.

Throughout most areas of the Gulf Coast, the Buckner anhydrite vertically seals the Smackover formation. But in this case the basement

FIGURE 1. Lower Haynesville trend

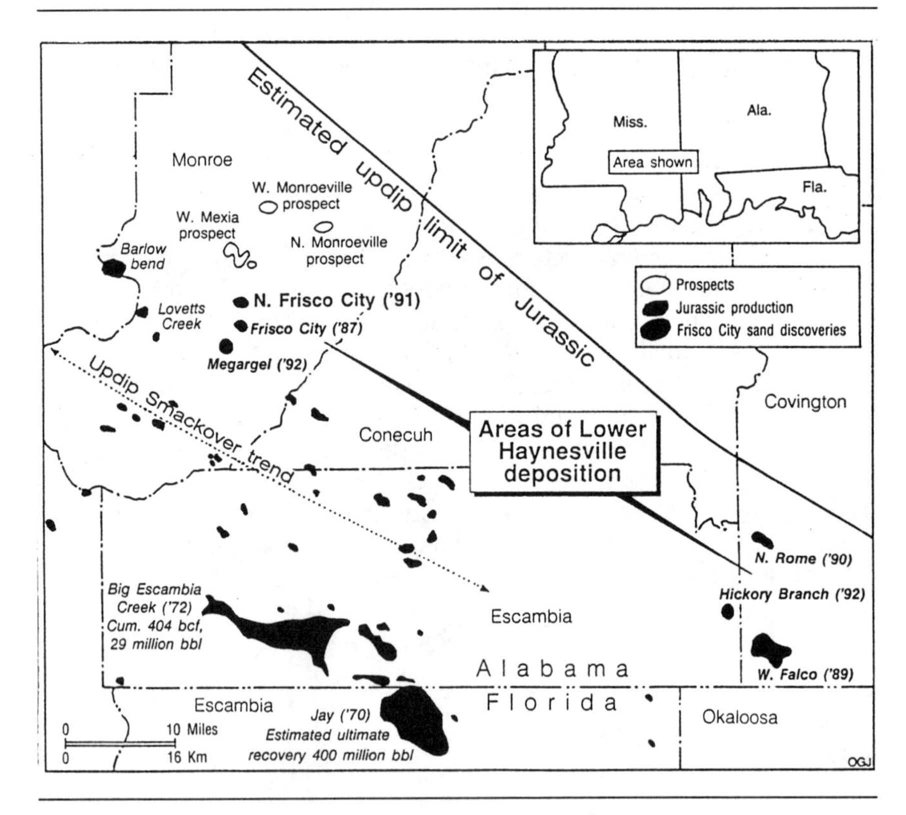

FIGURE 2. Hydrocarbon trapping model, Monroe County

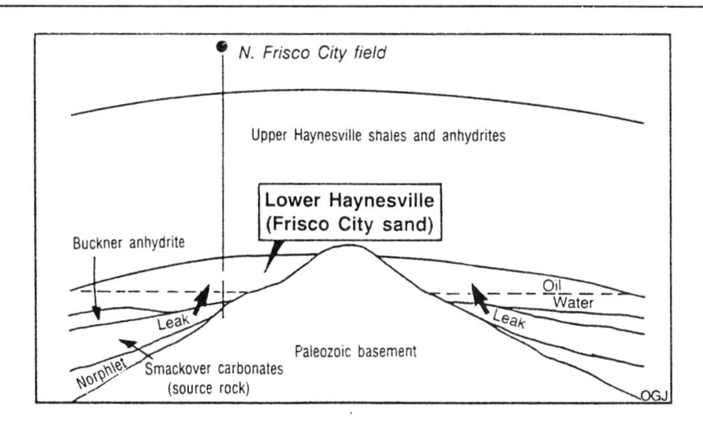

Throughout most areas of the Gulf Coast, the Buckner anhydrite vertically seals the Smackover formation. But in this case the basement rock rises above the depositional limit of the Buckner. The pinchout of the Buckner anhydrite seal allows hydrocarbons generated in the Smackover to migrate into the Frisco City sand. The overlying Upper Haynesville shales vertically seal this sand, providing the basis for the current play.

The Frisco City sand was deposited by braided streams sourced from an alluvial fan system. The sand is thin or absent on top of the basement highs and thick on the flanks (Figure 2). This resulted from the sand being deflected off the highs, creating a highly complex depositional pattern.

After the Frisco City field development, Paramount became the Haynesville play's most active prospect generator. Paramount discovered West Falco, North Rome, and North Frisco City fields in 1989, 1990, and 1991, respectively.

The West Falco and North Rome discoveries extended the play to a similar localized delta system 50 miles to the southeast. But North Frisco City field—the star performer—was found just two miles from the original Frisco City discovery.

Paramount generated the North Frisco City prospect off extensive knowledge of the region's geology and state of the art, proprietary 2D seismic shot through a 1984 well drilled on the lease (Figure 3). The 1984 well found oil shows in the Frisco City sand, but completion was not attempted because the area had no other production from that interval.

FIGURE 3. Structure produced manually from 2D seismic

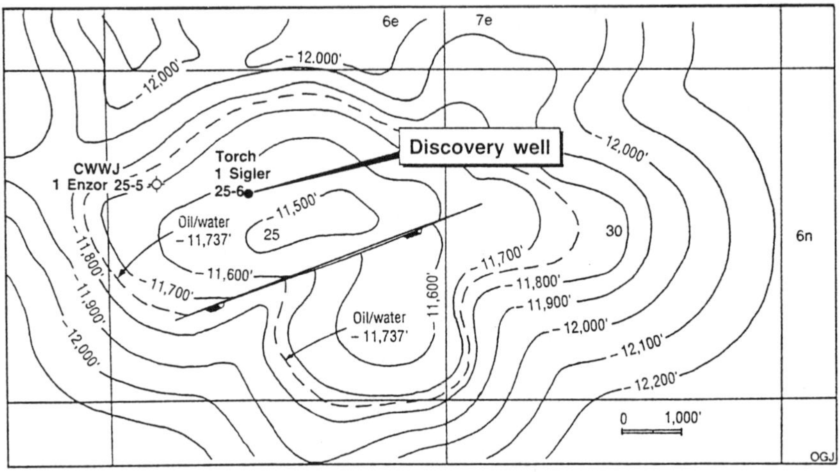

3D SEISMIC ACQUIRED

Torch's 1 Paramount Sigler 25-6 discovery well at North Frisco City was completed in March 1991. It encountered 92 net ft of oil pay, flowing 832 b/d of oil and 1.03 MMcfd of natural gas through a 14/64 in. choke with 2,410 psi flowing tubing pressure—good but not remarkable for the area.

The project's partners considered acquiring a small 3D survey over North Frisco City field. Factors driving the decision included the Haynesville's stratigraphic complexity, the highly irregular, unpredictable deposition of Frisco City sand, and dry holes encountered at nearby Frisco City.

Development drilling at Frisco City had thus far resulted in one additional producing well and three dry holes. The aggregate dry hole cost approached $1.5 million and had a significant impact on the small field's profitability.

The cost for onshore 3D seismic had been declining for several years. Permitting, shooting, and processing a 3D survey of approximately 3 sq miles at North Frisco City would now cost less than $200,000—less than half the area's typical $450,000–475,000 dry hole cost.

The numbers also compared favorably with 2D. Acquiring three 2D lines would cost about $120,000 for a small fraction of the information that would be obtained with 3D. Shooting 3D and using a CAEX workstation would allow the seismic data set to be treated as a solid earth volume rather than as individual lines.

This would allow arbitrary lines to be drawn throughout the survey area to quickly test interpreters' ideas at near-zero incremental cost. By comparison, when developing other Haynesville and Smackover fields, the project's partners had typically found it necessary to shoot additional 2D lines. They determined that at North Frisco City additional lines could push the total cost for 2D above $200,000 and add months to the field development process.

Permitting for the 3D survey began in June 1991 and took about four months due to the area's relatively dense culture. The survey was shot in October, processed the following month, and delivered to the partners at the beginning of December 1991.

CAEX TECHNOLOGY CHOSEN

Meanwhile, Paramount and Torch had yet to make a final decision regarding CAEX workstation technology. The two companies performed a manual interpretation using about 50% of the 3D data during an intensive two week period in December (Figure 4).

The location for an initial North Frisco City development well was then chosen based on the manual interpretation. The 1 McCall 25-7, spudded in mid-February and completed in early March, encountered 150 ft of net pay. It flowed 2,064 b/d of oil and 2.2 MMcfd of gas through a 24/64 in. choke with 2,087 psi flowing tubing pressure.

The partners selected a second development well location also based on the manual interpretation. However, Paramount and Torch were now nearing the end of their CAEX technology evaluation process.

FIGURE 4. Structure from CAEX Workstation, 3D Data—North Frisco City Field

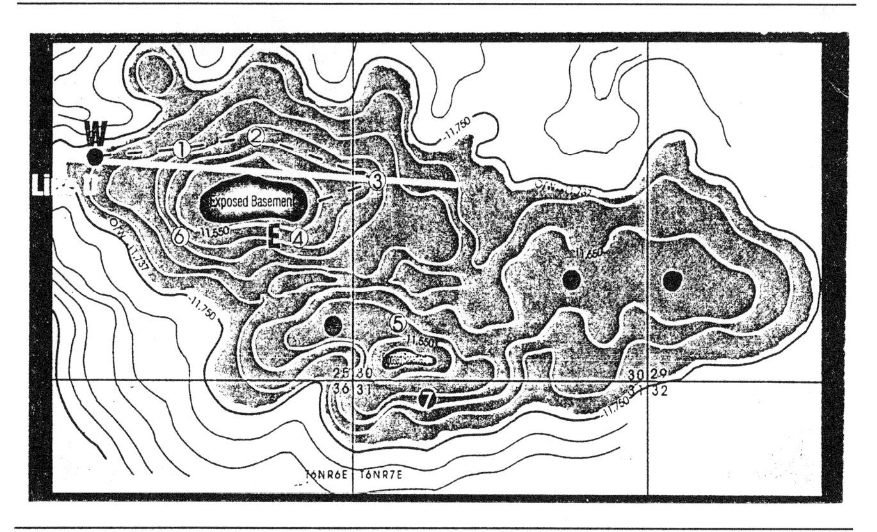

In mid-March, the partners provided a CAEX technology vendor with the North Frisco City 3D survey. As a demonstration, the vendor interpreted the 3D seismic, confirming the proposed well location.

The demonstration convinced Paramount and Torch to jointly acquire an advanced 2D/3D workstation system. Deciding factors included the system's ability to (a) speed the interpretation process and (b) make interpretations more thorough by allowing the use of arbitrary lines and taking into account well logs and other geological information.

Also, Paramount and Torch were by now convinced that 3D technology would be a pivotal factor in the ability of E&P companies to compete in the 1990s.

Therefore they saw the CAEX workstation purchase as an important strategic move as well as a tactical advantage for their current projects.

The well location confirmed by CAEX interpretation was drilled and completed in June. The 1 Lancaster 30-5 flowed 3,101 b/d of oil and 3.4 MMcfd of gas through a 34/64 in. choke with 1,855 psi flowing tubing pressure.

APPLICATION OF CAEX

Paramount and Torch took delivery of the CAEX workstation system in June 1992. It provided an immediate manyfold productivity improvement, enabling them to interpret the entire North Frisco City 3D data set in two days versus the one to two weeks required to manually interpret half the data.

Further, using the workstation improved the quality of the interpretation and significantly increased management's confidence in the proposed drilling locations (Figure 5).

As an example, the interpretation team was now able to generate arbitrary lines through the producing wells and proposed locations (Figure 5). It thus took only minutes to tie the seismic directly back to the wells, allowing comparisons with synthetic seismograms and the team's existing two dimensional models. Also, synthetics from the sonic logs were quickly and easily compared to the actual reflection responses in the same wells.

The first well interpreted entirely on the CAEX system, the 1 McCall 25-9, was the highest well drilled to date on the feature. Completed in July 1992, it proved 211 ft of oil column and came close to setting an Alabama state production record. The well flowed 3,559 b/d of oil and 3.4 MMcfd of gas through a 34/64 in. choke with 1,695 psi flowing tubing pressure.

The project's partners have since drilled two more North Frisco City wells based on the 3D data and CAEX. One, the 1 McCall 30-13, completed in August, flowed 3,060 b/d of oil and 3.9 MMcfd of gas through a 34/64 in. choke with 1,555 pounds of tubing pressure. The other, the 1 Sigler 25-11, completed in early October, flowed 3,007 b/d of oil and 3.3

FIGURE 5. Arbitrary line generated on CAEX workstation

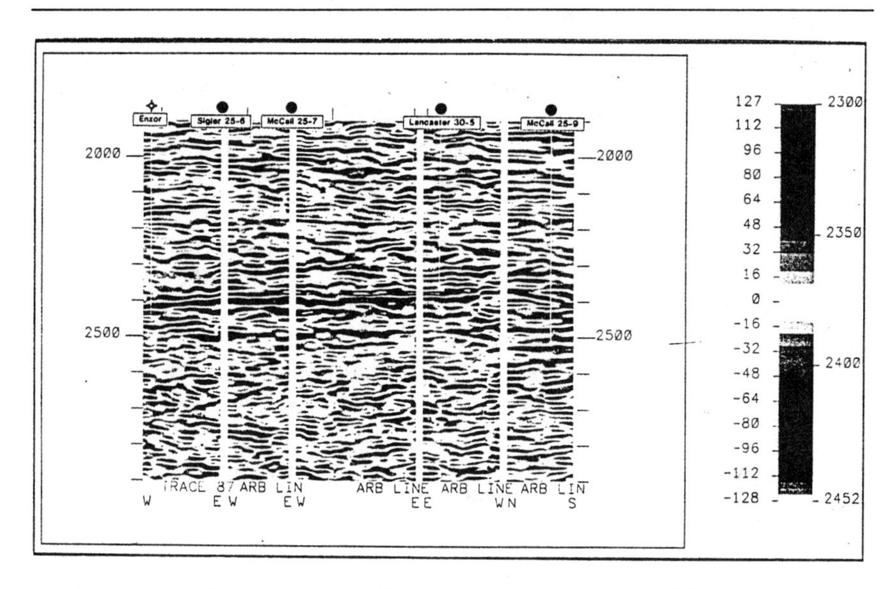

MMcfd of gas through a 26/64 in. choke with 1,740 pounds of tubing pressure.

The CAEX workstation added substantial value to the North Frisco City development in two ways, significantly improving the field's economics:

1. *Better drilling and production decisions resulting from higher quality interpretation.*

The CAEX workstation used all the available 3D seismic data and could optionally take into account other data such as well control. 3D plus CAEX produced very high resolution images providing substantial accuracy and detail. The Paramount-Torch interpretation team using the workstation was able to generate arbitrary lines, rapidly autopick horizons, and edit maps, enabling the team to quickly test ideas, refine the model, and improve its understanding of the play.

Impact on production and the field's economics:

The partners have achieved 100% development drilling success, compared with 40% in a nearby field without the use of 3D. High resolution images have enabled the partners to maximize recovery by gaining more oil column—confidently drilling higher on the structure while avoiding the dry, sand-devoid summit areas.

2. *Accelerated development of the field.*

Shooting 3D at North Frisco City has thus far prevented the need to acquire additional seismic data. The partners estimate that the "data acquisition to oil production" cycle has thereby been shortened by several months and that production has been "ramped up" much more quickly as compared with development based on 2D seismic. Also the CAEX workstation dramatically speeded the interpretation team's hypothesis testing and modeling, further accelerating development of the field.

Impact on production and the field's economics:

Paramount and Torch believe that if 2D rather than 3D had been employed at North Frisco City, development there would currently lag by at least two wells. This would translate into lost production at the wellhead of at least 40,000 bbl/month of oil and 40 MMcf/month of gas. Therefore the use of 3D has provided the project's partners with accelerated cash flow and increased return on investment and return on assets as compared with 2D. For the two partners that are public companies—Nuevo and Howell—such factors appear to have contributed to a rise in share price, increasing the companies' value to their investors.

CONCLUSION

In a project characterized by complex geology, an independent E&P company made its first use of 3D seismic and CAEX.

The combination significantly improved the project's development and production economics as compared with a nearby field in the same play where 3D technology was not employed.

As a result, Paramount has quickly acquired a second CAEX workstation, shot 3D surveys over two more properties in the play, and plans to expand its use of 3D technology as a competitive advantage throughout the 1990s.

ACKNOWLEDGMENTS

The authors express their appreciation to the following individuals who have contributed to the successful development of North Frisco City field: Sam Wilson, Torch Operating Co.; Bryan Richards and Dave Burkett, Howell Petroleum Corp.; Guy Joyce, Rimco; Jim Harmon, GEDD Inc.; John Bush, Paramount Petroleum Co.; and Bob Gaston, consulting geophysicist.

ABOUT THE AUTHORS

Mark A. Stephenson is a senior geologist with Paramount Petroleum Co. since 1990 and focuses on generating prospects. Before joining Paramount he was vice-president of Austin Production Co. He spent 10 years as a geologist with Union Pacific Resources Co. and Coastal Oil & Gas Corp. from 1981–89. He has a BS in geology from the University of North Carolina.

John G. Cox is a senior prospect generating geologist with Paramount since 1988. Before joining the company he spent a combined 10 years with Coastal Oil & Gas Corp. and Champlin Petroleum Co. as a senior geologist. Cox holds a BS from Milsaps College, Jackson, Miss., and a masters in geology from the University of Mississippi.

Pamela Jones-Fuentes is a senior geophysicist with Paramount since 1990 focusing on post stack analysis of seismic data using specialized software packages. She has a combined 13 years' experience as a geophysicist with Vulcan Exploration Inc., Allen Geophysical Consulting, GECO, and Texaco Inc. She holds a BS in mathematics from the University of Houston.

Cost-Effective 3-D Seismic Survey Design*

by **Richard D. Rosencrans**
Marathon Oil Company
Cody, Wyoming

As acquisition costs again begin to climb, designing cost-effective seismic surveys becomes an important consideration. This need is paramount for 3-D surveys due to the higher costs associated with the increased data volume. The ability to meet the management goal of prospect definition at a low cost can be attained only with a proper understanding of the 3-D acquisition parameters.

Most treatments of the basics of 3-D survey design concentrate on the theory and mathematics used for determining acquisition parameters. Consequently, they do not include examples that illustrate how varying field parameters affect data quality. Through the study of several 3-D surveys, I have developed some general guidelines to ease the burden of program design. With this paper, I strive to reduce some of the guesswork that goes into the task. The result, hopefully, will be cost-effective 3-D surveys. That is, the surveys will meet project objectives without being over- or underdesigned and overpriced.

To develop these guidelines, I first describe some typical project objectives and the field parameters that can influence them. Then, through two data examples, I show that alteration of two field parameters—bin size and migration aperture—affects the data quality and the final interpretation. Some project goals can be met with imperfect field parameters, yet other objectives suffer from marginal acquisition methods. Finally, I outline a cookbook flow for determining survey size (migration aperture) and bin size.

The factors affecting survey design, in the general order in which they should be evaluated, are: structural or stratigraphic azimuth; survey size; bin size; bin shape; interpolation; receiver and source line orientation; arrays; roll direction; fold and line spacing; recording equipment (master-slave or distributive system); and patch size (i.e., the number of channels). Unfortunately, it is beyond the scope of this paper to discuss all of these topics in detail.

Two 3-D surveys, both in the Rocky Mountain region, are used to illustrate the parameters discussed in this paper. Laredo, an exploration

*Reprinted, by permission, from *Geophysics: The Leading Edge of Exploration*, Vol. 11, No. 3, March 1992, courtesy of The Society of Exploration Geophysicists and the authors.

survey, was designed to oversample a stratigraphic target to ensure proper imaging. Previous 3-D surveys in this reef play did not properly image the anticipated reefs. This survey was designed to determine if the reefs were not imaged due to inadequate survey parameters.

The targeted reef was well imaged, so the Laredo data set was reprocessed in two ways to simulate field acquisitions of (1) a larger bin size and (2) a smaller overall survey. The original 5 mi^2 survey was shot with up to 5000 ft of migration aperture for the steep reef flank and an 80 × 160 ft subsurface bin, interpolated to an 80 × 80 ft bin. These data were first reprocessed to simulate a 160 × 320 ft bin. In two other versions, 1000 ft and 2000 ft of maximum migration aperture were simulated by reducing the overall survey size to 1 mi^2 and 2 mi^2, respectively. The purpose of this work was to determine the most cost-effective field parameters needed to adequately image the exploration target.

The second data example is taken from a production 3-D survey shot over a producing anticline in the Rocky Mountains. The field parameters satisfied the theoretical design guidelines. However, as interpretation of the data set proceeded, it became apparent that the field parameters were marginal for the reservoir characterization being attempted. Examples are taken from this data set to illustrate the need for correct determination of the field parameters.

This survey was shot using a 100 × 200 ft subsurface bin and 3500 ft of migration aperture for the steep flank. The project objectives included identifying very closely spaced minor faults on the 15–20° dipping limb, positioning the major bounding reverse fault, and mapping the 10–20 acre sinkholes on the top of one of the reservoirs.

EXPLORATION VERSUS DEVELOPMENT

Clear, concise objectives for any seismic survey are important for the project to succeed. Communication with the geologists or engineers is the first step, for they are the customers who will utilize the final geophysical interpretation. You must discern from these individuals what problems exist and how the 3-D seismic will provide answers to those problems. A general set of objectives is shown in Table 1 to demonstrate that there is a fundamental scale difference between exploration and development projects. This greatly affects survey design and the final project cost.

Consider the exploration objectives in Table 1. These objectives are usually met with a coarse grid of 2-D data shot with a 100–200 ft group interval. The interpretation is made by timing every fifth or 10th shot point. The final maps on the target are usually very subjective and a geologic model is used to finalize the interpretation. Exploration 3-D surveys can improve the sampling and minimize the reliance on a geologic model,

TABLE 1. 3-D survey objectives

Exploration	Development
Presence of target	Reservoir characterization
Size	Presence of compartments
Depth	Size of compartments
Dip magnitude & azimuth	Major & minor faulting
Formation thickness	Design of EOR program
Faulting	Production anomalies
Well location	Stratigraphic information
	Rock/fluid information

but a coarse image of the target often suffices. However, engineers and development geologists demand seismic resolution of 50 ft faults and subtle porosity and fluid changes. These are difficult demands, but some can be met if the reservoir is well imaged during the field acquisition.

The 3-D survey objectives must be directly related to the survey parameters. The first three exploration objectives in Table 1 are affected by survey size; the last four by bin size. Survey design must ensure adequate sampling and offset for the migration aperture to ensure complete collapse of the Fresnel zone. Proper imaging for all the development objectives will be adversely affected by incomplete migration and insufficient sampling.

SURVEY SIZE

When designing a 3-D survey, the first item to consider is size. Several factors, including structural dip, diffraction dip, fold build-up, and migration aperture should be taken into account. In addition, survey requirements, exploration versus development, and specific survey objectives must be predetermined.

An exploration survey requires a large fringe around the target area to allow for errors in positioning and size of the objective. If the target has been poorly imaged on 2-D data, then raypath geometry may be complicated, the anomaly may be smeared, and allowances must be made for the uncertainty in size and location. These uncertainties may have large margins of error and require compensation in the design. On the other hand, an exploration target may not need perfect imaging so full-fold and migration requirements can be minimized and the survey size reduced.

In contrast, a development objective requires more strict adherence to technical restrictions on survey size due to the need for greater resolution. The need for precision and detail requires a larger fringe area to build up fold and more accurately migrate the data. However, there may be a little uncertainty concerning the size and location of a production target, so the survey size can be minimized.

Once the initial survey objectives are defined, the geophysicist should estimate or measure structural and diffraction dip, and calculate the migration aperture necessary to properly record and migrate the data. In addition, some operators believe an additional one-half migration fringe width is necessary to allow for fold build-up and a better migration. In practice, this is usually unnecessary for exploration work.

Development surveys require greater resolution and, in practice, I have observed that pushing the theoretical limits will result in a less than ideal data set. If theoretical calculations indicate a 2000 ft migration fringe is needed, increasing it by an extra one-half migration fringe results in a better data set. An exploration target may be adequately imaged with less than the 2000 ft fringe.

The cost importance of properly determining the area of the migration fringe is demonstrated with a simple example. For a 2 × 2 mi^2 development survey with a critical dip in a north-south direction and a 2600 ft migration aperture, 1 mi^2 of data on either end of the survey is acquired solely for migration. At a cost of \$40,000 per square mile, \$80,000 of a \$160,000 survey, or one-half the survey cost, is spent on proper migration.

Proper design for 3-D surveys becomes economically critical and the desire to be frugal is strong. However, by shrinking the migration fringe, the interpreter may not achieve the project goals. Then the entire survey cost may be wasted and the viability of the tool could be in question.

To ensure that project goals are attained for a development survey, field parameters should meet or exceed design limits so proper imaging and adequate resolution are achieved. An explorationist may not need precise migration to image the target and can compromise the acquisition specifications to a certain extent.

Two data examples are used to demonstrate the effect of the migration fringe on data quality and interpretability. The first example (Figures 1a,b) is taken over a producing anticline in the Rocky Mountains. The stacked and migrated sections show what the migration fringe does to the data. In this case, field production starts right at the edge of usable data. Maximum cost savings are achieved, but there is no possibility of evaluating the change from reservoir to nonreservoir rock. On an exploration project, there would be no opportunity to determine the full size of the structure.

The second example is taken from a stratigraphic exploration play in the Rocky Mountains. The 3-D survey (Figure 2a) was shot over a pinnacle reef originally identified from 2-D seismic lines. Because there was uncertainty in the size and position of the reef, the 3-D survey was designed larger than necessary to ensure proper imaging of the target.

FIGURE 1. Diffraction energy and dipping beds cause data to be recorded beyond where the geologic horizon actually occurs. Migration moves the reflection data back to its proper position but can leave dead zones from where the data were moved. The theoretical bin size for the unimaged right flank is 150 ft. For an exploration prospect, either displayed migrated section would properly locate the structure. For a development project, neither section would provide sufficient detail for fully evaluating the structure.

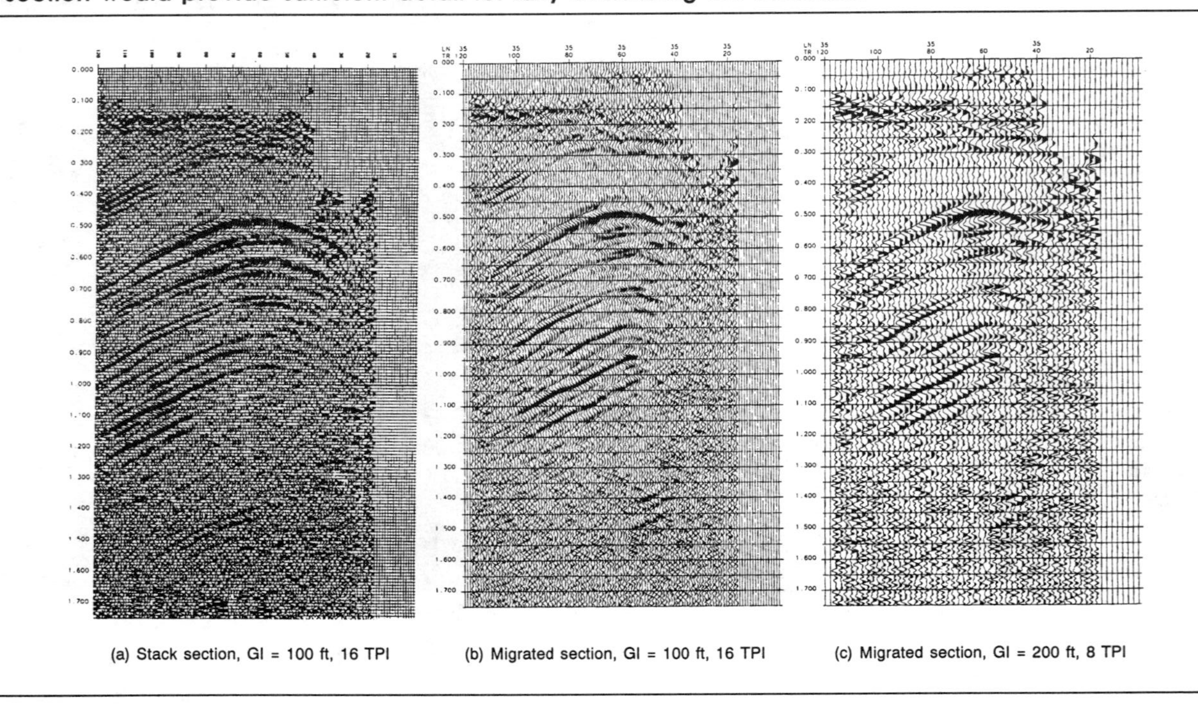

(a) Stack section, GI = 100 ft, 16 TPI (b) Migrated section, GI = 100 ft, 16 TPI (c) Migrated section, GI = 200 ft, 8 TPI

FIGURE 2.

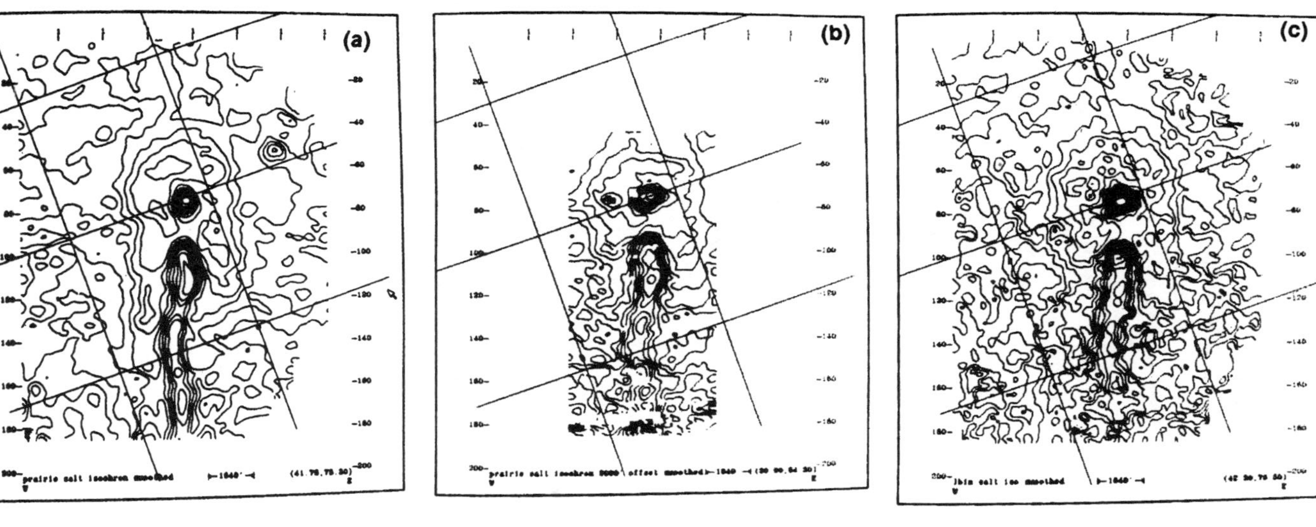

(a) THE SALT ISOCHRON OVER THE REEF WAS CONSTRUCTED USING THE ORIGINAL 3-D DATA SET WITH OFFSETS EXCEEDING 5000 FT AND A FIELD BIN OF 80 × 160 FT INTERPOLATED TO 80 × 80 FT.

(b) THIS SALT ISOCHRON WAS CONSTRUCTED USING THE 3-D DATA REPROCESSED TO SIMULATE ACQUISITION LIMITED TO A 2000 FT OFFSET BEYOND THE KNOWN EXTENT OF THE REEF.

(c) THIS SALT ISOCHRON WAS CONTRUCTED USING THE 3-D DATA REPROCESSED TO SIMULATE ACQUISITION WITH A 160 × 320 FT BIN INTERPOLATED TO 160 × 160 FT.

Stratigraphic dips on the reef flank varied greatly (from 12–60°, averaging 19°). The theoretical limit for the migration aperture was 3500 ft. The actual aperture attained over most of the survey was 5000 ft.

The reef data set was reprocessed to simulate a limited migration aperture. One line (Figure 3) is shown to demonstrate the effect of reducing survey size below the theoretical minimum. In this example, the maximum offset is reduced from 5000 ft to 2000 ft. Smearing of the reef increases as the offset is reduced below the theoretical limit of 3500 ft. Additional smearing is observed when the offset is reduced to 1000 ft.

Although the reef is imaged on all sections, the apparent detail is diminished. To adequately determine if the project goals are attained, a comparison of the final interpretations is made. Figures 2, 4, and 5 are several maps for the original survey and the 2000 ft maximum offset survey. From the isochron maps (Figures 2a,b) it can be seen that smearing occurs. A comparison of volumetrics calculated on each map would

FIGURE 3. Crossline 142. As the maximum offset available is reduced below the theoretical minimunm, resolution of the reef diminishes. Both data positioning and amplitude are smeared.

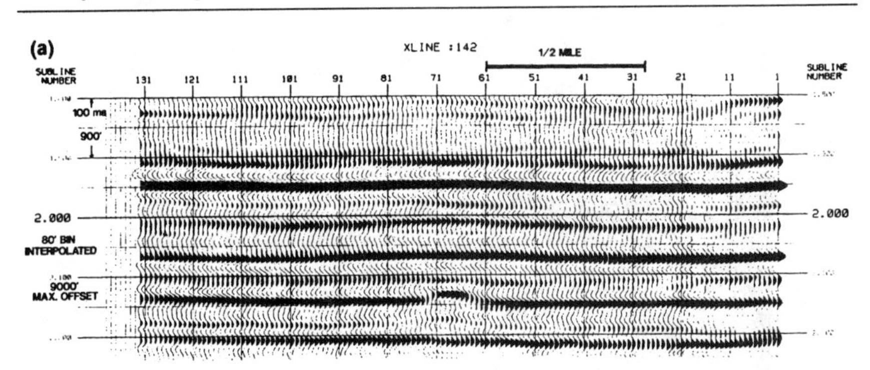

(a) ORIGINAL SURVEY WITH MORE THAN A 5000 FT FRINGE AROUND THE REEF.

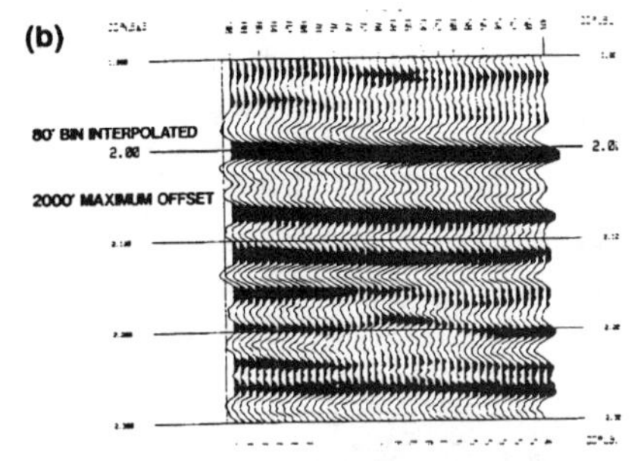

(b) SURVEY SIZE LIMITED TO A 2000 FT FRINGE.

FIGURE 4. Amplitude maps on the top of the Winnipegosis show deterioration in resolution as acquisition parameters are changed.

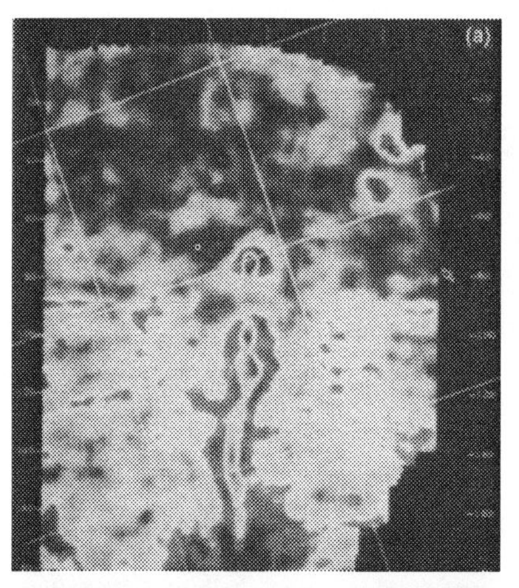
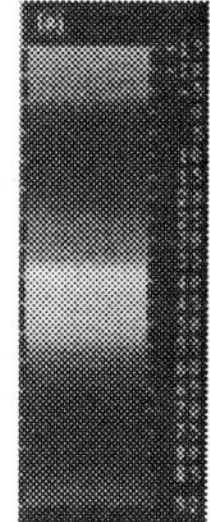

(a) ORIGINAL SURVEY WITH A 5000 FT FRINGE AND 80 × 80 FT BIN.

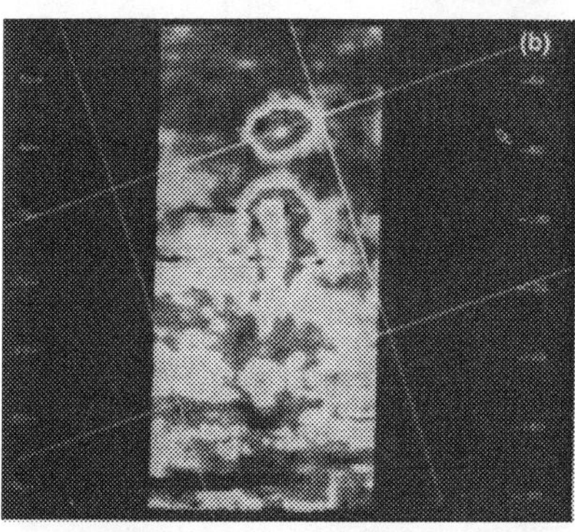
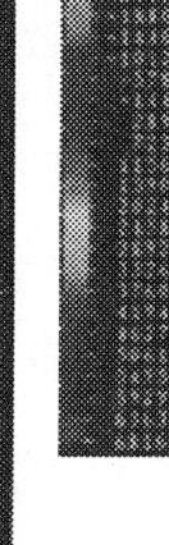

(b) SURVEY SIZE LIMITED TO A 2000 FT FRINGE AND 80 × 80 FT BIN.

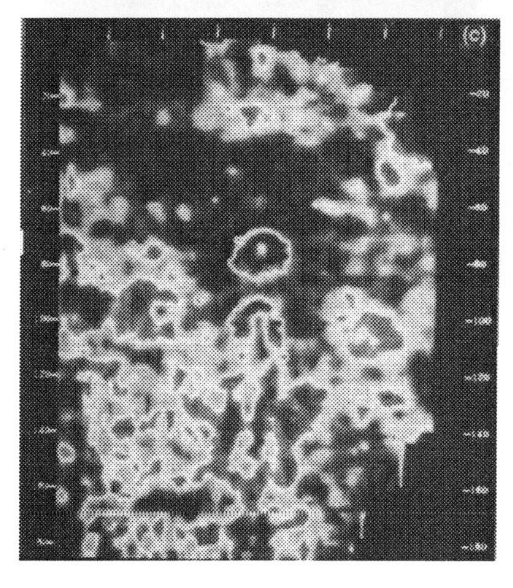
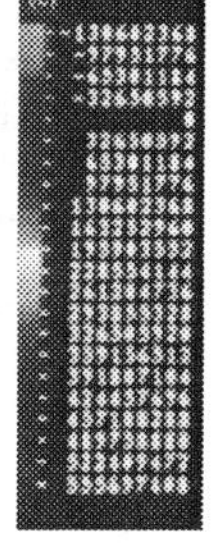

(c) ORIGINAL SURVEY SIZE WITH 160 × 160 FT BIN.

FIGURE 5. Perspective views of the reef show tremendous geologic and geomorphic detail. However, much of this detail is lost as field parameters are adjusted to minimize acquisition costs.

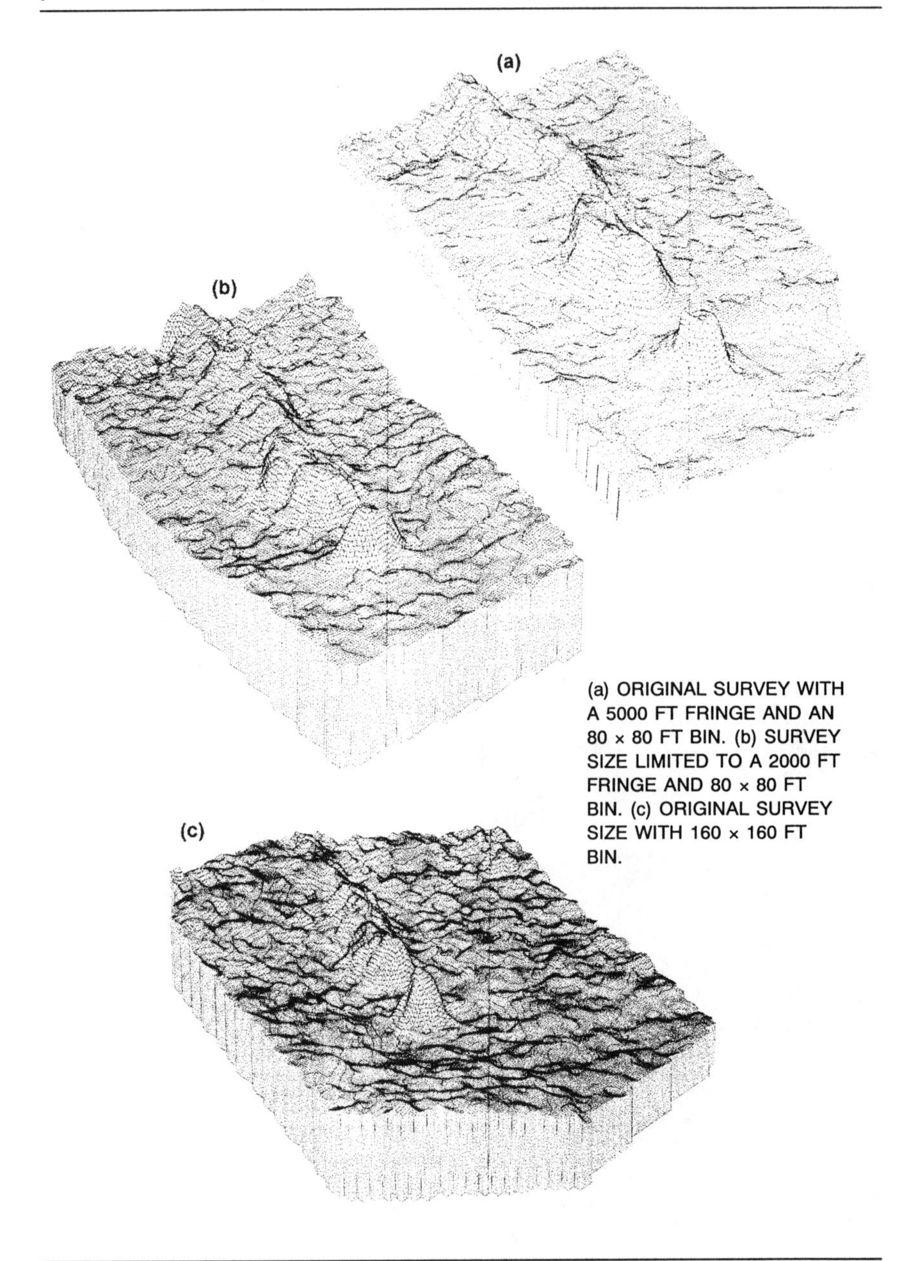

probably show a significant discrepancy. This discrepancy would affect both exploration and development economics, though not equally.

Seismic amplitudes are often important to the final interpretation on stratigraphic prospects. Because of the incomplete migration, a smearing of the low amplitudes also occurred (Figures 4a,b). This smearing could detrimentally affect bed thickness measurements made below tuning thickness on stratigraphic plays.

Frequently, a geologist will need to review the final geophysical interpretation to make facies predictions for a determination of the effective reservoir size and porosity. In this case, a comparison of the perspective plots of this reef (Figures 5a,b) demonstrates that much geologic detail can be attained, but also that much can be lost if geophysical requirements are not met. In this example, the geologist accurately predicted the reef facies that was encountered in the wellbore using this plot and a geologic model developed from all geologic information.

Tremendous detail can be seen in the perspective plot from the original survey. The reef rampart on the windward edge and a backwater or lagoonal area can be seen. A slump on the long sinuous reef can be seen just below the steep reef face. Dips ranging from 14° on the back of the northern reef to 58° on the windward edge can be calculated. Most of this fine detail is missing from the survey with a smaller than recommended migration fringe.

BIN SIZE

The bin size chosen for a survey can have a tremendous impact on the interpretability of the final product. Table 2 lists those factors which need to be considered when designing a survey. The restriction for optimizing all of these concerns with a smaller bin size is cost.

Spatial aliasing is the major concern in deciding on the appropriate bin size. Theoretical calculations are easily made and require an estimate of structural dip, diffraction dip, and the maximum usable frequency.

TABLE 2. Geologic parameters that determine bin size and shape

Structural dip	Fault spacing (imbricate faults)
Diffraction dip	Fresnel zone
Dip azimuth	Stratigraphic strike
Fault dip	Structural strike
Fault orientation	

Migrated structural dip is steeper than diffraction dip, so both stacking and migration velocities should be considered during calculations. The need to collapse Fresnel zones adequately to resolve a small stratigraphic target also requires a small bin.

An understanding of the ultimate objective of the project is critical to designing a cost-effective survey. An exploration project may require a less detailed sampling of a dipping layer. Consider the two seismic lines in Figures 1b,c. The theoretical estimate for the bin size on the steeply dipping right flank is 150 ft. The bin size in Figure lb is 100 ft and, on the line in Figure 1c, it is 200 ft. For exploration purposes, a 3-D survey shot with either spacing would adequately image the anticline. For development work, even the gently dipping flank in Figure lb may be inadequately sampled and the steep right flank, though theoretically imaged, cannot be seen. It appears unwise to push the theoretical limits too closely on a development project.

Subtle near vertical faults will always be difficult to detect on seismic data but can be inferred from dip magnitude and dip azimuth maps if there are sufficient data samples on both sides of the fault. On the anticline in the previous example, the geologist has mapped a series of nearly vertical faults spaced 200–300 ft apart. Faults were mapped using 10 acre well spacing. Bin size on this survey is 100 × 200 ft. That sampling density positions only one data point between faults, insufficient for mapping the subtle dip changes that can occur across faults. Dip magnitude and dip azimuth maps cannot show these geologically mapped faults. The theoretical limit for spatial aliasing was satisfied with a 150 ft bin, but not the interpretational limit for mapping faults 200–300 ft apart.

Stratigraphic resolution can be greatly enhanced using a smaller bin. This final example, taken again from the Winnipegosis reef survey, demonstrates how small bins and proper 3-D migration can collapse Fresnel zones so that detail much smaller than the Fresnel zone can be well imaged. This survey was reprocessed to simulate field acquisition with a 160 × 320 ft bin, double the actual bin. The data were then interpolated to a 160 × 160 ft bin. Compare the seismic line (Figure 6b) with the original data (Figure 6a). You can see the smearing of the reef in the large bin data. Again, the impact on the final interpretation should be the deciding factor in designing a cost-effective survey. A comparison of the isochron maps (Figures 2a,c) from the original and reprocessed versions shows that the smearing is extensive and the reserves will probably be miscalculated. A comparison of the amplitude maps (Figures 4a,c) again shows smearing and is an indicator of potential problems for those plays which weigh heavily upon amplitude analysis. Finally, a comparison of the per-

FIGURE 6. Crossline 78. As the bin size is increased from 80 ft to 160 ft, data quality deteriorates. Both amplitudes and data positioning are affected.

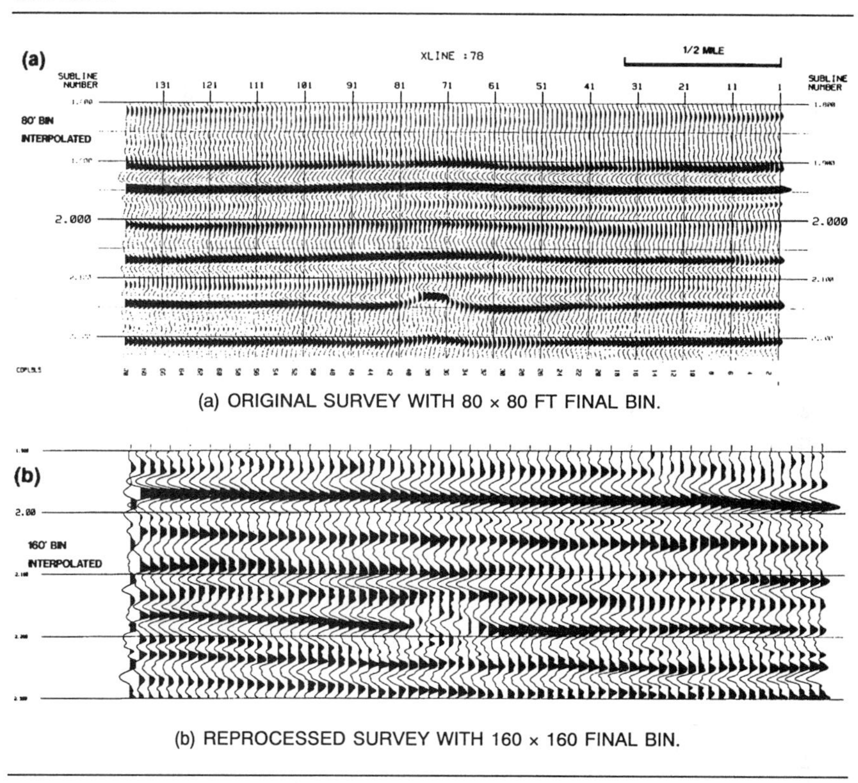

(a) ORIGINAL SURVEY WITH 80 × 80 FT FINAL BIN.

(b) REPROCESSED SURVEY WITH 160 × 160 FINAL BIN.

spective plots (Figures 5a,c) shows that the detailed geologic facies interpretation for the original survey cannot be easily made on the reprocessed data set.

3-D DESIGN FLOW

When designing a 3-D survey, consider following this flow path to ensure thoroughness in the design work. On a map, draw a line around the area that needs accurate imaging. From existing seismic data, determine the two-way time to the target horizon and the stacking and migration velocities for that event. Also, from existing seismic data or geologic cross sections, calculate the maximum structural dip and note its position and direction within the target area. If applicable, calculate the dip on the diffraction energy. Diffraction energy must also be adequately sampled to ensure proper migration. Again, from existing seismic, determine the maximum usable frequencies expected in the 3-D data set. Next using the theoretical equations (several commercial software programs are available), calculate the theoretical migration aperture and bin size. These calculations can be done in a spreadsheet format to obtain a range if there is uncertainty in any of the input. The theoretical answers you obtain are not necessarily the field parameters used.

Certain questions must be resolved at this point. What are the survey objectives? Is this an exploration or development survey? How much risk or uncertainty in the data set can the project economics tolerate? How precise a sampling of the reservoir is needed for the interpretation? These are only some of the questions you need to ask yourself. At this point, 2-D or 3-D computer modeling of the target may provide important information to guide the design work. Every incremental decrease in bin size and increase in survey size (migration aperture) results in an increase in survey cost.

The examples shown here suggest the theoretical results are sufficient for exploration work If economics dictate, these parameters can be pushed 10 percent (maybe more) and still attain reliable data. However, the opposite is true for development work. If ideal parameters include a 2500 ft migration aperture and 100 ft bin size, then consider at least a 10 percent adjustment to a 90 ft bin and 2750 ft aperture. This allows for uncertainty in the initial measurements and adjusts for practical versus theoretical results. If economically feasible, increase the migration aperture by 50 percent to build up fold in the aperture window. This should improve the migration quality (Figures 2–5).

Now draw a line on the map around the target for the migration aperture. If the maximum dip is only on one side of the survey, this fringe can be reduced. Typically, it's best to keep the fringe constant because it is also used to ensure fold build-up on the target.

You now have a survey size and bin size that can be used to estimate survey costs. If a square bin is too expensive, consider doubling or tripling the bin dimension in one direction. Again, you must look at the geologic dip and fault spacing to determine if it is constant in both directions or if one direction will be sampled adequately with a larger bin dimension.

The survey design work is not done at this point. The other factors listed in the introduction remain to be determined. This paper has covered two of the more important parameters. The rest start to fall naturally into place.

CONCLUSIONS

Survey size and bin size can dramatically affect the quality of a 3-D data set and its interpretability. We all recognize that larger surveys and smaller bins mean better data. However, perfect data acquisition is not always possible due to exploration and development budget restrictions, and our charge to economically find additional reserves. The geophysicist must decide what is and is not critical to the bottom line. In the reef examples, the ability to generate a detailed perspective plot to identify reef morphology would be useful to field development. However, a 100 ft bin would probably sufficiently sample the reef. For an exploration prospect, the 160 ft bin and 2000 ft offset would sufficiently define the target for a drilling decision.

In the anticline example, identification of the small faults could be critical to development but might not be critical in an exploration prospect. This anticline was probably inadequately imaged for field development. An 80 ft bin would have been more appropriate. This survey, as shot, would be overkill for an exploration prospect. A grid of 2-D data would suffice for detailing the structure for an exploration decision.

A better 3-D survey can be obtained by following a logical path during the design process. The flow should proceed by identifying the target, then determining dip, strike, and fault spacing. Through an evaluation of the survey objectives, theoretical calculations and modeling results, you must decide on the migration aperture and bin size. Double check the preliminary work and the final decision and then determine the other field parameters.

I believe that management has charged us with obtaining cost-effective seismic. With a small amount of effort in survey preplanning, we should be able to obtain data tuned to our particular project goals.

ACKNOWLEDGMENTS

The author thanks Marathon Oil Company for permission to publish this paper. Many thanks are extended to J.A. Allen, K.M. Kent, P.L. Fonda, J.V. Guy, and M.D. Greenspoon for their various contributions to this work.

ABOUT THE AUTHOR

Richard Rosencrans received an MS in geophysics (1983) from Pennsylvania State University. He then joined Marathon where he has worked in several Rocky Mountain basins. For the last three years, he has been involved in the design, acquisition, processing, and interpretation of land 3-D seismic surveys.

APPLICATIONS OF TOMOGRAPHY TO BOREHOLE AND REFLECTION SEISMOLOGY*

by Larry Lines
Amoco Research
Tulsa, Oklahoma

About ten years ago, a selected group of Amoco Research geophysicists was asked to participate in a study to predict the important geophysical problems of the next decade. Sven Treitel, the patriarch of Amoco geophysical research, went to the blackboard and wrote "v(x,z)." As a follow to this, Ken Kelly, the leader of Amoco's inversion group, wrote "v(x,y,z)." The two geophysicists essentially placed velocity estimation in two and three dimensions as the leading problems in exploration geophysics.

It is difficult to imagine any problem that is more fundamental to seismic data processing than the problem of seismic velocity estimation. The variation of velocity controls the traveltimes and amplitudes of the data which we record, process, and interpret. In this discussion, I would like to describe the applications of traveltime tomography, and will emphasize that tomography represents a general method for seismic velocity analysis.

TOMOGRAPHY

Seismic tomography provides a versatile tool for the velocity estimation from either borehole or surface reflection data. The word *tomography* is derived from the Greek *tomos* meaning cut or slice. The term literally means "graphing slices of an object." The goal of tomography is the imaging of material properties by using observations of wave fields which have passed through the body. The tomography method has been used in many fields including medical imaging, materials testing, civil engineering, and geophysical exploration. Here we will concern ourselves mainly with the estimation of seismic velocities by use of traveltime information. There are several current applications of traveltime tomography to seismic exploration and the method represents a viable approach to these problems, since tomography is based on a very general description of the earth's subsurface. Due to its generality and wide range of applications,

*Reprinted, by permission, from *Geophysics: The Leading Edge of Exploration*, July 1991, courtesy of The Society of Exploration Geophysicists and Larry Lines.

tomography has been used for many exploration problems involving both borehole and reflection data. The traveltime tomography method consists of three steps:

1. Identification of seismic traveltimes for various source-receiver positions—the data gathering step.
2. Ray tracing to formulate traveltime equations—the modeling step.
3. Solution of traveltime equations and the estimation of updated velocity models—the inversion step.

The first of these steps is generally the most time-consuming, in terms of a geophysicist's effort, since the picking of traveltimes essentially involves the interpretation of prestack seismic data. A typical interpretation session involves the identification of tens of thousands of arrival times on an interpretive workstation typically requiring 1–2 days of user time. For good data, automatic picking can substantially decrease the amount of time taken by the interpreter for this step. However, even with good data, traveltimes from automatic picking procedures should be inspected for possible errors. Since the traveltimes represent the data to be inverted in tomography, the interpretation step is a very important one. Although time-consuming, the traveltime picking phase allows the skilled interpreter to avoid noisy data zones and visually "filter out noise" by the interpretation process itself.

Step 2 involves the modeling of the traveltimes for some velocity model. Generally, ray tracing is used to compute the traveltimes for some velocity configuration. A recommended method for this ray tracing procedure is described by Langan, Lerche and Cutler (*Tracing rays through heterogeneous media*, 1985 GEOPHYSICS) and by Stork (*Ray trace topographic velocity analysis of surface seismic reflection data*, 1987 PhD thesis, California Institute of Technology). This ray tracing method honors Snell's law by curving rays in the presence of velocity gradients within cells. This velocity cell model allows for the modeling of velocity gradients in both the horizontal and vertical directions.

The third step in the procedure involves combining the results of Steps 1 and 2 and finding a solution of the traveltime equations. These equations express traveltimes for raypaths as the sum of distance × slowness products for the cells traversed. The slowness (or reciprocal velocity) is used rather than velocity in these equations since the system of equations is linear in slowness for the case of straight raypaths. The traveltime equations can be expressed as $t = \underline{\mathbf{D}}s$ where t is a vector containing the traveltimes of a raypath, s is a vector containing slowness values for each cell, and $\underline{\mathbf{D}}$ is a matrix of values, $\underline{\mathbf{D}}_{ij}$, which represents the distance traversed by the ith ray in the jth cell.

At first glance, this system appears to describe a linear relationship between traveltime and slowness. However, raypaths bend according to Snell's law and, therefore, this system is generally nonlinear since the distance values are themselves functions of slowness. Thus, as with most nonlinear system, we iteratively solve a series of linearized problems in order to obtain a solution.

In this nonlinear problem, a solution for updating the velocity or slowness model arises from starting with some "best guess" initial velocity model and iteratively solving $\mathbf{D}\boldsymbol{\delta}\,\mathbf{s} = \boldsymbol{\delta}\,\mathrm{t}$, where $\boldsymbol{\delta}\,\mathbf{s}$ is the adjustment in the slowness vector and $\delta\,\mathrm{t}$ is the vector containing differences between picked traveltimes and the traveltimes of the starting model for each iteration.

The system of traveltime equations is large, sparse, and illposed. The **D** matrix has n rows and p columns where n = the number of picked traveltimes and p = the number of velocity cells. Typically both n and p are of the order 10^4 in size. **D** is usually a large sparse matrix. That is, most of the members of **D** are zeros since a typical ray samples only a small fraction of the number of cells. Therefore, sparse equation solvers are very useful in solving the traveltime system of equations. Since **D** is a rectangular matrix for this ill-posed system, solutions to the traveltime equations have traditionally involved least squares solutions. However, least squares methods suffer from the problem of bad picks or outliers in the data. Fortunately, we are essentially able to use the iterative reweighted least squares method (IRLS) to find the least absolute deviation (LAD) of traveltime differences between data and model response, rather than the sum of squares. The LAD method is less sensitive to large errors than the least squares method. As shown by Scales, Gersztenkorn, and Treitel (*Fast LP solution of large, sparse linear systems: Application to seismic traveltime tomography*, 1988, *Journal of Computational Physics*), this robust estimation can be invoked by solving the reweighted least squares equations such that the LAD criterion is satisfied. The IRLS solution in this mode essentially downweights the large traveltime picking errors in finding the optimum solution. The paper of Justice et al. (*Acoustic tomography for monitoring enhanced oil recovery*, February 1989 *TLE*) has also advocated the use of IRLS in tomography following extensive comparisons of different traveltime inversion methods. The continued use of IRLS to update the velocity model generally produced improved fits to the data in the traveltime inversion process.

BOREHOLE SEISMIC TOMOGRAPHY

Borehole tomography often involves the modeling of rays which are transmitted without reflection from a known source position to a known receiver position. As shown by Figure 1, transmission tomography is typically used in cross-borehole profiling and vertical seismic profiling (VSP). The cross-borehole case has sources in one borehole and receivers in another borehole, while the VSP case has sources on the earth's surface and receivers in the borehole.

FIGURE 1. Two typical transmission tomography problems include the cross-borehole and surface to borehole (VSP) problems.

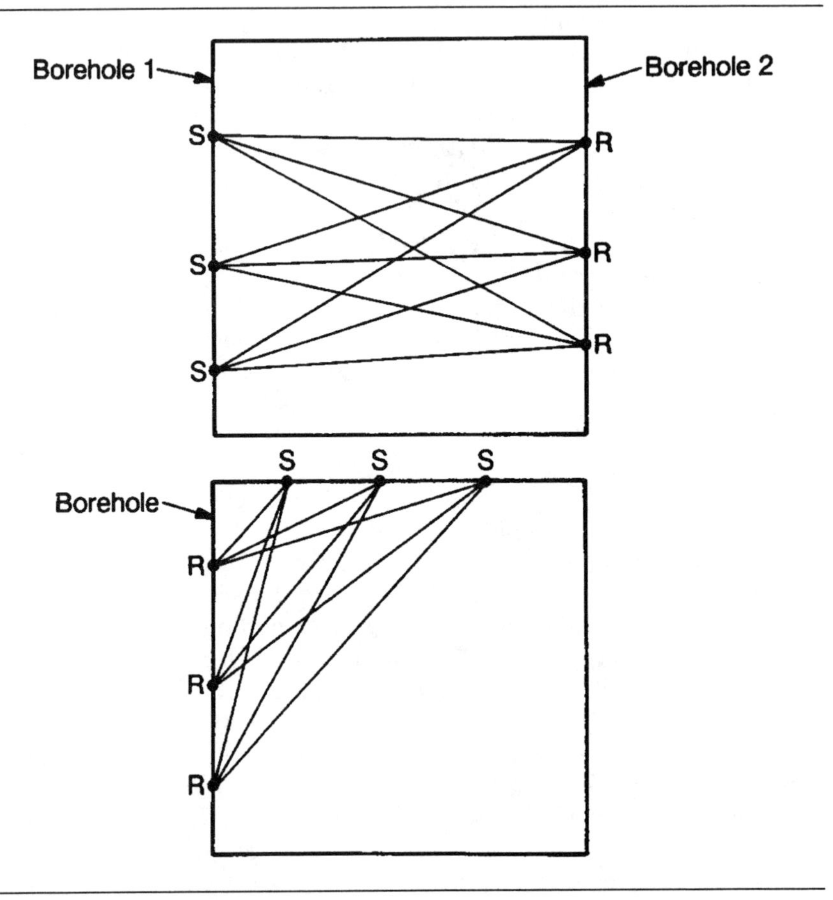

One of my first experiences with tomography resulted from a VSP imaging problem as described by Whitmore and Lines (*Vertical seismic profiling depth migration of a salt dome flank*, 1986 GEOPHYSICS). Velocity estimates were obtained by inversion of transmitted and reflected arrival traveltimes from VSP records. This velocity model allowed for an improved depth migration image of a salt dome flank. Figure 2 shows reflected rays for the final salt dome model. Interestingly enough, this VSP project used the tools of tomography and depth migration which were later used in many subsequent projects involving reflection and cross-borehole data.

The cross-borehole problem has become especially interesting in recent years because the advent of borehole sources has allowed large

FIGURE 2. Model of VSP reflections off a salt dome flank. The flank position was obtained by use of tomography and depth migration (from Whitmore and Lines, 1986 GEOPHYSICS).

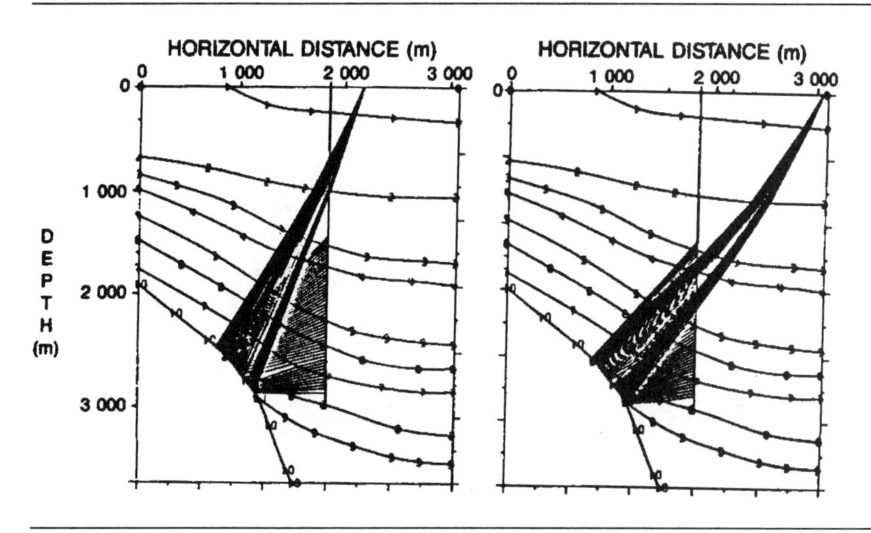

amounts of cross-well data to be gathered. The applications by Justice et al. (cited above) are excellent examples of using cross-borehole tomography for steamfront mapping in enhanced oil recovery. This tomography application was primarily based on the fact that acoustical velocity in tar sands often decreases substantially (25–30 percent) as temperature increases from 20° C to 120° C. Repeated use of cross-well tomography monitors the decrease of P-wave velocity as a function of temperature increase and thereby provides a method of monitoring steamfronts. The successful mapping of steamfronts by reflection traveltime methods has also been demonstrated by Pullin et al. (*3-D seismic imaging of heat zones at an Athabasca tar sands thermal pilot,* December 1987 *TLE).* Traveltime tomography methods have been frequently used in mapping steamfronts.

Figure 3 shows another successful application of cross-well imaging to an oil recovery prospect as presented by Lo et al. *(McKittrick cross-well seismology project, Part 2,* 1990 SEG abstracts). In two cross-borehole surveys, labelled A and B in the figure, tomography produced interesting images of the McKittrick thrust fault and associated faults. It is believed that velocity images in Figure 3 may allow reservoir engineers to locate zones of bypassed production.

Another example of borehole imaging comes from a paper by Harris et al. (*Cross-well tomogrophic imaging of geological structures in Gulf Coast sediments,* 1990 SEG abstracts). Figure 4 shows that cross-borehole tomography can be used in the detection of fault structures and porous

FIGURE 3. Cross-borehole tomograms from McKittrick field in California indicate rocks with different *P*-wave velocities (from 1990 SEG abstract by Lo et al.).

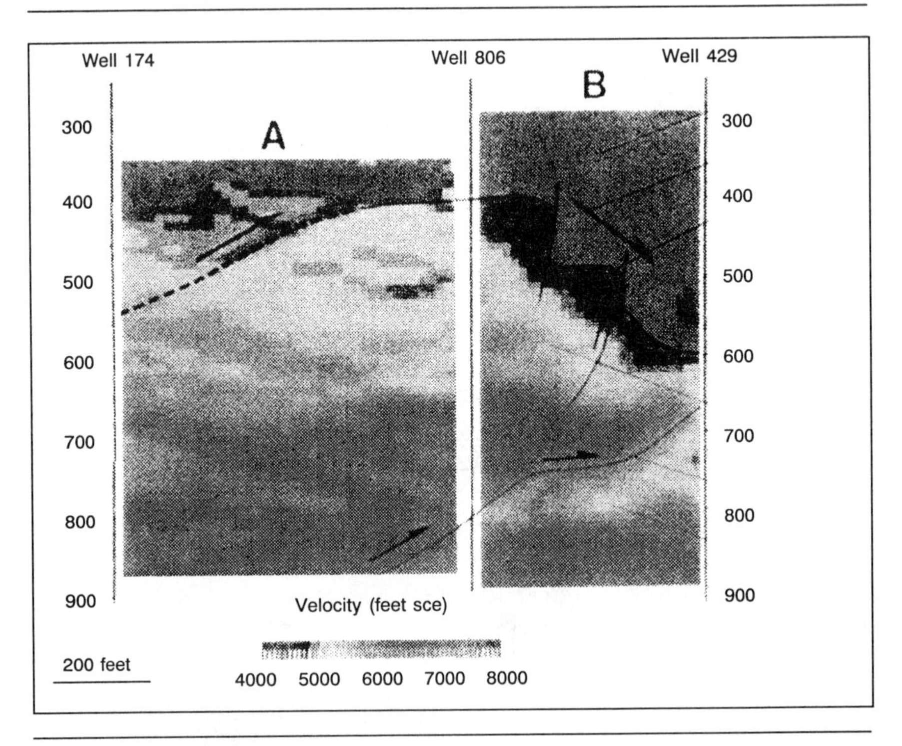

sandstone beds between wellbores. The tomogram was obtained from a joint Amoco-Stanford cross-borehole survey of about 6700 traces from the onshore Gulf Coast region. The velocity tomogram is shown along with well logs for wells along the edges of the tomogram. The left side of the tomogram is the position of the receiver well and the heavy line represents the position of the well containing a piezoelectric source. The log curves are responses of conductivity logs (CON), spontaneous potential (SP), and density from borehole gravity meter (DEN). Two faults, denoted as F1 and F2 in Figure 4, were detected due to the contrast in acoustical velocity. Beneath the faults, there are a series of low-velocity, low-density, porous sandstone layers labeled as M9, M10, and M10A. These observations are based on a joint interpretation of SP response, density variation, and of the tomogram. The velocity image was obtained by using two different tomographic methods applied by Amoco and Stanford research groups, and it was encouraging that the two methods gave very similar results. The continued advance in borehole seismic sources and receivers will undoubtedly be followed by more examples of cross-borehole tomography applications.

FIGURE 4. Cross-borehole tomogram from Harris et al. [1990 SEG meeting] indicates two faults, F1 and F2, at depths 3200–3400 ft and 2780–2880 ft respectively. The tomogram and well logs also show porous sandstone layers—M9, M10, and M10A—beneath the faults. These layers, shown in black on tomogram, are at approximate depths of 3380, 3500, 3630 ft. Also shown: conductivity log (CON), spontaneous potential log (SP), and borehold gravity meter density profile (DEN).

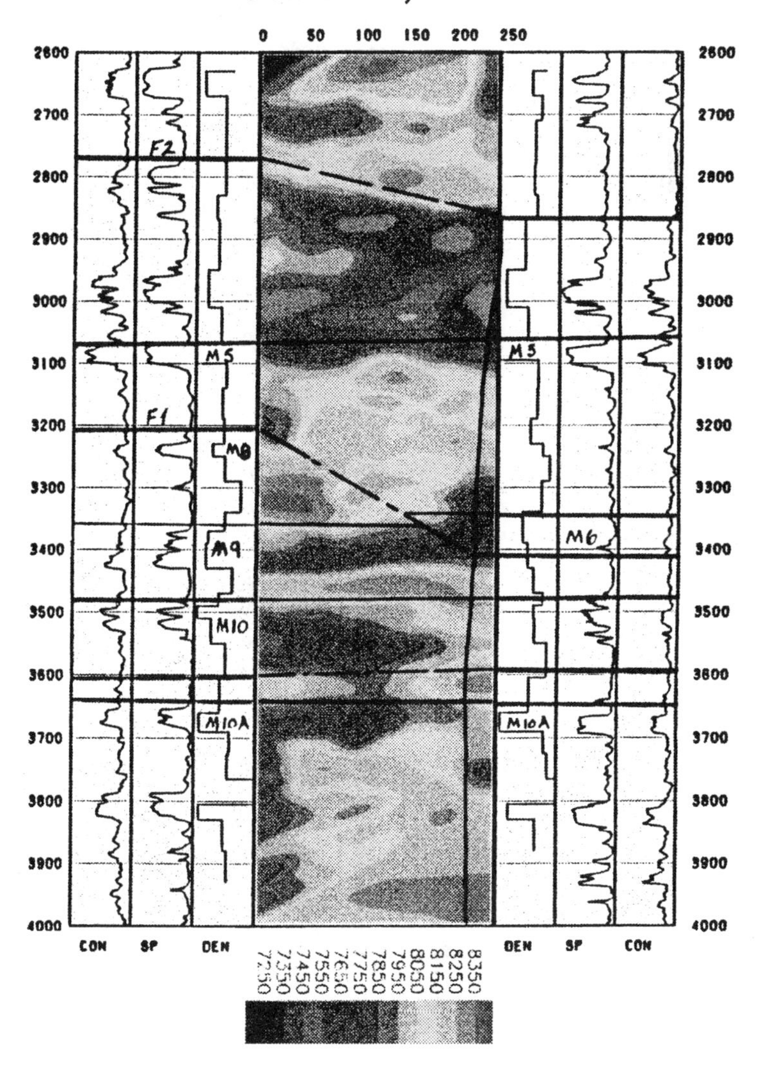

SURFACE SEISMIC TOMOGRAPHY

Figure 5 from Bording et al. (*Applications of seismic travel-time tomography*, 1987, *Geophysical Journal*) outlines the method of reflection tomography, which was made popular among many geophysicists following the publication of a number of papers by Gulf researchers at the 1984 SEG Annual Meeting. A summary of this pioneering research effort was presented by Bishop et al. (*Tomographic determination of velocity and depth in laterally varying media*, 1985 GEOPHYSICS). In the Gulf procedure, reflection tomography was used to estimate reflector positions and velocity variation. The problem of reflection tomography essentially involves two-way transmission, since seismic waves propagate to a reflecting interface and are then reflected upward to receivers. Reflection tomography is more difficult than transmission tomography due to the problem of defining reflector position. The matching of ray traced traveltimes to picked traveltimes produces reliable estimates of interval velocities, and such velocity estimates can be used in producing reliable depth migrations.

Depth migration and tomography are complementary processes. Depth migration requires reliable interval velocity estimates, which can be supplied by tomography, in order to produce a reliable depth image. Tomography requires good reflector definition from depth migration, in order to produce reliable velocity estimates. The two processes can be iteratively used in order to produce reliable depth images, since both processes are sensitive to velocity-depth variation. Following a suggestion by Gor-

FIGURE 5. Illustration of reflection tomography geometry.

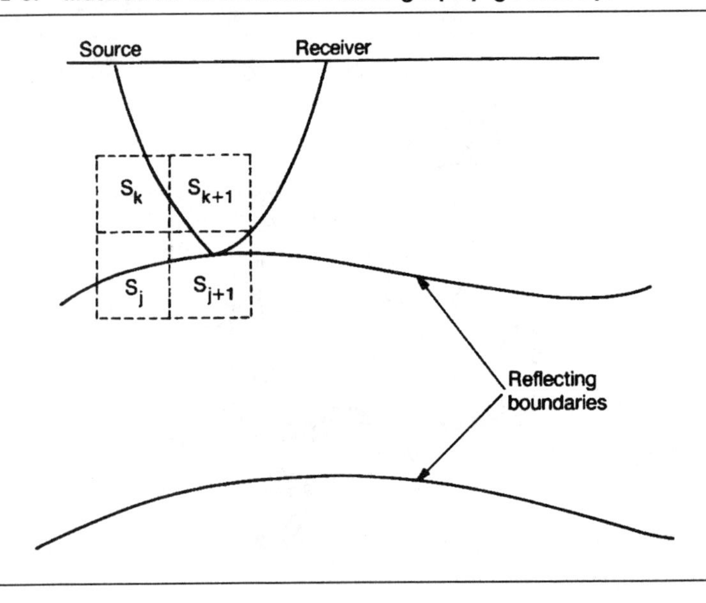

don Greve, our color plots of velocity tomograms are often superimposed on the depth migrated seismic data. The tomograms generally respond to gradual velocity changes, since traveltimes represent integrals of the earth's slowness over raypath. Migrated seismic sections respond to sudden velocity changes since seismic reflections are usually proportional to velocity differences in the earth. Although these two methods respond to different spatial variations in velocity, we often see agreement in the two images. In our experience, the iterative use of tomography and depth migration has been successfully used in imaging many different geologic features including overthrust folds, carbonate reefs, salt structures and sandstone lens features.

Some of these reflection tomography examples have been given in previous issues of THE LEADING EDGE. The cover of the December 1989 issue showed a Wyoming overthrust image obtained by our use of iterative tomographic migration. The result was similar to results obtained by extensive conventional velocity analysis, and it provided encouragement for many further applications. A sandstone lens image from Amoco's research efforts was shown by Sherwood (*Depth sections and interval velocities from surface seismic data*, September 1989 *TLE*). This tomography application outlined a high-velocity lens feature beneath a dominantly low velocity shale section.

Figure 6a shows an example of a salt wedge migration based on tomographic velocities. (This image resulted from the joint efforts of Amoco geophysicists Adam Gersztenkorn, Sam Gray, Phil Johnson, John Scales, and myself.) The high-velocity salt feature, with velocities between 13 500 and 14 800 ft/s, is surrounded by sediments of much lower velocity. The top and bottom of the salt feature, as shown by Figure 6b, were accurately estimated by tomographic depth migration. The interpreted reflector depths on the migrated section were later confirmed by a well drilled to the base of the salt feature. Depth estimation errors for the salt body in Figures 6a and 6b were less than 2 percent.

Figure 7 shows a schematic diagram of a classical reef problem in which the geophysicist attempts to detect the transition from a low-velocity basinal shale to a reefal carbonate of higher velocity. Figure 8 shows that iterative tomographic migration reliably imaged a velocity change in the transition from a low-velocity shale to high-velocity carbonate and thereby detected the edge of a carbonate reef. This tomogram was based on a data set of over 40 000 traveltimes interpreted by Frank Ariganello of Amoco's Houston Region. In the figure, the reef edge could be outlined by a line from (x, z) coordinates (11 600, 8000) to coordinates (21 500, 6500). Both the velocity images and the depth migration images coincide in this method and agreed with well data. This example exhibits the ability of reflection tomography to estimate lateral velocity variation.

FIGURE 6a. A velocity (superimposed on depth migrated traces) tomogram illustrates a contrast between high-velocity salt and low-velocity sediments. Dimensions of the tomogram are 63 000 ft wide by 16 000 ft deep.

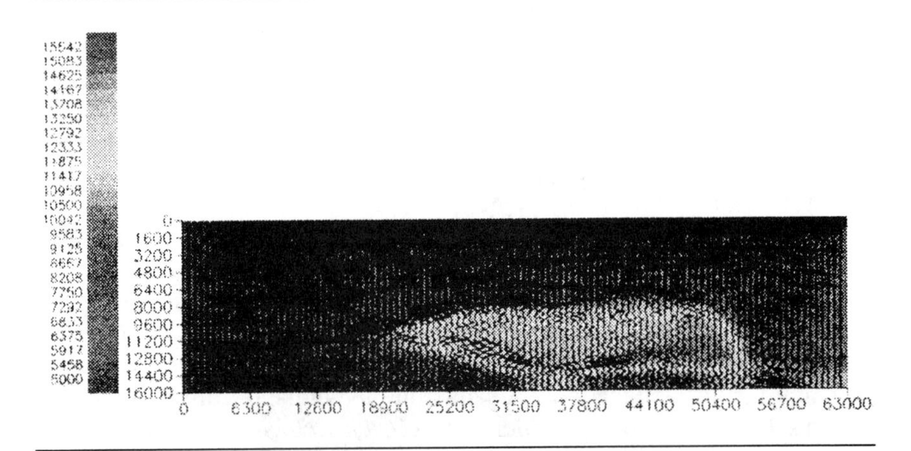

FIGURE 6b. Depth migration based on a velocity tomogram. The top of salt reflector and the bottom of salt are indicated below. Depth estimation errors for a well drilled near x = 25 000 were less than 2 percent. Dimensions of the section are 63 000 ft wide by 16 000 ft deep.

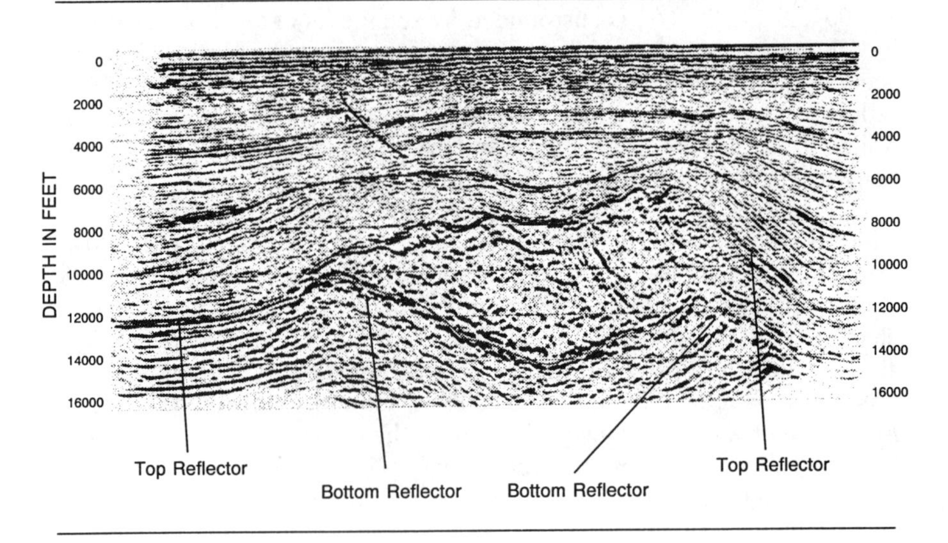

FIGURE 7. Schematic diagram of a possible carbonate reef enclosed by shales. For the present problem, typical shale velocities range from 11 000 to 14 000 ft/s, while carbonate velocities range from 14 000 to 20 000 ft/s.

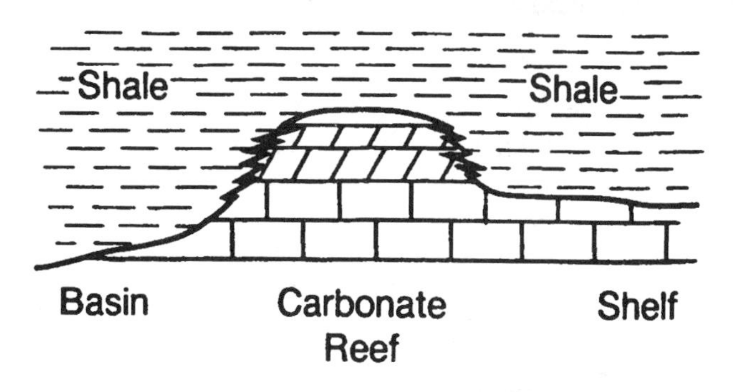

FIGURE 8. Tomographic migration for carbonate reef. Dimensions of tomogram are 40 000 ft wide by 10 000 ft deep.

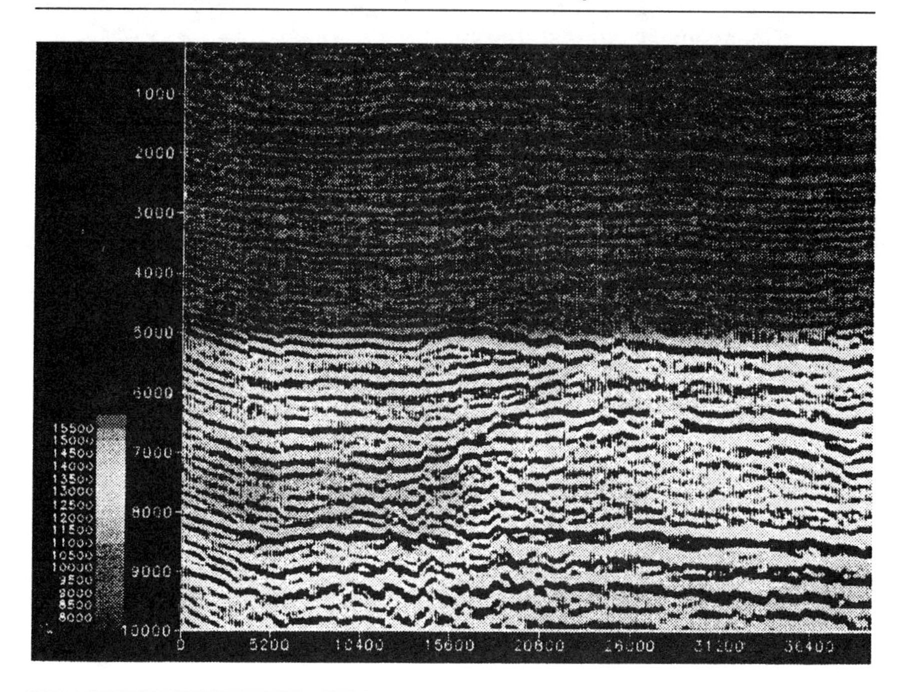

DISCUSSION

Experience has shown that tomography is useful in imaging both borehole and surface reflection data. In the case of cross-borehole imaging, tomography is the method of choice, whereas in reflection seismology, tomography is one of a number of possible methods of velocity estimation which can aid in depth migration.

In recent years, the emphasis in tomography has focused on borehole problems rather than surface reflection problems. It is believed that this trend will continue for the following reasons. In terms of user effort, first arrival traveltimes for cross-borehole data are generally easier to pick than the arrivals on surface reflection records. Also, problems with reflection tomography can occur due to the reflector depth-velocity ambiguities. Such ambiguities are problematic for small offset to reflector depth ratios in a seismic survey. Moreover, while tomography appears to be the main method in borehole velocity analysis, several techniques (including iterative prestack migration) exist for reflection velocity analysis. Finally, much cross-borehole data has recently become available for tomographic applications, making cross-borehole tomography a viable tool for crosswell velocity estimation. The traveltime tomography method will undoubtedly see many applications in the future.

ACKNOWLEDGMENTS

This tomography project would not have been possible without the cooperative efforts of a large number of fellow researchers. The author wishes to thank: Frank Ariganello, Phil Bording, Paul Docherty, Adam Gersztenkorn, Sam Gray, Jerry Harris, Phil Johnson, Ken Kelly, Ed LaFehr, Joe Lee, Jim Myron, Mike Sabroff, John Scales, Christof Stork, Henry Tan, Paul Thomas, Sven Treitel, Dan Whitmore, and the many other Amoco support personnel who contributed to the project.

ABOUT THE AUTHOR

Larry Lines received a BSc and an MSc in geophysics from the University of Alberta and a PhD in geophysics from the University of British Columbia. In 1976 Lines joined Amoco Canada where he worked as an exploration geophysicist. In 1979 he transferred to the Amoco Research Center in Tulsa where he is presently a research associate. During 1980–82 he was an adjunct professor in geosciences at the University of Tulsa. His main interests are in geophysical modeling and inversion. Lines is a member of SEG, IEEE, EAEG, CSEG, GST, and is a registered professional geophysicist in the Canadian Province of Alberta.

Success Through a Multidisciplinary Approach: A Case History*

by T.S. Dickson, A.P. Ryskamp, and W.D. Morgan
Amoco Canada Petroleum Company Ltd.
Calgary, Canada

Seismic imaging provides a powerful tool for a variety of development applications. But do seismic images accurately reflect the shape of the reservoir? While defining an injection well location in a gas cycling pool (Upper Devonian Nisku formation, west central Alberta, Canada), our project team was confronted with this question. The problem was recognized and solved by carefully integrating the seismic data with geologic and engineering information to produce a superior reservoir model.

GEOLOGY

At the beginning of Nisku time, a shallow water shelf extend over the study area depositing fossiliferous, argillaceous platform carbonates (Figure 1). Colonies of corals and stromatoporoids began to establish themselves throughout the area. At the end of this depositional cycle, an abrupt rise in relative sea level caused a lateral diversification of facies. A shallow water carbonate bank (shelf) deposited argillaceous carbonates in the southeast. Reef growth along the bank margin marked an abrupt transition to a basinal environment. In the shallow part of this basin, numerous pinnacle reefs grew in a broad northeast-to-southwest trend (Figure 2). The basin then filled with calcerous shale, followed by a resumption of carbonate bank sedimentation across the study area.

The Nisku formation was affected by numerous diagenetic events. Most significantly, dolomitization and leaching created abundant porosity in the reef facies. The pore system is dominantly vuggy in nature, although variable amounts of intercrystalline and fracture porosity provide communication between the vugs. Early cementation maintained the competency of the reef rocks during burial, allowing differential compaction to occur with respect to the less competent basinal shales.

The overlying shelf carbonates and siltstones provide an effective top seal for the reef reservoirs. The pinnacle reefs are sealed laterally by

*Reprinted, by permission, from *Geophysics: The Leading Edge of Exploration*, Vol. 11, No. 3, March 1992, courtesy of The Society of Exploration Geophysicists and the authors.

FIGURE 1. Facies differentiation during the Upper Nisku interrupts regional carbonate bank deposition.

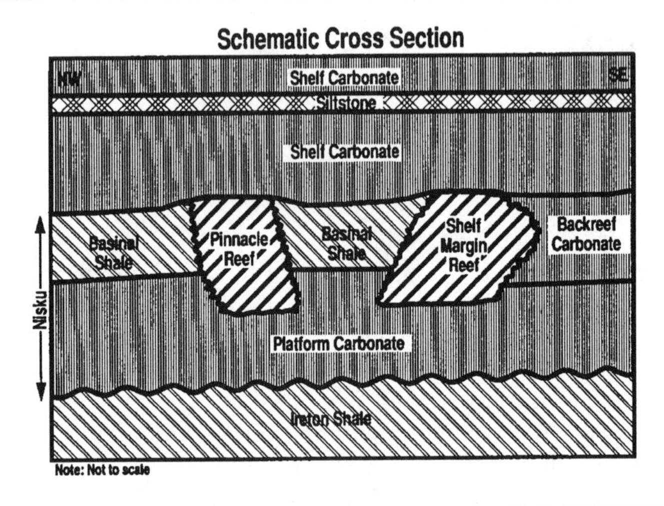

FIGURE 2. During the Upper Nisku, reefs developed in the basin and along the shelf (bank) margin.

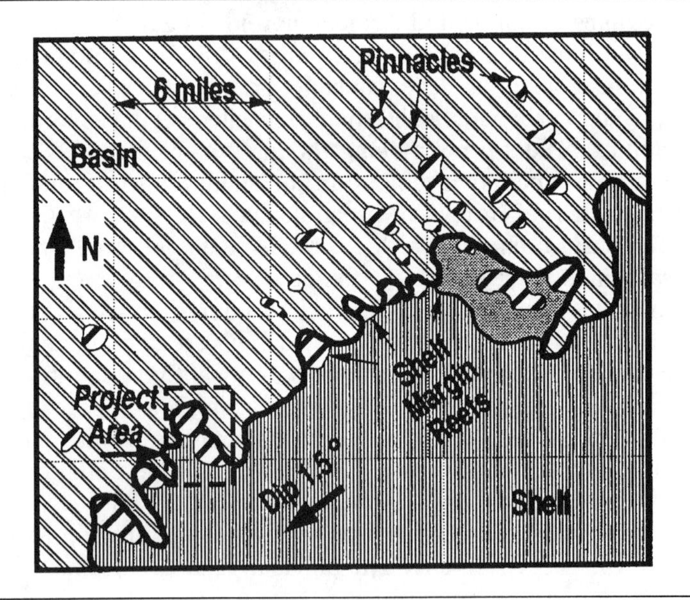

the basinal shale. Minor tilting of the basin toward the southwest at 1.5°, combined with the irregular morphology of the bank margin, allows hydrocarbons to be trapped in some bank margin reefs. Pools vary in size from 0.5–30 million bbl oil or 10–100 billion ft^3 gas, depending on the size of the reef buildup and the effectiveness of the seal.

GEOPHYSICS

Figure 3 shows the seismic response of the Upper Devonian in this area. The bank and shelf carbonates have the highest velocity of the facies present (about 6000 m/s) and the shales have the lowest (about 4700 m/s). Reef velocities vary between these values depending on porosity. The increase in velocity at the top of the shelf carbonates generates a trough

FIGURE 3. Reefs are characterized by moderate Nisku amplitude and thin Wabamun-to-Nisku isochron.

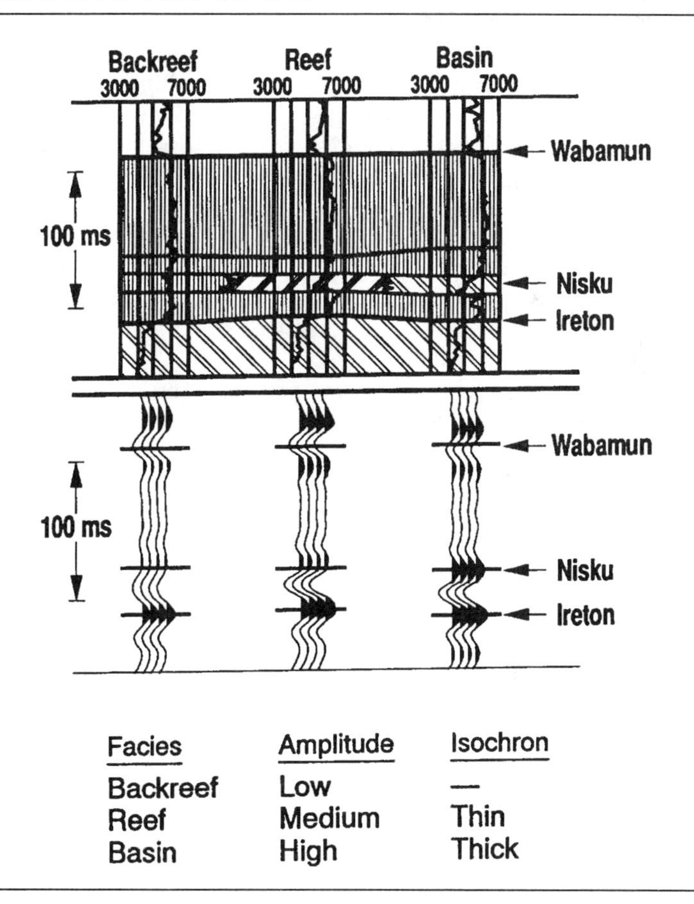

Facies	Amplitude	Isochron
Backreef	Low	—
Reef	Medium	Thin
Basin	High	Thick

(Wabamun reflection) on the polarity shown. The top of the underlying shale causes a peak (Ireton reflection). These regional marker events are present throughout the study area.

Seismic attributes at the Nisku level are related to facies. Where tight carbonate is present, there is little or no velocity contrast so no reflection is generated. Both porous carbonate and basinal shale have lower velocities than the surrounding rock, causing a peak to develop near the top of the interval (Nisku reflection). The basinal response usually has higher amplitude than the reef response because of its lower velocity. Also, the basinal event is normally about 5 ms stratigraphically lower because of early compaction of the basinal shale. The stratigraphic position of the event is measured by mapping the Wabamun-to-Nisku isochron and noting terminations of weak events immediately above the Nisku.

ENGINEERING

Natural gas from the Nisku reservoirs in this area is rich in natural gas liquids, condensate, and hydrogen sulphide. At original reservoir temperatures and pressures, all the components form a single phase gas mixture. Under conventional production practices, as hydrocarbons are produced and reservoir pressure declines below the dew point, the heavier hydrocarbons condense (Figure 4). Once in a liquid state, they become much less mobile and cannot be efficiently recovered.

To improve recovery from these reservoirs, gas cycling is used (Figure 5). Dry gas is injected into the reservoir to replace wet gas volumes produced. No condensation occurs because pressure is maintained above

FIGURE 4. Primary production in a wet gas reservoir causes abundant hydrocarbon liquids to remain at abandonment, compared to a dry gas pool.

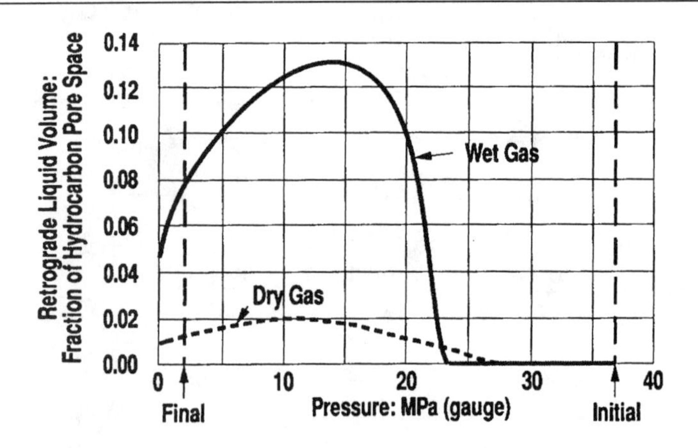

FIGURE 5. Gas cycling improves liquid recovery by displacing wet gas with dry gas at pressures above the dew point.

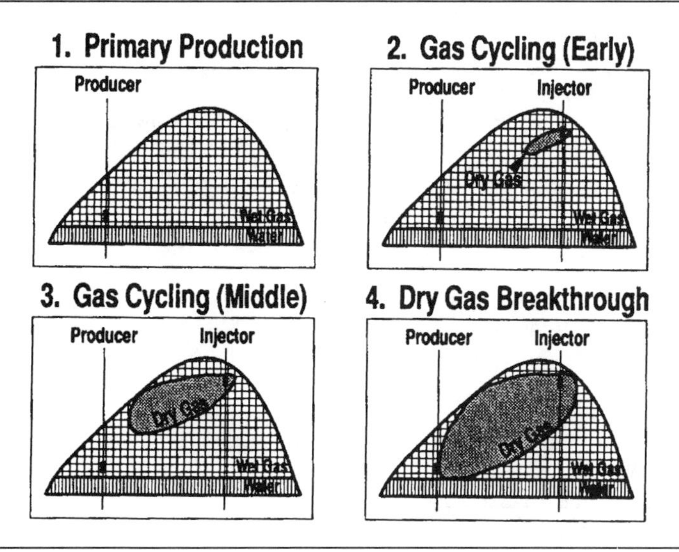

the dew point. The dry gas front moves gradually toward the producing well until dry-gas breakthrough occurs. The injection well is then recompleted as a producer and the reservoir is again drained conventionally (blowdown).

The improvement in liquid recovery through this technique is governed by the size of the swept region which, in turn, is a function of three interacting factors. First, the greater the pressure differential, the greater the tendency to draw injected gas directly toward the production perforations. Secondly, because dry gas is less dense, it tends to rise to the top of the reservoir above the wet gas. Finally, recovery is affected by reservoir heterogeneity, which can reduce the swept volume by channeling the dry gas between injection and production points.

PROJECT HISTORY

The project team set out to drill an effective injection well in a single well pool that was approaching dew point. The following criteria were chosen to optimize the location:

- Injection well must be in pressure communication with the producer.
- Both wells should be as far apart as possible to improve horizontal sweep.
- Injection well should penetrate the highest point of the reservoir to optimize the vertical sweep.
- Injection well should have good reservoir quality.

FIGURE 6. The production/pressure history of the discovery well suggests an OGIP of 71 billion ft³.

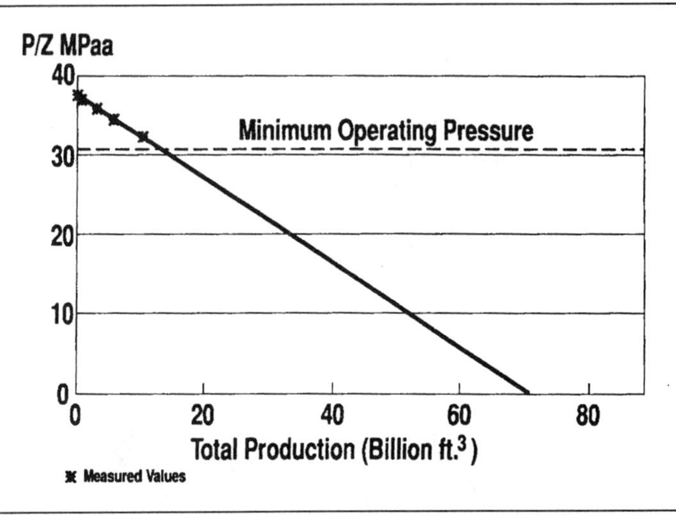

The pool discovery well had produced about 12 billion ft^3 gas with 1.2 million bbl condensate and 152 000 tons sulfur at daily rates of about 10 million ft^3 gas with 1200 bbl liquids. Pressure performance (Figure 6) shows the linear characteristics of volumetric reservoir behavior common to the pools in the area. Extrapolating the P/Z line to the axis yields an original gas in place (OGIP) estimate of 71 billion ft^3. Gas cycling should increase condensate recovery by 900 000 bbl or 12 percent.

Figure 7 shows the location of the pertinent wells. The discovery well encountered 44 m of Nisku reef with 23 m of gross pay over water. Using a three percent porosity cutoff, net pay equals 19 m and the average porosity over this interval is 9.5 percent. By contrast, the backreef well to the west encountered only 1 m of net porosity.

The seismic maps in Figure 7 were prepared from an irregular 1/2 × 1/4 mile grid of data acquired by various crews during the late 1970s and early '80s. The Wabamun-to-Nisku isochron is thick (80–84 ms) on the eastern side of the map and thins toward a plateau around the discovery well (72–76 ms). The west side of the map has no contoured data because the Nisku event is not present in this area. The amplitude of the Nisku event generally increases toward the east.

Figure 8 shows an east-west line which passes by the wells. The high Nisku amplitude and thick Wabamun-to-Nisku isochron at the east end of this line are characteristic of the basinal response. The change to a lower amplitude and the termination of the event above the Nisku mark the reef edge. Near the discovery well, typical reef attributes are seen. The low Nisku amplitude at the west end of the line indicates tight backreef facies. A parallel line south of the wells shows the same character-

FIGURE 7. Well locations, key seismic lines, and seismic attributes are highlighted.

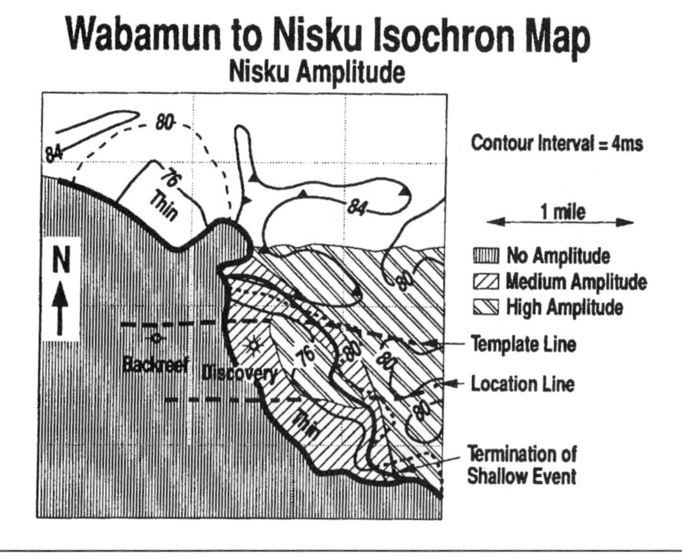

FIGURE 8. This line confirms the seismic signature of the reef.

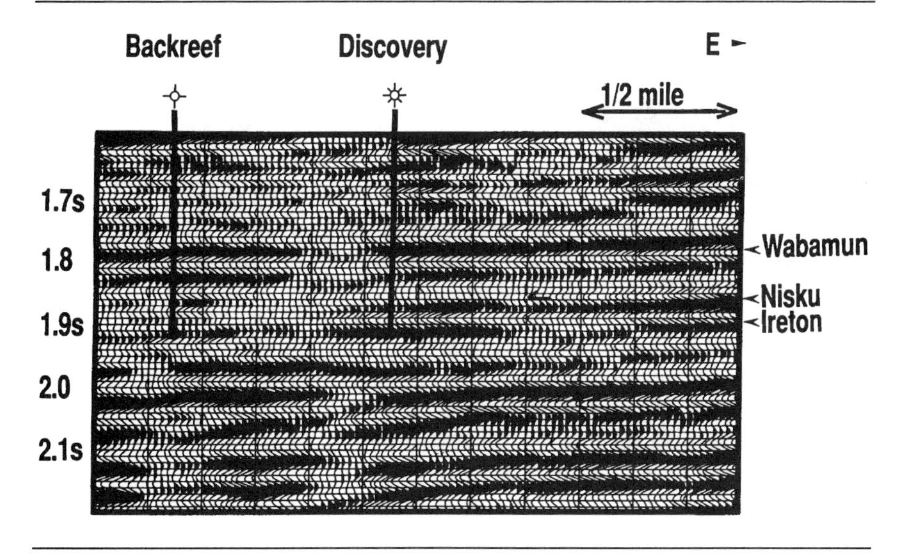

istics (Figure 9). A well location was suggested based on analogy with the discovery well.

The team prepared a pay map (Figure 10a) to confirm that this reservoir interpretation was accurate. No net pay was assigned to the low-amplitude backreef and interpreted basinal areas. The area with thin iso-

FIGURE 9. **The initial location was chosen for its similarity to the discovery well.**

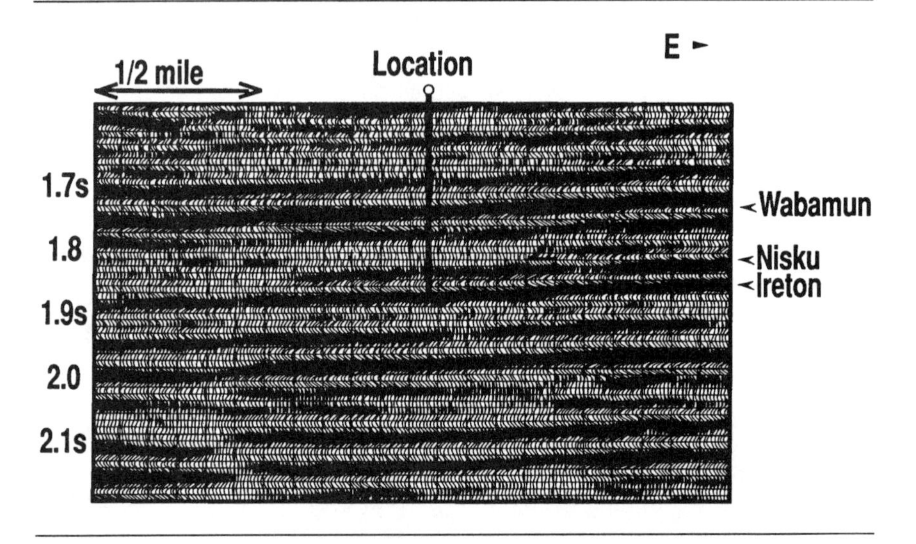

FIGURE 10. **(a) Inital reservoir interpretation assumed that the reef had the same shape as the isochron map. (b) Second reservoir interpretation assumed a tabular reef morphology with a steeper easter margin. (c) Final reservoir interpretation assumed that the isochron slope area contained maximum reef development similar to the reef to the northwest. For each isopach, contour interval is 10 m and porosity is 9.5 percent.**

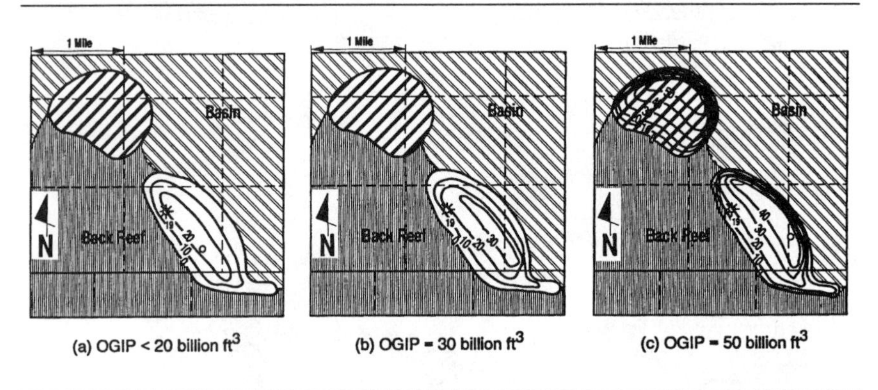

chron values was assigned reservoir parameters consistent with the discovery well. The eastern slope of the reef conformed with the shape of the isochron map and regional dip was superimposed on the model. This concept agreed with the available seismic and well data. However, the OGIP that could be accommodated in this estimate was under 20 billion ft^3 of less than one-third of the reserves suggested by pressure data.

The second reservoir map (Figure 10b) was based on reef slopes being steeper than those in the first model. The pool outline was kept constant but the eastern slope was compressed to add volume to the reservoir. This map accommodates an OGIP of about 30 billion ft^3, still considerably lower than the volume suggested by pressure data.

A review of other reefs in the area suggested that reef isopachs tend to increase and average porosity values improve values toward their eastern margins. This may be related to an easterly prevailing wind that caused the biota to flourish and reef growth to climax along the high-energy windward margin.

From these observations, a third reservoir model was mapped (Figure 10c). The reservoir outline and well data were unchanged from the first two models. Reef thickness was increased over the area previously interpreted as reef slope. The volumetric reserves were recalculated to be about 50 billion ft^3 from this model. This was a much more acceptable fit with the pressure data, especially since even more reserves could be accommodated by increasing the porosity estimate for the windward margin.

The injection well location was based on the updated reservoir model. Its seismic tie is marked on Figure 11. The amplitude and isochron values at this position lie between the known reef attributes and basinal values. It was difficult to believe that these intermediate seismic characteristics suggested maximum reservoir thickness.

FIGURE 11. The location was moved to the position indicated. Compare to Figure 9.

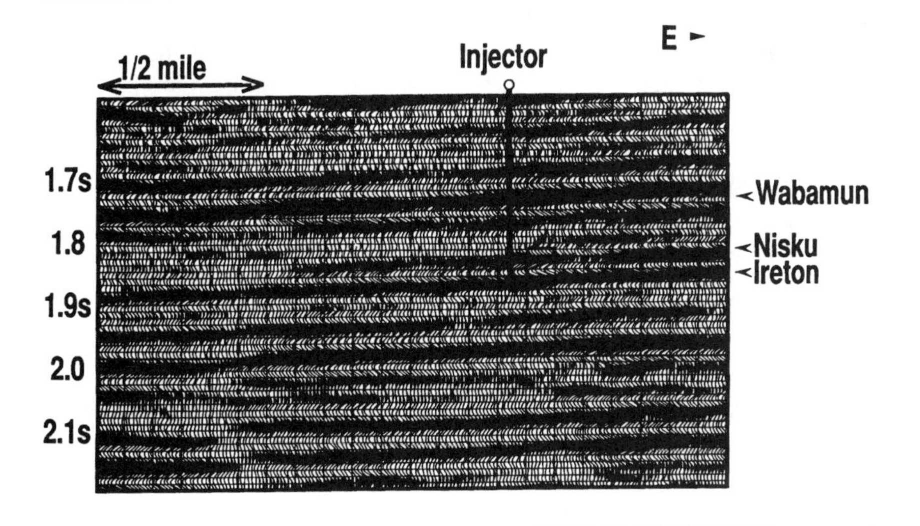

FIGURE 12. Synthetic traces verify that the seismic attributes of thick reef are between thin reef and basinal values.

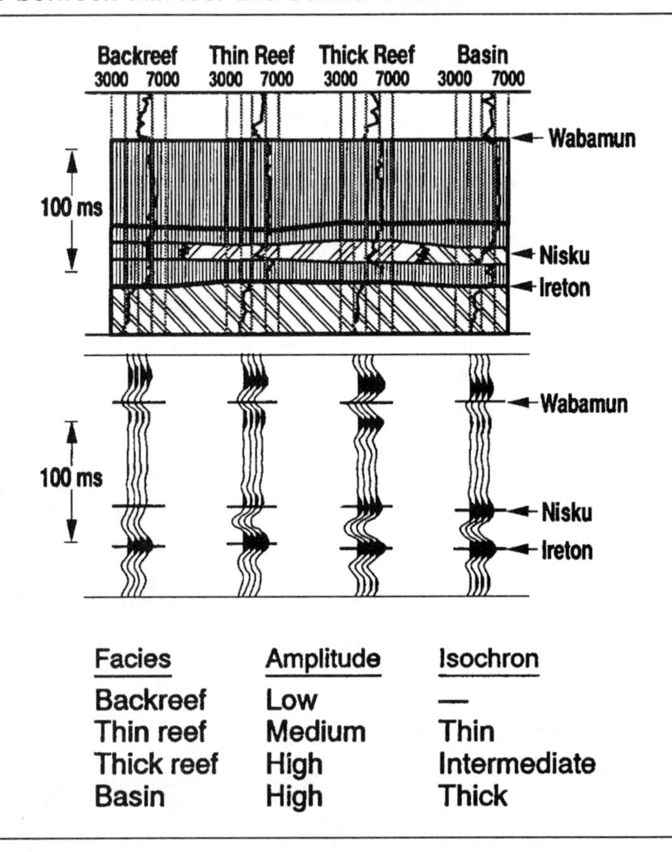

Facies	Amplitude	Isochron
Backreef	Low	—
Thin reef	Medium	Thin
Thick reef	High	Intermediate
Basin	High	Thick

However, the well results confirmed the third reservoir model. The injection well encountered 41 m of net pay with an average porosity of 11 percent (as compared to 23 m and 9.5 percent in the discovery well). Figure 12 shows that the seismic signature of thick, highly porous reef is, in fact, intermediate between the typical reef and basinal responses.

SUMMARY

The seismic image of this reservoir did not define the distribution of reservoir characteristics in the subsurface. Geologic interpretation was impaired by insufficient well control. Engineering data contained no information about the location of the reserves. These difficulties were overcome by combining the reservoir outline from seismic with a geologic shape and the volume suggested by engineering to produce an accurate working description of the reservoir. These data exist for most reservoir development projects. The challenge is to integrate the information effectively.

ACKNOWLEDGMENTS

The authors wish to acknowledge Rob Kostash for his contribution to this project; Mike Kusman and Lynn Peacock for their work in preparing the figures; and the managements of Amoco Canada Petroleum Company, Esso Resources Canada, and Encor Energy Corporation for permitting publication.

ABOUT THE AUTHORS

Tom Dickson is currently a staff geophysicist with Amoco Canada. After receiving a BSc in geophysics from the University of Saskatchewan (1980), he joined Hudson's Bay Oil and Gas. Since then his work has focused on development seismic interpretation as well as seismic acquisition and processing in the Western Canada basin.

Phil Ryskamp joined Hudson's Bay Oil and Gas in Calgary in 1980. Successive merges have moved him to Dome Petroleum and Amoco Canada where he is currently an area geologist. Ryskamp holds a BS in geology from the University of Calgary. His career has been focused on exploration and development geology in the Western Canada basin.

Bill Morgan is a senior staff production technologist at Amoco Canada. He has 12 years of petroleum industry experience in production and reservoir engineering. He also holds a diploma in chemical and metallurgical engineering from the British Columbia Institute of Technology.

Geoscience in Reservoir Development—A Sleeping Giant*

by Gordon M. Greve
Amoco Production Company
Houston, Texas

As the search for new hydrocarbon reserves becomes more and more difficult, attention in the industry is turning to the economic recovery of products from existing oil fields. In many cases, these reserves have been exploited with little help from the geoscience community; so it is not surprising, considering the desire to better characterize reservoirs, that the formation of geoscience/engineering teams is an industry trend. One would expect the utilization of geoscience in reservoir development would pay handsome dividends.

The ultimate purpose of using geoscience technology in hydrocarbon production is to improve the profitability of the resource under production. From what I have seen, no project is undertaken that cannot be economically justified. Although this is also true in exploration projects, the parameters there are much more loosely defined. In reservoir development, when one applies geoscience technology, the direct result of that action must be profitable. There is a balance between technology and profits.

Leaving profits aside for the moment (but never entirely out of the picture), let's turn to geoscience. Geoscience applications in reservoir development can be placed into four categories with a fifth one (reservoir engineering) providing information that further constrains the geoscience solutions to reservoir characterization. The four categories are geology, geophysics, geochemistry, and petrophysics. These form an interlocking geoscience network and the key to profitability.

Integration of these categories is an important area of current development. We are also seeing the data and programs to analyze the data coming together on geoscience workstations. It is obviously necessary to have all the information about an oil reservoir available at the fingertips of those who are studying it. Many of these workstations were demonstrated at the exhibits of the last SEG Annual Meeting. However, to be useful, the workstations need to be applied to a specific task.

---

*Reprinted, by permission, from *Geophysics: The Leading Edge of Exploration*, June 1992, courtesy of The Society of Exploration Geophysicists and Gordon M. Greve.

These tasks seem to fall into three areas: to determine rock properties in well bores; to extrapolate these properties and their containing surfaces between wells; and to monitor enhanced recovery projects.

ROCK PROPERTIES

The determination of rock properties in well bores is accomplished primarily by means of wireline logs and core analysis. But what are these rock properties? Porosity, pore throat size, permeability, and rock type have to rank high on the list, but other important properties are mineralogy and fluid types (their character and saturations). However, this list is not exhaustive and as I study the science of reservoir characterization, I find there is a seemingly unending list of other rock properties.

Porosity is certainly one of the key characteristics. I recommend reading the October 1991 AAPG *Bulletin* in which Bob Ehrlich and his colleagues at the University of South Carolina report on their work involving the characterization of porosity, its relation to pore throat size, and permeability. They determine all these parameters quantitatively from a mathematical analysis of thin sections.

Many of the larger scale rock property features (such as fracture porosity) can more easily be seen now that the Formation Micro-Scanner (FMS) has been introduced. Its high resolution makes detailed features of the well bore wall visible. This allows a better interpretation of other logs since one can determine if the log measures a value characteristic of the stratigraphy or just an anomalous point.

Another area where great strides have been made in the determination of rock properties is the adaptation of continuous core drilling from the mining industry to the petroleum industry and the attendant full-scale measurement of core properties. With these compact rigs, a whole core is continuously taken. In order to obtain information from this core, special equipment must be assembled and located at the drill site to prevent the shipping of immense amounts of core around the world. Thus, the equipment must be housed in portable modules. Rock properties that are typically measured are density (both grain and saturated), porosity, *P*- and *S*-wave velocity (at elevated pressure), shear-wave birefringence, magnetic susceptibility, permeability, total organic content, quantitative mineralogy, and compressive strength. The output from the modules can be displayed in typical log curve form and are also retained in digital form for filing and manipulation. One of the many advantages of continuous core measurement is that the reservoir rock as well as shales, anhydrite or any other type of lithology is included. In normal coring operations, usually only the reservoir rock is recovered. This is only half of the information that a geophysicist needs to calculate a reflection coefficient. Continuous coring and the measurement of rock properties from the core is a development that is only now emerging, even though the continuous core rigs have been available for some time.

EXTRAPOLATION OF SURFACES

The second area of geoscience applications in reservoir characterization is the extrapolation of surfaces and rock properties. Table 1 lists many of the technologies employed in this task. Some, such as well log and paleo correlations, I will not discuss. Others, like geologic models (which are extremely useful, especially where seismic data is very poor) and sequence stratigraphy (which is useful where seismic data is very good) require more extensive discussion than is possible here.

There is no doubt that the advent and use of 3-D seismic surveys has been a major factor in reservoir development. Refer to the article by E.O. Nestvold elsewhere in this issue for the latest on the use of this technology. However, I would like to add one comment regarding the spread of 3-D seismic surveys to land use in the United States. Here the major barrier is cost. Not only are the surveys expensive but the surface and subsurface permitting costs escalate to unbearable levels. Certainly, something has to change. Either less expensive methods have to be employed or land and mineral owners will have to lower their permitting costs. Since a reduction in permitting costs seems unlikely, less expensive methods for land 3-D surveys will have to be developed.

Two less expensive techniques are shown in Figure 1. One is to use the old box technique where receivers as laid out along the perimeter of the area to be surveyed and the sources proceed around the perimeter recording into all receivers. This technique results in two-fold coverage with low permit costs. Additional fold can be obtained by rolling the box along. Surface permits need only be obtained along the perimeter. If the field owner is conducting the survey in his own behalf or on behalf of the other owners, the subsurface permit fees are also waived. Another method is the cross technique. Here the receivers are placed on one line and

TABLE 1. Extrapolation of surfaces & petrophysical properties

- Well log correlations
- Geologic models (including sequence stratigraphy)
- Paleontological correlations
- Seismic
- Stochastic modeling
- Borehole tomography
- VSPs
- Pressure analysis

FIGURE 1. Inexpensive land 3-D.

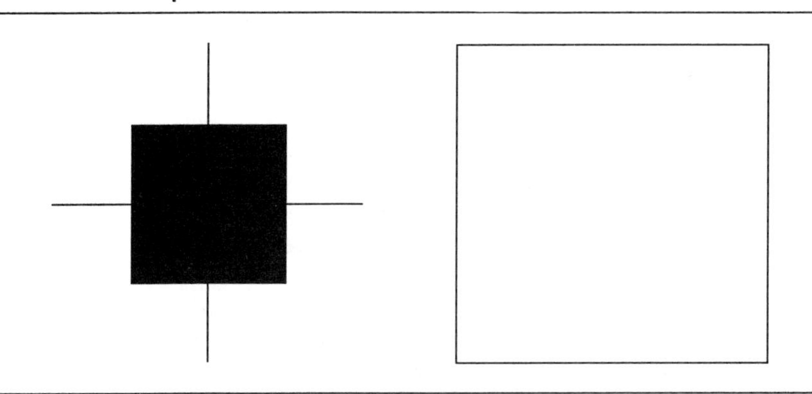

the sources on another. Other methods are already known and not practiced, or are yet to be discovered.

Another technique, which is expanding rapidly, is the use of geostatistical methods to extend a surface or rock property away from points of control. Often, from either 2-D or 3-D seismic surveys, a surface can be reliably mapped or the gross features between horizons inferred, However, relating the fine detail to the gross features calls for geostatistical techniques. For those of you unfamiliar with this term, I should mention that the prefix geo does not stand for geologic but for geographical. (It may also be beneficial to study up on this technology.) Geostatistics is the use of statistical concepts to infer spatial relationships. At the 1991 EAEG meeting, there were many papers on the use of geostatistics to assign measured rock properties using sequence stratigraphy models. Some discussed how geologic modeling techniques are used to predict a sequence which is then assigned rock property values taken from outcrop and well data. In cases where seismic controls is available, this geologic information can be integrated into the model using geostatistical techniques as the integrating tool. I expect we will see much greater utilization of geostatistics incorporated in workstations.

The application of geostatistical techniques yields many different realizations of the subsurface, all fitting the input control. As more control is obtained, the number of realizations may decrease but I doubt one will ever get enough control to result in the real answer. We simply cannot afford to sample at the density required. This implies that geostatistics will continue to be employed on an ever broader scale. We should expect improvements in both the basic mathematics and in computer programs to apply these concepts.

Another tool that can be used for both the extrapolation of surfaces and rock properties as well as being useful in EOR monitoring is borehole tomography and its relative—the VSPs. This is an area where development has involved consortiums because of the relation between cost and the perceived economic benefits. Perhaps this consortium approach to research is a hallmark for future research efforts in our industry.

Key to the success of this technology is the development of borehole sources and receivers. There are at least eight different types of sources either under development or available. Maximum range varies from 300–2500 ft although none of them is really ideal. In the receiver development area, there are four commercial tools available with others under development. So far most of these tools suffer from either bandwidth problems or number of receivers that can be deployed in the hole.

Most important to borehole geoscience are the applications which are possible with this equipment. Figure 2 shows four different schemes which are commonly used in borehole work. The scheme on the upper left shows the source and receiver in the same well. As indicated, this would be useful to determine distance from salt domes, reflective faults or other near-vertical phenomena. On the lower right is a VSP where the sources have been placed in the hole. On the upper right is borehole-to-borehole tomography where time, amplitude, or some other characteris-

FIGURE 2. Borehole seismic technology.

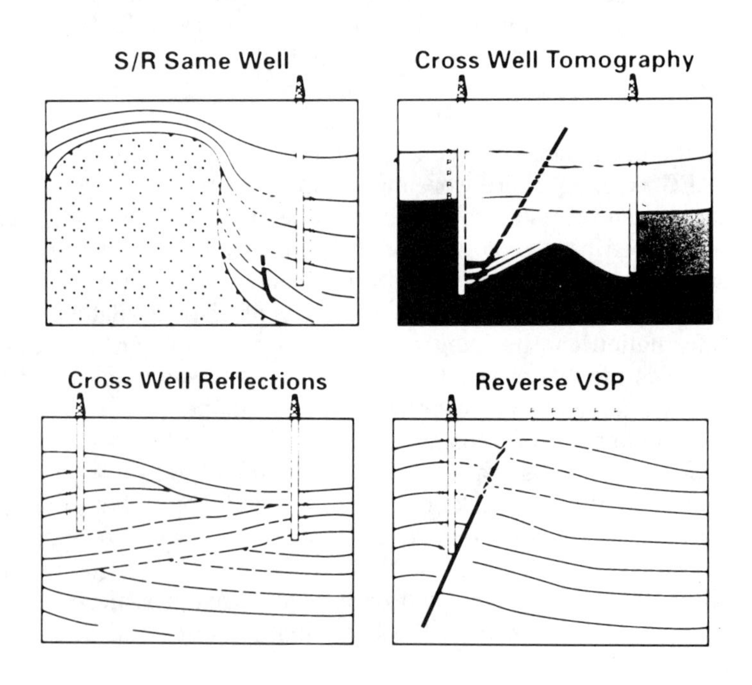

tic of the direct arrival is measured. This scheme is useful to interpolate surfaces or rock properties between wells. Examples of borehole-to-borehole tomograms appear in many papers; thus far transit-time tomograms seem to be the prevalent output, but I expect we will soon see absorption tomograms made by measuring first arrival amplitudes. The scheme on the lower left is cross-well reflection where the direct arrivals *and* the later reflected arrivals are both measured. This technique can be coupled with the borehole-to-borehole technique to look for other exploratory targets between wells or for deeper targets below wells.

EOR MONITORING

This is the area where borehole tomography has been extensively used. Most of the applications involve the monitoring of heavy oil movement caused by some sort of heat source. The application of heat lowers the velocity of the rock which is monitored by measuring seismic traveltime delay. Another method is to monitor hydrocarbon movement by surface measurements. Both methods rely upon the difference between two sets of measurements which greatly enhances the ability to detect subtle changes. Certainly, the surface methods are in their infancy and should see much growth in the next few years.

SOME ODDS AND ENDS

One area that I have not touched so far is reservoir engineering. The data collected while the reservoir is under production can be important in deducing its geologic features. Measured pressures in a reservoir, taken over a period of many years, can yield important information about the reservoir's size. This information is incorporated into the geoscience-calculated volumes expected in the field and the discrepancies rectified. Long term pressure measurements can also lead to subtle pressure barriers present in the field but previously undetected.

I can't close without mentioning the location problem. All this great geoscience is useless if the control points are not where thought to be. Surveying, especially in old fields, can be primitive. Maps and map projections were sometimes poorly controlled. Errors can occur either in the surveying or when unknown map coordinate system was used. There have been cases where more than one coordinate system was used over the years with no correction for previous coordinate systems made. When working on these old fields and in some of the newer fields where the surveying may not be the best, the locations should be resurveyed using the Global Positioning System of orbiting satellites.

SUMMARY

The important points to remember are:

- Application of geoscience in hydrocarbon development is driven by economics.
- The work of Ehrlich and his colleagues is important because of their quantitative approach to porosity, permeability, and pore throat radius.
- The FMS has dramatically improved our picture of well bore phenomena.
- The development of continuous core measuring technology will vastly increase our knowledge of nonreservoir rocks.
- Although 3-D seismic has been a dramatic improvement in extrapolating and measuring surfaces, less expensive means need to successfully extend the technology to large scale deployment in land surveys in the United States.
- Geostatistics will find its way into more and more applications.
- Borehole tomography is ready to explode onto the scene over the next few years as a new technology for both interpolation between well bores and EOR monitoring.

ACKNOWLEDGMENTS

I would like to thank the many geophysicists and geologists who developed the technology which made my presentation possible. Special thanks go to those geoscientists at Amoco Research who provided the figures.

ABOUT THE AUTHORS

Gordon M. Greve received a BS in electrical engineering, an MS and PhD in geophysics all from Stanford University. He began his career with Amoco Production Company in 1960. In 1986 he was promoted to manager of geophysics for Amoco's worldwide activities. He has served SEG as 1st Vice-President on the 1990–91 Executive Committee, as chairman of the Student Sections/Academic Liaison Committee, and as Vice-Chairman of the 1984 Annual Meeting in Atlanta. He is also a member of AAPG and EAEG.

CHAPTER 14

CONCLUSIONS

For all that modern science adds to petroleum technology, exploration and development of hydrocarbons still involve considerable art as well as applied science. A prospect may begin as a geologist's dream, but an economic success may depend on the contributions of the geophysicist and the engineer, and each one's ability to properly interpret the data at his disposal and integrate the results. The skill and experience of the interpreter will influence the utility of the data. Failure of any of these to correctly apply his own knowledge, or to integrate the expertise of one of the others with his own, may result in an economic disaster.

How can we, as earth scientists, prevent such a disaster? By communicating with each other to take full advantage of the expertise of each discipline. We must learn each other's terminology and educate one another so that we understand the value, potential, and limitations of all the disciplines and can communicate effectively. Then we can all work together toward one goal; that is, to discover and recover more hydrocarbons.

The value of seismic surveying to exploration geology is well known. Engineers, however, have been skeptical that seismic data have sufficient resolution to aid in reservoir exploitation. This may have been true at one time, but advancements such as Vertical Seismic Profiling, 3-D Seismic Surveys, Tomography, and vastly improved data processing are changing this notion. Monitoring enhanced recovery is one important application of seismic-to-oil production.

The future of the oil industry will lie not only in discovering new reserves, but also in recovering the most from the present reserves. All the geological, geophysical, and engineering information should be integrated to recover all possible reserves from the ground. The key is to understand each other's terminology and establish intelligent communication to close the gap between these disciplines.

Cross-discipline training and communication should be the first step. Engineers should start conveying their needs to geologists and geophysicists by teaching them engineering terminology, what tools are available, and their limitations. Geophysicists should always convey the latest techniques in the field of geophysics and what can be done to help the engineer better understand the reservoir. Geologists should be able to communicate to the other disciplines about the rock properties and how they relate to production and reservoir lithology.

With today's unstable oil prices and low level of wildcatting, minimizing risk in exploration is a must. Reservoir geophysics will play an important role in aiding oil finders to enhance their methods of oil recovery by better describing the reservoir properties using borehole seismic measurements.

In summary, better communication and cross-discipline training will open a more intelligent line of communication between earth scientists of different disciplines. If we can accomplish this, we will get more answers in a shorter period of time. We will also recover more hydrocarbon from the ground and probably at a lower cost than finding new reserves. What has been left in the ground is probably more than we can find by future exploration. By finding and mastering better ways to recover it, monitoring the enhanced oil-recovery projects, and increasing our efficiency of recovery by only 10%, we may be able to double the proven recoverable oil.

APPENDIX A

To understand F-K migration, it is important to understand migration in the depth domain. Chun and Jacewitz (1981) explained this method in a very elegant and clear manner.

a. Depth Domain Migration

Figure A–1a illustrates the vertical earth model ($\theta_a = 90°$).

Consider a seismic source A with a signal recorded in the same point; then the only energy that can be recorded at A is the horizontal path in the ray theory approach. Any nonhorizontal traveling wave will be reflected downward and will not return to A. Mapping the distance of the horizontally traveling path in the (X,Z) plane in the Z direction is shown in Figure A–1b.

Since AO = AC, the dip angle reflector in the record section is equal to 45°. Thus, for a 90° reflector the reflection takes place only at a point on the surface, and the recorded reflections are mapped along a 45° line on the depth plane as A moves along the surface.

Next, consider a dipping earth model as in Figure A–2a. Assuming that the source and receiver are at point A, the wave from A will be reflected at C′ and will be recorded at A. The travel distance is thus AC′ = AC. When the earth model in (a) is superimposed on (b) we can see that:

$$\sin\theta_a = AC'/AO = AC/AO = \tan\theta_b$$

This equation describes the relationship between migrated angle (θ_a) and recorded angle (θ_b). Since point C maps to C′ under migration, this process moves data up-dip.

b. Diffraction Concept

The concept of diffraction is required to understand migration properly. Diffractions are normally associated with discontinuities, and reflections may be considered as a superposition of diffractions.

So the general process of mapping a reflector or diffractor to the earth model will be described as a diffraction process.

Migration proceeds from a record to the earth model. Diffraction proceeds in the opposite direction, from earth model to record section.

FIGURE A–1. A 90-degree reflector model (modified after Chun and Jacewitz, 1981)

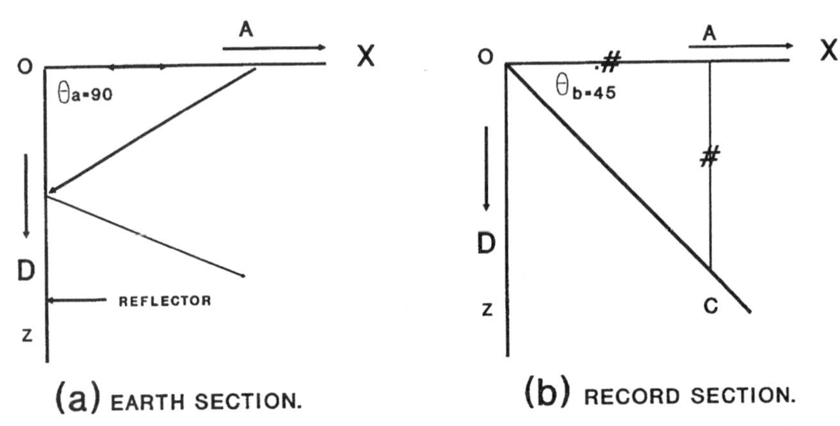

THE ONLY ENERGY THAT CAN BE RECORDED AT A IS FROM THE HORIZONTAL PATH IN THE RAY THEORY APPROACH. ANY NONHORIZONTALLY TRAVELING WAVE WILL BE REFLECTED DOWNWARD AND WILL NOT REACH POINT A. MAP THE TRAVEL DISTANCE OF THIS RAY ON X–Z PLANE OA = AC. THE DIP ANGLE θ_b OF THE RELECTION IS 45 DEGREES. MIGRATION MAPS THE 45 DEGREE RELECTION IN (b) ONTO THE 90-DEGREE REFLECTOR OF (a).

FIGURE A–2. A dipping reflector model

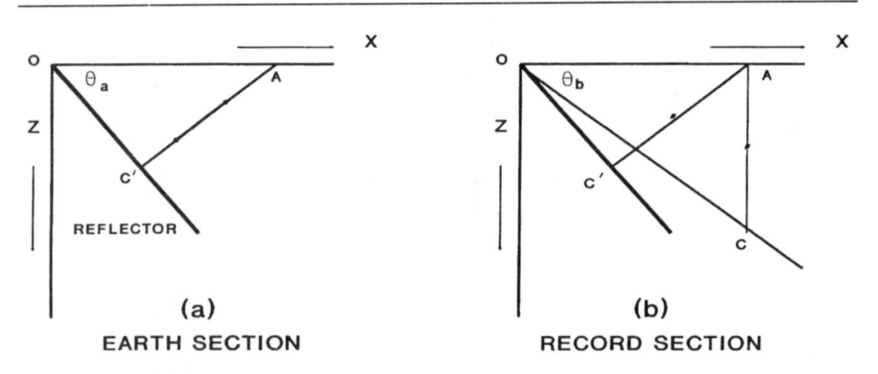

THE WAVE FROM A WILL BE REFLECTED AT C′ AND WILL BE RECORDED AT A. THE TRAVEL DISTANCE AC′ IS MAPPED VERTICALLY ON (X–Z) PLANE AS IN (b) TO SEGMENT AC. THE TRAVEL DISTANCE IS THUS AC′ = AC. THE EARTH MODEL IN (a) IS SUPERIMPOSED ON (b), WE CAN SEE

$$\sin\theta_a = AC'/OA = AC/OA = \tan\theta_a$$

THIS EQUATION DESCRIBES THE RELATIONSHIP BETWEEN THE MIGRATED DIP ANGLE (θ_b) AND RECORDED ANGLE (θ_b). SINCE C MAPS TO C′ UNDER MIGRATION, THIS PROCESS MOVES DATA UPDIP.

APPENDIX B

1. Design of Maximum Offset (non-dipping case)

Figure B–1 explains the calculations of maximum offset for a non-dipping target horizon. One can see that the maximum offset is a function of the depth and angle of incidence. Note that the maximum offset is from source to reflector to receiver point, and it is indicated by $2X$.

2. Design of Maximum Offset (dipping case)

Figures B–2 and B–3 show how to calculate the maximum offset of a dipping event.

The offset is a function of depth in unit distance, angle of incidence, and dip angle.

Note: Figure B–2 is to calculate X, which is the offset distance from source S to the reflection point.

Figure B–3 is to calculate Y, which is the distance from the reflecting point to the receiver. The sum of X and Y is the maximum required offset for the case of a dipping target.

To compute the shot line spacing (see Figure B–4), one must first compute shot interval.

$$\text{Shot line spacing} = \frac{\text{number of receivers per receiver line times G.I.}}{\text{number of fold in the direction of shot line}}$$

Example: Suppose that the required fold in the receiver line direction is four, and the number of fold in the shot line is six. For maximum fold, $4 \times 6 = 24$ fold. Number of receivers per receiver line is 60 and group interval is 55 feet.

The shot line spacing = $60 \times 55 / 6 = 550$ feet. Need to space the shooting lines at an increment of 550 feet.

FIGURE B–1. Maximum offset—nondipping case

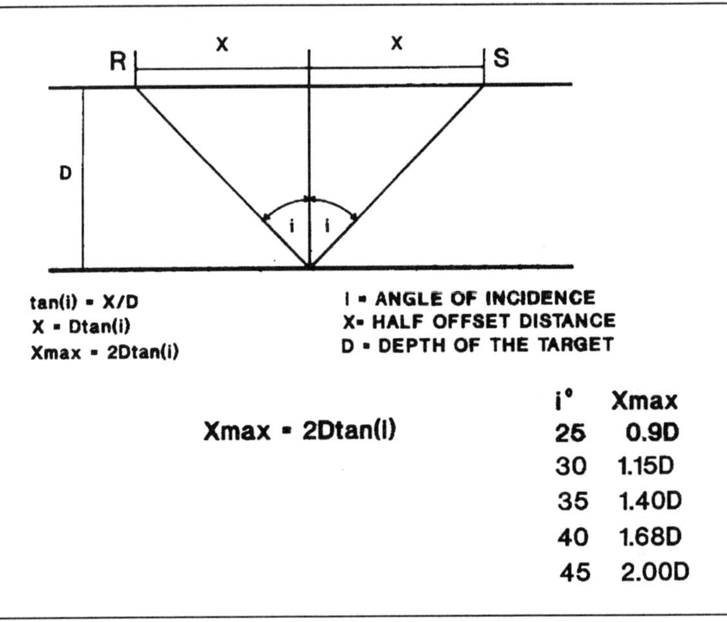

Xmax = 2Dtan(I)

i°	Xmax
25	0.9D
30	1.15D
35	1.40D
40	1.68D
45	2.00D

FIGURE B–2. Design of maximum offset—dipping case

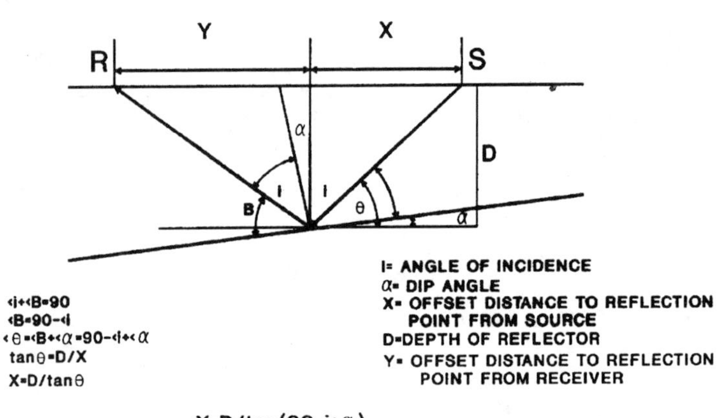

<i+<B=90
<B=90−<I
<θ=<B+<α=90−<I+<α
tanθ=D/X
X=D/tanθ

I= ANGLE OF INCIDENCE
α= DIP ANGLE
X= OFFSET DISTANCE TO REFLECTION POINT FROM SOURCE
D=DEPTH OF REFLECTOR
Y= OFFSET DISTANCE TO REFLECTION POINT FROM RECEIVER

X=D/tan(90−i+α)

FIGURE B–3. Design of maximum offset—dipping case

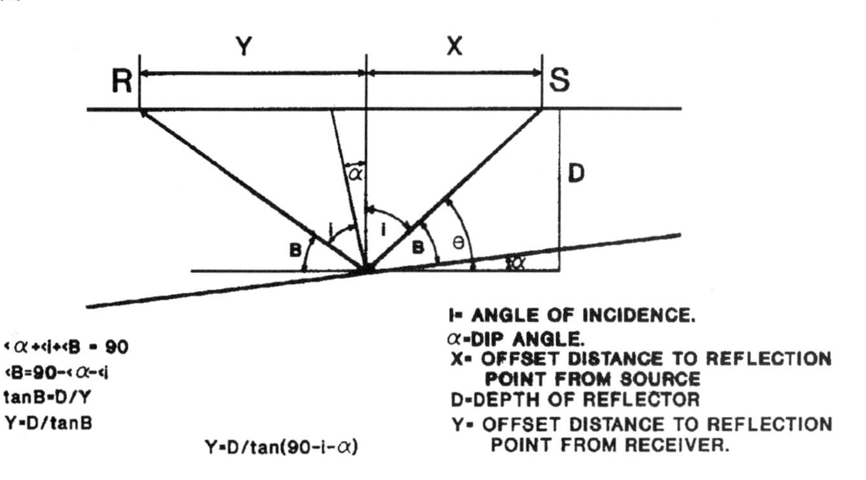

FIGURE B–4. Computation of shot line spacing

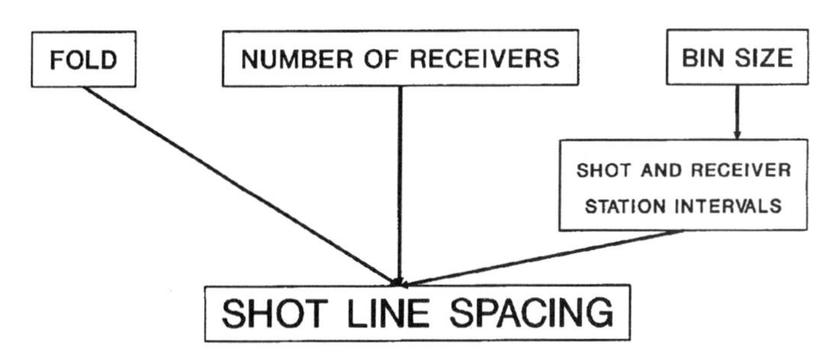

TO COMPUTE THE SHOT LINE SPACING,THESE ABOVE MENTIONED PARAMETERS ARE NEEDED. THERE MAY BE LATER MODIFICATIONS.

APPENDIX C

ANSWERS TO SELECTED EXERCISE QUESTIONS

CHAPTER 3

Question 3:

Angle of travel in the lower layer = 43.16 deg.
Critical angle = 30 deg.
Travel time = 3.293 sec.

Question 5:

a. $\lambda = 2000/50 = 40$ m.
b. $\lambda = 6000/25 = 240$ m.

CHAPTER 4

Question 3:

Period (T) = $1/f = 1/50 = 0.02$ sec.
Wavelength = $V/f = 2000/50 = 40$ m.

CHAPTER 5

Question 1:

a. A: (–2,3,14,3,–2)
   B: (–1,2,–1)
b. A: 14
   B: 2

Question: 3

C: (–4,2,3,–5,1,3)

Question 5:

A: 1– 1.00 sec.   2– 1.248 sec.   3– 0.248 sec.
B: 1– 1.16 sec.   2– 1.223 sec.   3– 0.063 sec.

Question 9:

a. Ray A: $P = 0 = \sin\theta_1/1.25 = \sin\theta_2/1.875$
So
θ_1 and $\theta_2 = 0$

Ray B: $P = 0.333 = \sin\theta_1/1.25 = \sin\theta_2/1.875$
$\theta_1 = \sin^{-1}(.416) = 24.59°$
$\theta_2 = \sin^{-1}(.624) = 38.64°$

Ray C: $P = .167 = \sin\theta_1/1.25 = \sin\theta_2/1.875$
$\theta_1 = 12.05°$
$\theta_2 = 18.25°$

b. Ray A: $\Delta h_1 = \Delta Z_1 \tan\theta = 0$
$\Delta h_2 = \Delta Z_2 \tan\theta = 0$
Half-offset $= \Delta h_1 + \Delta h_2 = 0$

Ray B: $\Delta h_1 = (3{,}300)\tan 24.59 = 1{,}510$ ft.
$\Delta h_2 = (4{,}950)\tan 38.64 = 3{,}957$ ft.
Half-offset = 1.035 miles

Ray C: $\Delta h_1 = (3{,}300)\tan 12.05 = 704$ ft
$\Delta h_2 = (4{,}950)\tan 18.25 = 1{,}632$ ft
Half-offset = 0.44 mile

c. Ray A: $\Delta t_1(0) = \Delta Z_1/V_1 = 3{,}300/(1.52 \times 5{,}280) = 0.500$ sec.
$\Delta t_2(0) = \Delta Z_2/V_2 = 4{,}950/(1.87 \times 5{,}280) = 0.500$ sec.
Vertical one-way traveltime $= \Delta t_1 + \Delta t_2 = 1.00$ sec.

Ray B: $\Delta t_1 = \Delta t_1(0)/\cos 24.59 = 0.55$ sec.
$\Delta t_2 = \Delta t_1(0)/\cos 38.64 = 0.64$ sec.
Slant one-way traveltime $= \Delta t_1 + \Delta t_2 = 1.19$ sec.

Ray C: $\Delta t_1 = \Delta t_1(0)/\cos 12.05 = 0.511$ sec.
$\Delta t_2 = \Delta t_2(0)/\cos 18.25 = 0.526$ sec.
Slant one-way traveltime = 1.037 sec.

CHAPTER 7

Question 1:

a. $R1 = .688$, $R2 = .161$, $R3 = .143$, $R4 = .150$, $R5 = .174$
b. $T1 = .527$, $T2 = .974$, $T3 = .980$, $T4 = .978$, $T5 = .97$

Question 3:

a. The composite wavelet (C) : (1.6,1.4,–3,–2.4,3,3.2,–2.6, –3.2,1,1.8,.4,–.8,–.4)
b. The dominant frequency = 50 Hz. This example shows the tuning effect destructive interference.

CHAPTER 10

Question 1:

a. $R = (\lambda Z/2)^{0.5} = (100 \times 2000/2)^{0.5} = 316.2$ m.
b. $R = (V/2)(t/f)^{0.5} = (3000/2)(2/20)^{0.5} = 474$ m.

GLOSSARY

Abnormally high-pressure zone
High formation pressure that is higher than the expected hydrostatic pressure at a certain depth. The normal static pressure gradient is 0.5 psi/ft. Reservoirs that have abnormal pressure usually show low seismic velocity.

Abnormal pressure
Formation fluid pressure that differs from the normal hydrostatic pressure, which is the pressure produced by a column of fluid extending to the surface.

Absorption
A process that converts seismic energy into heat while passing through a medium and causes attenuation in the seismic amplitude.

Acceleration
The velocity gradient—rate of change of the velocity with time.

Acoustic impedance
Seismic velocity multiplied by density.

Aliasing
Frequency ambiguity due to signal sampling process.

Alias Filter
A filter used before sampling to remove undesired frequencies. It is also called antialias filter.

Ambient noise
Noise associated with surrounding environment.

Amplitude
The maximum swing of wave from an average value.

Anomaly
A deviation from uniformity in physical properties. In seismic usage, generally synonymous with structure. Occasionally used for unexplained seismic event.

Autocorrelation
The cross-correlation of a wavelet with itself.

Azimuth
The horizontal angle specified clockwise from true North. In 3-D survey the Azimuths are calculated for each recording line as a part of acquisition parameters.

Bandwidth
(1) The frequency band required to transmit a signal. (2) The range of frequencies over which a given device is designed to operate within specified limits.

Bin
A bin consists of cells with dimensions one-half the receiver group spacing in the inline direction and nominal line spacing in the cross-line direction (same as CMP). In data processing, sorting data in cells is called binning.

Binary gain
A gain control in which amplification is changed only in discrete steps by factor of 2.

Blowout
Unexpected flow of liquids or gas due to abnormal high-pressure reservoir. The reservoir pressure exceeds the pressure of the drilling fluid.

Bow-tie effect
The appearance of a buried focus on a seismic section. It is two intersecting seismic events with apparent anticline below it.

Bright spot
A large increase in the amplitude due to a decrease in the acoustic impedance from the overlaying shale to the sandstone reservoir saturated with gas. As little as 5% gas saturation can cause this amplitude anomaly. It is called a direct hydrocarbon indicator for gas sands.

Buried focus
For zero offset and constant velocity, buried focus occurs if a reflector's center of curvature lies beneath the recording plane.

Cable feathering
Cable drifting at an angle to the marine seismic line due to crosscurrent.

Caliper log
A wireline logging tool that records the borehole diameter.

Calibration
A check of equipment readings with respect to known values.

Carbonate
A rock formed from calcium carbonate. Limestone and dolomite are carbonate rocks; they are potential reservoir rocks.

Casing
Tubes used to keep a borehole from collapsing (caving in).

Cementing
Pumping wet cement to fill the space between casing and walls of the well to protect the well from collapsing.

Convolution
Change in the wave shape as result of passing through a linear filter. It is a mathematical operation between two functions to obtain a desired function.

Constant angle stack
Commonly used in amplitude versus offset analysis to check the variation of the amplitude with the angle of incidence. Certain values of angles are chosen within the CMP before stack, then a stack is generated.

Closure
The vertical distance from the apex of a dome to the lowest closing contour. Areal closure is the area contained within the lowest closing contour.

Common cell gather (CCG)
In 3-D seismic surveys, data are sorted in common cell gathers same as CMP in 2-D surveys.

Common depth point (CDP gather)
The set of traces that have a common depth point. Each trace is from different source and receiver. It is used where there are horizontal reflectors.

Common mid-point (CMP gather)
The set of traces that have a common depth point. Each from different source and receiver. It is used where there are dipping reflectors.

Common shot point
Seismic data are recorded from same source and different detector stations on the ground.

Compressional waves
A p-wave or waves travel (propagate) in the same direction of motion; it propagates through the body of a medium.

Core
A rock sample cut from a borehole.

Critical angle
An angle of incidence for which the refracted ray travels parallel to the surface of contact between two types of rocks of different velocities.

Correlation
Measurement of degree of linear relationship between traces, or degree of how two traces are alike.

Cross-correlation
A measure of similarity between two functions. In a mathematical sense, it is a cross-multiplication of sample values, addition, shift one sample and so on. Zero cross products is indication of no similarity between the functions.

Deconvolution
A process designed to enhance the vertical resolution of the seismic data by attenuating the undesirable signals such as short period multiples. It is also called inverse filtering.

Demultiplex
Is to rearrange the recorded digital field data in such a way that all the samples belong to every trace in the record are together in one channel. It is called trace sequential form.

Density
Mass per unit volume, usually measured in gram per cubic centimeter.

Diffraction
The phenomenon by which energy is transmitted laterally along a wave crest. When a portion of a wave is interrupted by a barrier, diffraction allows waves to propagate into the region of the barrier's geometric shadow.

Digitizer
An instrument to sample curves, seismic traces, or other data recorded in analog form.

Dim spot
A lack of seismic amplitude. It is caused by abnormally low reflection coefficient. Shale overlaying a porous or gas saturated reef can cause this amplitude anomaly.

Dispersion
Variation of velocity with frequency. Dispersion distorts the shape of a wave train; peaks and troughs advance toward the beginning of the wave as it travels.

Dispersive wave
Wave that has changed in its shape because of dispersion. Surface waves usually suffer very large dispersion due to the near-surface velocity layering.

Dip moveout (DMO)
The change of the arrival time of a reflection due to the dip of the reflector. DMO process corrects for the reflection points smear that results when dipping reflectors are stacked.

Downgoing wave
A seismic wave where the energy hits the detector at the top. It is widely used in the VSP terminology.

Down time
The time during which the drilling operation stopped to conduct different operations such as logging, VSP, or fishing.

Dynamic correction
Normal Move Out (NMO) is a type of dynamic correction. Dynamic correction depends on the distance from the source and the time of the seismic event. Normally, the deeper (more time), the less the dynamic correction because velocities increase with depth.

Elastic
The ability for a material to return to its original shape after the distorting stress is removed.

Electrical conductivity
The ability of a material to conduct electrical current. Units are sisms per meter.

Electrical method
By which measurements at or near the earth's surface of natural or induced electric field is investigated in order to locate mineral deposits.

Enhanced oil recovery (EOR)
Techniques used for maximizing the oil production after primary recovery.

Evaporite
A sedimentary rock layer such as salt formed after water was evaporated. Other evaporite such as Gypsum and Anhydride.

Exploration
The search for commercial deposits of useful minerals such as hydrocarbons.

Exploitation
The development of petroleum reservoirs; wells are drilled to optimally drain the reservoir and have fairly low risk. The development wells may have several producing wells in the nearby drilling or spacing units.

Extrapolation
Projection or extension of unknown value from values within known observations or interval.

Fault
Discontinuity in a rock type due to a break caused by tensional forces which cause normal faults or compressional forces which cause reverse faults.

Fermat's principle
See bow-tie effect (buried focus).

First break times
The first recorded signal from the energy source. These first breaks from reflection records are used to obtain information about the near-surface weathering layers. According to SEG polarity standard, an initial compression usually shows as a downkick.

Flat spot
A horizontal seismic reflection due to an interface between two fluids such as gas and water.

Fold (Structure)
An arch in a rock layer. A fold is usually formed as a result of deformation of rock layers by external forces. Folds include anticlines, cynclines, and overturns, etc.

Fold (seismic)
The multiplicity of common mid-point data.

Formation
A distinctive lithological unit or rock type.

Formation velocity
The speed by which a certain type of wave travels through a particular formation.

Floating point (recording)
A number expressed by the significant figures time a base raised to a power. This prevents the loss of a very small or a very large number. Computer uses bases that are a power of two rather than the base ten.

Frequency
A periodic waveform repetition rate in a second. Measured in cycle per second or Hertz.

Frequency-wavenumber (F-K)
A domain in which the independent variables are frequency (f) and wavenumber (k), yield two-dimensional Fourier transform of a seismic section. k is the reciprocal of the wavelength.

Gate
Also called window, it is a time interval where certain process or function is performed.

Ghost
Type of multiple in which seismic energy travels upward and then reflects downward as occurs in the base of the weathering layer or at the surface.

Gravity
A method by which the subsurface geology is investigated on the basis of variations in the earth's gravitational field which are generated by a difference in rock densities. It is measured in milligal.

Ground roll
Surface-wave that travels along or near the surface of the earth. It is considered as noise on the seismic record. It is also known as Rayleigh wave.

Heterogenous
Lateral and vertical variations of the rock properties.

Homogenous
Constant properties through the rock material.

Horizontal resolution
How closely two reflecting points can be situated laterally, yet can be recognized as two separate points. It is known as first Fresnel zone.

Horst
A block formed by the upthrown sides of two normal faults.

Huygen's principle
Each point on a wave front can be considered as a secondary source.

Intrabed multiple
Also called pegleg multiple. A multiple generated due to successive reflections between two different interfaces and then reflect back to the surface. Intrabed multiple has irregular travel path.

Interpolation
Determining values at location where there was no measurement. It is performed between two measured values.

Inversion (seismic)
Is to derive from the observed field data a model to describe the subsurface. Also can be used to calculate the acoustic impedance from the seismic trace.

Iteration
Procedure that repeats with improved output until some conditions are satisfied.

Least squares
An analytic function that approximates a set of data such that the sum of the squares of the distances from given points to the curve is a minimum.

Limestone
Is a sedimentary rock composed of mainly calcium carbonate. It is an important type of reservoir rock. Its matrix density is 2.7 gm/cc, and its matrix velocity is about 23,000 ft/sec.

Log
A record of measurements, especially those made in a borehole (e.g, resistivity, sonic and density logs).

Love wave
A surface seismic wave characterized by horizontal motion perpendicular to the direction of propagation of the seismic wave with out vertical motion. It is near surface shear wave. Its counterpart in the p-waves is the ground roll or rayleigh wave.

Magnetic method
A method by which the subsurface geology is investigated on the basis of variation in the earth's magnetic field. It's measured in gamma.

Matrix
A rectangular array of numbers called elements. An mxn matrix A has m rows and n columns. If m=n, it's called a square matrix.

Migration
Is a process to move the dipping seismic reflection, refraction to their proper subsurface positions to obtain better imaging of the structure and stratigraphic picture.

Migration aperture
Is the length that should be covered in a seismic survey of a geologic feature. It is usually larger than the actual lateral extent of the feature especially dipping ones.

Mud weight
Density of drilling mud (mass divided by volume) and expressed in pounds per gallon. The heavier the mud weight, the greater the pressure and may cause loss of circulation. Light mud weight may cause blowout if formation pressure exceeds the pressure of the mud column.

Multioffset VSP
Is a survey where a string of geophones is laid out around the well while a VSP survey is conducted. It is used to investigate the subsurface from

the borehole. It is a good tool for stratigraphic implication such as sand channel mapping or delineating small faults.

Multiple
Seismic energy that reflects more than once from the same horizon. Multiple reflections may mask out stratigraphic and structure details and is one of the undesirable signals to be attenuated.

Multiplex format
Is a way of digital field recording where the first sample of channel 1 is recorded, followed by the first sample of channel 2, then the first sample of channel 3 etc until all first samples of all given channels are recorded followed by the second sample of channel 1, then the second sample of channel 2 etc. SEG set some standard formats for digital field recording such as SEGA, SEGB, and SEGC. Some manufacturers of recording systems have modified versions of these formats.

Mute
To exclude part of the seismic data. Normally, it is applied in the early part of the traces that contains first arrivals and body waves and is called front end mute. Mute can be performed over a certain time intervals to keep ground roll, airwaves, and noise out of the stack section. This process is called surgical mute.

NMO stretch
Increase in the period (lower frequency) due to the application of normal-moveout correction to offset traces. It is noticed on the far traces within the seismic record or CMP.

Noise pattern (seismic)

Any seismic signal but the primary reflection. This includes multiple reflection, ground roll, airwaves, source generated noise, and ambient seismic noise.

Noise test
A test or set of tests conducted in the field to analyze the noise patterns in an area to design the optimum recording parameters that will yield good signal-to-noise ratio seismic data.

Permeability
Ability of the rock to transmit fluid; it is measured in millidarcies.

Petrophysical properties
Are physical aspects of the reservoir such as porosity, permeability, and fluid content.

Pilot flood
A small waterflood or enhanced oil-recovery project that is run on small portion of a field to determine its efficiency.

Pilot signal
A signal has a predetermined frequency range, commonly used in vibrioses system. This signal is cross-correlated with the recorded signal to produce the seismogram record.

Poisson's ratio
It is an elastic constant and defined as the ratio between transverse contraction to longitudinal extension when a rod is stretched. In seismic method, it is a function of p-wave and s-wave velocities.

Polarity (seismic)
The condition of the amplitude being positive or negative referred to a base line.

Porosity
Pore volume divided by bulk volume.

Quadratic fit
A second order approximation to get the best fit to a set of data points.

Random noise
Undesired signals without a uniform pattern; it can be attenuated by the stacking process.

Ray
A line normal to the wave front.

Ray Tracing
Determining the arrival time at detector's locations.

Rayleigh wave
See ground roll.

Reconnaissance survey
Is a survey to determine an area's main geological features. It is done to delineate an area of interest to focus on it.

Recovery factor
A percentage of what can be recovered from the reservoir fluids or gas. It varies from field to field and depends upon the geological setting of the area.

Reflection method
A technique used to investigate the subsurface by analyzing the seismic response of waves reflected from rock interfaces of different velocities and densities (acoustic impedances).

Reflectivity series
Represents reflecting interfaces and their reflection coefficients as a function of time, usually at normal incidence.

Refraction method
A technique used to map the subsurface structure by analyzing waves that enter high-velocity medium near the critical angle of incidence to the interface. It travels in the high-velocity material parallel to the refractor.

Reservoir (petroleum)
Rock containing hydrocarbon accumulation.

Residual statics
Statics remain after applying elevation statics. They are due to the near-surface formation and velocities irregularities and cause seismic distortion. They are solved by applying refraction base or short period statics in surface consistent manner.

Seismic resolution
Ability to separate two features close together.

Root-Mean-Square velocity (VRMS)
RMS velocity approaches the stacking velocity that is obtained from velocity analysis based on the application of the normal moveout correction as the offset approaches zero. The assumption is that the velocity layering and the reflectors are parallel and that there are no changes within the layers (isotropic).

Saturation
Percentage of pore space of a certain rock filled with a particular fluid (water, oil, or gas).

Seismic detector
A device that detects seismic signals. Geophone on land surveys, hydrophone in marine surveys.

Seismic marker
A continuous distinguished seismic character on a seismic section.

Seismic record
A plot or display of seismic traces from a single source point; a seismogram.

Seismic section
A display of seismic data along a line. The horizontal scale is in distance units and the vertical scale is usually two-way time in seconds or sometimes in depth units.

Seismic signature
A waveform generated in a certain medium by a seismic source.

Seismic trace
"Wiggle trace" is the response of a single seismic detector to the earth's movement due to seismic energy. Each part of the wiggle trace has some

meaning, either reflected or refracted energy from a layer of rock in the subsurface, or some kind of noise pattern. Excursions of the trace from a central line appear as peaks and troughs; conventionally peaks represent positive signal voltages, and the troughs negative signal voltage.

Seismic velocity
The speed by which a seismic wave travels in a particular medium. It is measured by unit distance per unit time.

Shadow zone
A portion of the subsurface from which reflections are not present because its raypath did not reach the detectors.

Shear wave (S-wave)
Is a seismic wave; its particle motion is perpendicular to the direction of propagation. Approximately, the velocity of the *S*-wave is one-half the velocity of the *P*-wave.

Shot point
The location where an explosive charge is detonated. Also used for the location of any source of seismic energy.

Signal
A part of a wave that contains desired information.

Slowness
Reciprocal of velocity (1/V).

Snell's law
It shows the relationships between the incidence and reflected wave.

Snell's law of reflection
States that the angle of incidence (angle between the incident ray and the normal to the interface) equals to the angle of reflection (angle between the reflected ray and the normal to the interface).

Snell's law of refraction
States that the sine of the angle of incidence divided by the velocity in the upper layer equals to the sine of the refraction angle in the second layer divided by the velocity of the second layer. The velocity in the second layer is higher than the first layer.

Sparse system
A large system and has about 1% nonzero values (e.g., sparse matrix).

Split spread
Arrangement of geophone groups in relation to the source point. In this case, the source point is in the middle between the geophone groups.

Stack
A composite record made by combining traces from different records. It is done in order to improve the signal-to-noise ratio and reveals the subsurface geology.

Static correction (statics)
A correction applied to the seismic data to correct the irregularities of the surface elevations, near-surface weathering layer, and weathering velocities, or reference to a datum.

Stratigraphic column
A chart where the rock units are arranged from bottom (older) to top (younger) chronologically.

Swath shooting
A method by which 3-D data are collected on land. Receiver cables are laid out in parallel lines (inline direction), and shots are positioned in a perpendicular (cross-line direction). Also called multiline shooting.

Synthetic seismogram
A seismic trace generated from the integration of the sonic and density logs by calculating the reflectivity series. This series is filtered with the same filter of the seismic section for better correlation. It is a man-made seismic trace and one of its applications is to transfer lithology to the seismic section.

Transmission coefficient
The ratio between the amplitudes of the transmitted ray and the incident ray.

Transit time
Is a measure of the sonic velocity of the rock layers and is obtained by sonic tool. Transit time is measured in microsecond per foot and varies with rock type, porosity, and fluid content.

Trap
A shape of rocks that is able to confine fluids such as oil. A trap should have a cap rock in order to prevent fluids escape. A stratigraphic trap can be formed by permeability termination.

Tomography (seismic)
The word is derived from the Greek words Tomos (section) and graphy (drawing). It is a method for obtaining models that adequately describe seismic data observations and show the effect of rock properties on the seismic wave propagation.

Tuning effect
Interference resulting from closely spaced seismic reflectors. It can cause enhancing or smearing to the individual reflection.

Unconformity
Is a buried erosional surface. It separates older rock from younger overlaying rock. Unconformities are normally good seismic markers and hydrocarbon accumulation occurs above or below the surface of the unconformity.

Upgoing wave
It is a seismic wave that hits the detector from the bottom after it reflects from horizons.

Velocity pull-up
A pull-up of a reflection due to abnormal high velocity of a material such as salt.

Velocity survey
A series of measurements in a well to determine the average velocity as a function of depth. Sometimes it applies to sonic log or vertical seismic profiling (VSP).

Vertical resolution
The ability to separate two features that are very close together. Maximum vertical resolution is one-quarter of the dominant wavelength.

Vertical seismic profiling (VSP)
Is seismic survey in which a seismic signal is generated at the surface close to a well and recorded by geophone placed at various depths in the borehole.

Vertical stack
A process combining seismic records from several sources at nearly the same location without correcting static or offset differences.

Wave equation
An equation that relates the lateral (spatial) and the vertical (time) dependance of disturbances that can propagate as waves.

Wave front
A circle of equal travel time or a leading edge of a waveform.

Wave length
Velocity in unit distance per second times the period (time in seconds between two peaks or two troughs of a seismic wavelet); it's measured in unit distance. Also can be expressed as velocity divided by frequency.

Well logging
Is done by a borehole tool where a particular device can measure a variable that can be used to determine a rock property such as porosity, saturation, lithology, and formation boundaries.

Well prognosis
Prediction of geological targets before drilling.

Wildcat (well)
A well drilled in a newly explored area where hydrocarbon accumulations are not discovered commercially.

REFERENCES

Hyne, N.J. *Dictionary of Petroleum Exploration, Drilling & Production.* Tulsa, Oklahoma: PennWell Books, 1991.

Sheriff, R.E. *Encyclopedic Dictionary of Exploration Geophysics.* Tulsa, Oklahoma: SEG, 1991.

INDEX

E

F

G

H

I

K

L

M

N

O

P

R

T

U

V

W

Z